Forest-people interfaces

Understanding community forestry and biocultural diversity

edited by:
Bas Arts
Séverine van Bommel
Mirjam Ros-Tonen
Gerard Verschoor

Wageningen Academic
Publishers

ISBN: 978-90-8686-193-4
e-ISBN: 978-90-8686-749-3
DOI: 10.3920/978-90-8686-749-3

First published, 2012

To Freerk Wiersum
On the occasion of his retirement 1 March 2012

Table of contents

Part III. Biocultural diversity

Preface

This *Festschrift* is a tribute to Dr Freerk Wiersum, Associate Professor in Forest and Nature Conservation Policy at Wageningen University and Research Centre, the Netherlands, on the occasion of his retirement on 1 March 2012. During his career he was involved for about 15 years in forestry-related development cooperation projects in Asia and Africa and then for about 25 years in high-quality academic research and teaching in the field of forest-people interfaces. In the latter period, he produced over 200 publications (journal articles, book chapters, books, reports and conference papers). His most cited work deals with sustainability in forestry, forests and rural development, social forestry, small-scale forest ownership, non-timber forest products, rural livelihoods, forested landscapes and community forestry.

There are two ways of writing a *Liber Amicorum*. One is to make the person the subject, which has value only in terms of collegiality and friendship. Such books generally consist of personal tributes, short essays or even poems. The second type is no less an expression of collegiality and friendship, but also of the wish to contribute to the field of expertise of the laureate. This book is of the second type. It contains 18 academic chapters from both junior and senior researchers from all over the world, divided into four parts on non-timber forest products and agroforestry, community-based natural resource management, biocultural diversity and forest governance. The intention was not only to provide a state-of-the-art review of this academic field, but also to elaborate on a number of concepts embraced by Freerk Wiersum during the past few years: biocultural diversity, institutional bricolage and governance. Our aim was to produce a book that contributes to the current academic debates on the forest-people interface.

We wish Freerk all the best in his post-academic life and hope to continue collaborating with him in the years to come. Above all, we thank him for his boundless inspiration and valued cooperation, past and present.

The editors

1. Forest-people interfaces: from local creativity to global concerns

Bas Arts[1], Séverine van Bommel[2], Mirjam Ros-Tonen[3] and Gerard Verschoor[4]
[1]Wageningen University, Forest and Nature Conservation Policy Group, Droevendaalsesteeg 3, 6708 PB Wageningen, the Netherlands; [2]Wageningen University, Department of Social Sciences, Communication and Innovation Studies Group, Hollandseweg 1, 6706 KN Wageningen, the Netherlands; [3]University of Amsterdam, Amsterdam Institute of Social Science Research (AISSR), Nieuwe Prinsengracht 130, 1018 VZ Amsterdam, the Netherlands; [4]Wageningen University, Rural Development Sociology Group (RDS), Hollandseweg 1, 6706 KN Wageningen, the Netherlands; bas.arts@wur.nl; severine.vanbommel@wur.nl; m.a.f.ros-tonen@uva.nl; gerard.verschoor@wur.nl

1.1 Introduction

This book takes the reader on a journey through four major themes that have dominated research on the people-forest interface since the UN Food and Agriculture Organization (FAO) published its *Forestry for Rural Development* paper and launched its Programme on Forestry for Local Community Development in 1976. This was the prelude to the FAO VIIIth Forestry Congress entitled 'Forestry for People', organised two years later, which drew attention to the role of forests in meeting people's livelihood needs. These events marked the emergence of social forestry as a new approach to forest management that aimed to increase community participation in the development and management of forest resources (Arnold, 1991; FAO, 1976; Wiersum, 1999). In the 1980s social forestry marked a shift away from an exclusive focus on industrial, timber-oriented forestry to participatory and cooperative management schemes (Colchester *et al.*, 2003). In the same period, the Canadian forester John Bene (Bene *et al.*, 1977) coined the term 'agroforestry' for the practice of integrating trees, food crops and/or animals in a combined production system compatible with the cultural practices of the local population. Bene played an important role in the establishment of the International Council for Research in Agroforestry (ICRAF) in Nairobi in 1997 (King, 1987). This is now known as the World Agroforestry Centre and has regional offices in India, Indonesia, Kenya, Malawi and Mali.

In parallel, but in a relatively separate development, another paradigm shift occurred in conservation circles. The publication of the World Conservation Strategy by the International Union for the Conservation of Nature (IUCN), the United Nations Environmental Programme (UNEP) and the World Wildlife Fund (WWF) created a basis for what is now commonly accepted: that conservation and development are intrinsically linked and that one is impossible without the other (Adams, 2008). The strategy gave rise to several community-based conservation approaches, such as transition zone management, integrated conservation and development projects, and community-based conservation, which all aim to enhance 'participation' in conservation efforts and combine conservation and development objectives (Ros-Tonen *et al.*, 2005). The 'non-timber forest product' strategy, which emerged at the end of the 1980s and has since figured prominently on the research agenda, can be seen in a similar vein. Together with agroforestry, it is the first theme of this book.

Both developments merged in an approach and a discipline that is known as community-based natural resource management (CBNRM), and which forms the second theme of this book. CBNRM is about the co-existence of people and nature. The term has a long history that dates back to debates in the US in the 1920s between proponents of a preservation (or 'hands off') approach to the management of natural resources and those who favoured the 'wise' or sustainable use of natural resources. The first group was concerned about the ravages of unfettered commercial exploitation and its environmental impact; the latter argued for restrictions on the misuse by 'big business' of federal property, while allowing small-scale use for local or communal purposes. The debate lingered on for decades (and goes on to this day, see Wilshusen *et al.*, 2002 for a summary of the positions taken) and it was not until the early 1990s that the idea that communities are the best caretakers of the natural resources in their immediate surroundings really took hold in multilateral organisations, non-governmental organisations (NGOs) and state agencies worldwide. This was a result of a coalition of arguments involving, amongst others, grass-roots approaches to development (e.g. Chambers, 1993), environmentalism, and human rights and indigenous peoples movements. Today, through the assimilation of a range of interests and concerns, CBNRM has become something of a grab bag that cannot be defined simply. However, as a summary approximation, CBNRM refers to a variety of practices that differ in kind, but which ideally are carried out 'by, for, and with local communities' (Western and Wright, 1994: 7). In one sense, it denotes a wide range of activities practised in various corners of the world that directly or indirectly lead to the sustainable use of resources, such as buffer-zone protection of parks and reserves, natural resource use by farmers in rural areas, and policies geared to benefiting the relationship between communities and their environment. In yet another sense, CBNRM refers to those evaluative frameworks or discourses that serve to justify specific calls for action and which, implicitly or explicitly, address the question of what should be conserved or managed, who should do so and where, and how results should be monitored and evaluated.

The common thread in the aforementioned approaches is the aim of combining the goals of biodiversity conservation and rural development, with different degrees of focus on either one. The concept of biodiversity was first introduced in 1986 at the conference 'The National Forum on BioDiversity', which was held in Washington DC (Jeffries, 2005; Wilson, 2006). This conference dealt not only with the richness of life on earth and the threat to and extinction of species, but also with the economics, functions, values and conservation of biodiversity. These subjects demonstrate that the concept has never had a purely biological connotation. The topic subsequently became the subject of international negotiations that culminated in the Convention of Biological Diversity (CBD) adopted at the United Nations Conference on Environment and Development (UNCED) in Rio de Janeiro in 1992. The CBD acknowledged the dependence of indigenous and local communities on biological resources and the need to protect and encourage customary and sustainable use of biological resources in accordance with traditional cultural practices. However, a countermovement emerged that called attention to the 'inextricable link' between biological and cultural diversity (Posey, 1999 cited in Cocks, 2006), embodied in the notion of biocultural diversity. It resulted in a third body of literature and research in community forestry that focuses on cultural and religious values in the use of forest resources. This is the subject of the third set of papers in this book.

Governance is the fourth theme addressed in this book. It emerged as a research theme in the 1990s and was described in a seminal paper by Rod Rhodes (1996: 652-653) as 'a *new* way of governing, a *changed* condition of ordered rule, or the *new* method by which society is governed'. In relation to forests it is the process of implementing and monitoring the allocation of forest land and resources and of making the relevant policy (Macqueen and Bila, 2004). It encompasses decisions on how and to what ends forests are managed, who are involved in these decisions, and what is done to enforce forest laws and policies on the ground. It also refers to the regulatory and institutional frameworks for the conservation, use and trade of forest resources and the principles that guide the interactions between those taking part in the design and implementation of different kinds of arrangements. In contrast to centralised government by the nation state, governance also includes governments at local levels as well as private and civic actors, such as companies, communities and NGOs (Ros-Tonen and Kusters, 2011). Through such decentralisation processes, governance is a theme of eminent importance to CBNRM and research at the forest-people interface. Below we look to the four building blocks of this book in more depth.

1.2 Non-timber forest products and agroforestry

The use and sale of non-timber forest products (NTFPs) has received much attention in community forestry research and practice as a way to simultaneously generate livelihood benefits and enhance participatory and sustainable forest management. Chapter 2 presents an overview of this 'NTFP debate', while approaching the use of NTFPs as a productive bricolage process. Analogous with the institutional bricolage concept addressed in Chapter 17, productive bricolage encompasses the flexible and dynamic crafting together of various livelihood options. It resembles the concept of livelihood diversification, but differs from it by considering the effects of the bricolage process on the landscape. It thus offers an approach to examining livelihoods, day-to-day decisions on natural resource use and landscape change in an integrated manner. It also provides a framework for reinterpreting Wiersum's work on the evolutionary continuum of forest-people interactions (Wiersum, 1997a) and the co-domestication of forests and trees (Wiersum, 1997b). The chapter makes it clear that local decisions on forest and natural resource are not always a matter of choice. Productive bricolage can also be driven by necessity and is becoming increasingly dependent on factors at other scales. Ros-Tonen therefore proposes analysing productive bricolage processes from a political ecology perspective that takes account of ecological, economic and political-institutional factors at multiple scales.

Chapters 3 and 4 apply the productive bricolage concept to gum and resin production in the Ethiopian drylands and to enset-coffee homegardens in Ethiopia. Both chapters make it clear that the respective production systems (and with them the landscapes in which they are embedded) are undergoing changes resulting from individual choices by farming households in response to internal and external pressures and opportunities. In Chapter 3, Woldeamanuel shows how livelihood portfolios, multilevel institutional arrangements and characteristics of resin and gum production affect landscape diversity and dynamics in the Ethiopian lowlands. He argues that the relative importance of NTFPs in people's livelihood portfolios and the economic and locational choices inherent in the productive bricolage process influence the landscape, but are not a sufficient basis to explain the degree of management and domestication of gum and resin trees. The chapter makes it clear that for an in-depth understanding of natural resource management and its effects on landscape diversity, it is also important to know whether NTFP use is compatible with other livelihood components (such as crop cultivation) and how access to resources and markets is regulated. As regards the latter, both statutory and customary institutional arrangements and macro policies considerably affect the management status of gum and resin-producing resources. In Chapter 4, Abebe and Bongers stress that the dynamics of the agroforestry homegarden of Southern Ethiopia ('a productive bricolage process on the micro scale') may jeopardise a production system that is ecologically and socially sustainable. These dynamics result from increasing population pressure and market opportunities for certain cash crops. This not only causes a further uniformisation of the landscape, but also threatens food security, biodiversity and other environmental services that the homegardens provide. Both chapters are illustrative of the argument that local-level household choices regarding livelihoods and natural resource management are entwined with landscape change, but that a broader approach is needed to unravel the influence of factors emanating from broader institutional and economic contexts.

The last chapter in this part of the book highlights the importance of woodfuel (fuelwood and charcoal) to the livelihoods of people in the Democratic Republic of Congo (DRC). As explained in Chapter 2, fuelwood has long been excluded from the NTFP debate as it did not fit the conservation paradigm very well. In Chapter 5, Schure confirms the ambiguous position of woodfuel in the NTFP debate. However, it is becoming increasingly clear that fuelwood is by far the most important forest product in terms of its contribution to people's livelihoods. Indeed, the empirical data from Schure's study in the DRC confirms that the sale of woodfuel not only contributes to poverty mitigation (by providing basic needs), but also to poverty elimination (by providing capital for investment in other economic activities). The latter implies that woodfuel revenues create opportunities for livelihood diversification. However, in contrast to the NTFP systems highlighted in Chapters 3 and 4, the production of woodfuel in the DRC is not sustainable and negatively affects the resource base on which it depends. This may eventually turn 'diversification by choice' into 'diversification by necessity', with an impoverished landscape being the result.

1.3 Community-based natural resource management

As outlined above, CBNRM refers to the two different domains of practice and discourse, as well as their interconnections. In this section, we take the reader through these domains. We start with the global concerns evidenced in discourse and slowly zoom in to focus on the creativity of local actors engaged in the management of their resources.

In Chapter 6, De Jong introduces the reader to the discursive domain by exploring the idea of 'community forestry', a concept introduced and promoted by the FAO in the 1970s. In his account, De Jong identifies a major shift in discourses about community forestry. This shift takes us from an initial 'forest resource scarcity' discourse, via one that emphasises 'tropical forest conservation', to a discourse in which pride of place is given to 'community enterprise'. Each discourse is presented in terms of main arguments, main protagonists and specific public manifestation. The author then goes on to link wider societal concerns to the shift in the community forestry discourse, showing how the forest resource scarcity discourse emerged shortly after concerns about the world's limited natural resources and the need to meet future economic growth and related consumption. Similarly, he shows how the tropical forest conservation discourse can be related to international concern about tropical deforestation and, finally, how the community enterprise discourse is closely linked to a mix between neoliberalism and a focus on poverty reduction in international development assistance. De Jong concludes his contribution by predicting that the present apprehension about climate change will likely affect the future community forestry discourse through the incorporation of climate change mitigation elements.

Chapter 7, by De Vette, Kashululu and Hebinck, also deals with the discursive realm of CBNRM, but focuses on how a specific (neoliberal) discourse is translated into specific state interventions. In a case study about a conservation initiative in Namibia the authors show the way in which a global conservation discourse subtly (and oftentimes indirectly) creates fluid, overlapping and constantly changing boundaries between communal institutions and their memberships. By unpacking the working of technologies of rule in conservancies and its concomitant making of subjects, the authors effectively problematise what constitutes a 'community' and go on to argue that the institutionalisation of conservancies forms the basis of new, parallel power relations in communal areas that compete with existing traditional hierarchies. This effectively curtails local capacity development through increased reliance on state regulatory agents and external expertise about tourism or human-wildlife conflict mitigation. Furthermore, this shift has resulted in an unfair distribution of benefits. Conservancy committees, for example, serve as a vehicle for the emergence of new local elites closely associated with the modern global commodity economy. In conclusion, the authors confirm statements that the present neoliberal CBNRM discourse is clouded by various romantic notions of 'community', which in turn leads to a range of paradoxes and ambiguities that have problematic, if not outright negative outcomes at community level. While decentralisation is the discourse, re-appropriation and recentralisation seems to be the practice.

In Chapter 8, Margaret Skutsch further disentangles the coils of the discourse/practice skein by addressing the implications of a global discourse on climate change for communities. As international REDD+ policy is being developed under the auspices of the United Nations Framework Convention on Climate Change (UNFCCC), many countries are proposing community forest management as a major element of their national REDD+ strategies. In the light of these developments, Skutsch first explores which elements of REDD+ can probably be tackled successfully by community forest management. She concludes that its major niche will be in promoting forest enhancement rather than in reducing emissions from deforestation, and that although it is likely also to reduce emissions from forest degradation, carbon credits for this will be difficult to claim as it is almost impossible to construct a reliable baseline for past forest degradation. She then goes on to consider the incentive systems being used to promote community forest management under REDD+ national plans. Although at the international level emission reductions achieved under REDD+ will be rewarded on an output basis (per tonne CO_2), it is questionable whether this same system is appropriate within national programmes. The author makes the case for input-based incentives, which will be easier to administer as they have much lower transaction costs. Moreover, she argues, an input-based incentive system will result in less conflict and higher participation. In other words, Skutsch makes the argument that even if a market system is necessary at international level to encourage investment in REDD+, at the local level more traditional stimuli will be more efficient.

In the final chapter of the CBNRM part (Chapter 9), Musalem and Verschoor turn more explicitly to the impasse between conservation and development goals. Advocacy and support for CBNRM initiatives, they write, may offer a possible solution to this impasse. This has resulted in a mushrooming of communal initiatives, yet the dynamics underlying these efforts remain a blind spot in the literature. The authors shed light on these dynamics through the description and analysis of the rise and demise of a Mexican community-based producer organisation involved in the commercialisation of *Chamaedorea* fronds used in flower arrangements in export markets. Following the organisational practices of key actors involved in the project, Musalem and Verschoor argue that the ultimate collapse of the initiative had less to do with inefficiencies in market strategies or lack of planning but, surprisingly, more to do with a struggle over control of local political and social networks.

1.4 Biocultural diversity

Over the past two decades, biocultural diversity research has emerged as a new field in the academic world. Interest in this topic grew through the 1990s, drawing from a variety of sources in the natural, social, behavioural and policy sciences, as well as in the humanities and applied sciences. It conveys the idea that many of the planet's areas of highest biological diversity are inhabited by indigenous and traditional people, providing an 'inextricable link' between biological and cultural diversity.

Chapters 10, 11 and 12 aim to gain attention and appreciation for the complexity and important role of humans and cultural values in forested landscapes. The point of departure of these chapters is that humans have maintained, enhanced and even created biodiversity through culturally diverse practices of management of 'wild' resources and the cultivation of domesticated species, leading to landscapes with a mosaic of vegetation types. Although the term biocultural diversity is increasingly used, there has been little critical reflection on what it precisely refers to. Often it is used with particular reference to 'indigenous traditional' people in rural areas in developing countries (Cocks, 2006). Chapters 10 and 11 challenge the idea that the concept is only relevant to poor indigenous communities in non-Western countries. In Chapter 10, Cock and Dold show that the amaXhosa rural (not always poor) people in South Africa are still intrinsically connected to their natural environment. Access to places and spaces within the natural environment provides a sense of wellbeing, a link to ancestral spirits, a location for religious rituals, plus a wealth of culturally-inspired uses of specific species and sacred places in the landscape. In their chapter, Cock and Dold call for institutional support for rural communities' attachment to local landscapes by incorporating local cultural values into the formal conservation process. In Chapter 11, Elands and Van Koppen challenge the idea that biocultural diversity is limited to non-Western society. They move away from the static picture of the ecologically noble savage and introduce the notion of 'biocultural creatives'. These are people or groups that, out of an engagement with society and nature, come up with new cultural models and practices of interaction with biodiversity. With this notion, the authors show that biocultural diversity values and practices are also present in Western society. In Chapter 12, Gerritsen uses the Sierra de Manantlán Biosphere Reserve in Western Mexico as a case study to provide a dynamic perspective of the concept. He concludes that a 'resource diversity' perspective might be more productive in gaining insights into the possible role of farmer participation in ecosystem management in these areas than the 'biocultural diversity' perspective.

In recognising the human and cultural values in biodiversity conservation, the question that then arises is how biocultural diversity can be incorporated into policy and management in order to protect biodiversity and secure livelihoods. In Chapter 13, Van Oosten and Hijweege discuss the governance of biocultural diversity in a mosaic landscape. They argue that current governance structures do not do justice to the biocultural diversity of most forested landscapes. Instead, they argue for a 'spatialisation' of governance, as a means to reconnect governance to landscape and build upon inhabitants' identities and sense of ownership to manage their biocultural diversity and govern their place. Spatial thinking and policy learning are crucial elements of constructing such landscape governance, which is strongly embedded in local practice and scaled up to higher politics of scale.

Chapter 14 reflects on this new scientific field by discussing the onto-politics of biocultural diversity. Van Bommel and Turnhout argue that classification systems do not merely mirror the world but also create a certain reality. By reflecting on the work of Freerk Wiersum on biocultural diversity, they conclude that his classifications create a world that consists of various natural, cultural and mixed categories which (1) privileges local/indigenous communities to manage the nature-culture mixtures, (2) creates a nature-culture continuum to allow for coordination across the nature-human divide, and (3) creates a network of scientists and practitioners from diverse disciplines that can arrive at a division of labour in research and management of the identified biological, human and cultural categories.

1.5 Forest governance

Within a short time, forest governance has become a very popular concept, both among scientists and practitioners, and both in the global North and the global South. In its broadest sense it refers to governing or steering society towards sustainable forest management by whatever institution, but the most common interpretation is of new modes of governing forest issues that go beyond traditional government: policy networks, certification schemes, social corporate responsibility, participatory forest management, markets for ecosystem services, public-private partnerships and the like. The field is therefore broad and complex. Arts and Visseren present a state of the art review of forest governance in Chapter 15.

Participatory forest management in particular has become a prominent approach in this field. The core idea is that local management of forests, or co-management of forests by state departments and local communities, is more cost-effective and efficient in conserving and sustainably using forest resources than the 'old' state forest institutions. This section contains two chapters with examples of participatory forest management from different countries: Tanzania (Chapter 16) and Bolivia and Ecuador (Chapter 17). The chapter by Katani and Babili shows that participatory forest management can be successful. In Tanzania, it has contributed to an improvement in forest condition in various areas, although the income premium for local communities has been less than expected. However, De Koning and Cleaver show that in Bolivia and Ecuador some of the participatory forest management initiatives have been more problematic. Many communities have not accepted the new forest rules because they did not feel that these took their interests into account, did not want to lose their autonomy, or were too poor (or their forests were too poor) to make participatory forest management a viable option.

However, forest governance is not only about involving communities, but also about involving markets. Most common in the forest field today are forest certification schemes and payments for environmental services. Forest certification entails a market-based mechanism that, through independent labelling and monitoring, guarantees to both consumers and producers that wooden products originate from sustainably managed forests. Examples are the Forest Stewardship Council (FSC) and the Programme for the Endorsement of Forest Certification (PEFC). Payments for environmental services, secondly, involve appointing and trading monetary credits for the provision of services such as water regulation through forest conservation or climate control through carbon sequestration. An example is REDD+. Whereas certification schemes have now become quite well established, payments for environmental services schemes are still in their infancy. Moreover, we are not sure whether such markets will work or not. There is still much uncertainty and lack of knowledge about how and whether markets can contribute to environmental public goods or not. Therefore, in Chapter 18, Hoogstra goes into the nature of forest markets by discussing various conceptualisations of markets, using forest certification schemes as an illustration.

Theoretically, the section on forest governance includes both mainstream and critical approaches to the topic. One mainstream approach is neo-institutionalism, which is founded mainly on the work of Elinor Ostrom. The main thesis is that conservation and sustainable use of forest resources can be achieved by designing robust institutions that regulate, among other things, access, use, management, benefit sharing, decision-making and conflict resolution at the local level. Chapter 16 on Tanzania is a good example of such an approach. However, Ostrom's neo-institutionalism has also been criticised. In Chapter 17, De Koning and Cleaver show the many unexpected ways people can respond to newly introduced institutions; they may recombine, alter, or even reject them. They call this process of reshaping institutions institutional bricolage (and their approach 'critical institutionalism'). In Chapter 15, though, more radical critics on forest governance are articulated as well: political ecology and governmentality. The former questions the new governance claims on livelihoods, empowerment and democracy; the latter considers governance as a continuation of governmental control 'at a distance'.

The mainstream approach to market governance is neoclassical economics, in which the market is considered to be an abstract allocation mechanism that guarantees efficiency. The question, though, is whether this theory holds for governing forests through markets. Hoogstra, in her contribution in Chapter 18, is sceptical about the neoclassical claims and turns to practice theory to analyse how markets actually work on the ground.

1.6 Conclusion

Reviewing the four major themes in research on the forest-people interface, the main conclusion that comes to the fore is that the naïve optimism associated with the introduction of new themes and regimes should be tempered. The papers show that in spite of expectations regarding the possibility of reconciling economic and political empowerment and environmental aims, local landscapes and communities are not as 'makeable' as was often assumed.

Part I on agroforestry and NTFPs makes that clear by introducing the productive bricolage concept as a hybrid concept that links people's livelihoods to landscape change. Although there is a link between local natural resource use and landscape change, it also becomes clear that livelihood decisions and landscape change at local level are embedded in developments that emerge at other scales of activity or decision-making.

Part II on community-based natural resource management shows that there is both an intimate connection and a serendipitous relationship between global concerns and local practices. When universal blueprints for the sustainable use and management of finite resources are passed on to the local level, one should be prepared for the unexpected: the creativity of local actors far outstrips the boundaries of what can be 'planned'.

The concept of biocultural diversity addressed in Part III stemmed from a romanticised idea about the link between indigenous cosmovisions and keeping biodiversity intact. The authors de-romanticise this notion by lifting the biodiversity concept beyond the rural development and indigenous management debate, and showing that it applies not only to the poor, not only to the global South, and not only to indigenous people.

Finally, Part IV introduces the concept of governmentality in order to enable a critical review of governance. It allows us to come to grips with shifts in governance, such as the shift from centralised state control to state control at a distance (through networks and markets) in which governance produces state-related disciplinary practices in new forms. However, local agencies are not 'docile bodies' that slavishly follow the new governance blueprints and rules. The chapters in Part IV therefore also emphasise other shifts: from institutional design to institutional bricolage and from governance models to governance practices. These shifts, again, illustrate the serendipity of local creativity regarding global and national blueprints and regimes.

References

Adams, W.M. (2008). Green development: environment and sustainability in a developing world. Routledge, London, UK, 449 pp.

Arnold, J.E.M. (1991). Community forestry. Ten years in review. Food and Agricultural Organization (FAO), Rome, Italy, 32 pp.

Bene, J.G., Beall, H.W. and Côté, A. (1977). Trees, food, and people: land management in the tropics. International Development Research Centre, Ottowa, Canada, 52 pp.

Chambers, R. (1993). Challenging the professions: frontiers for rural development. Intermediate Technology Publications, London, UK, 143 pp.

Cocks, M. (2006). Biocultural diversity: moving beyond the realm of 'indigenous' and 'local' people. Human Ecology 34(2): 185-200.

Colchester, M., Apte, T., Laforge, M., Mandondo, A. and Pathak, N. (2003). Bridging the gap: communities, forests and international networks. Synthesis report of the project 'Learning lessons from international community forestry networks'. CIFOR Occasional Paper No. 41, Centre for International Forestry Research, Bogor, Indonesia, 72 pp.

Food and Agriculture Organization (FAO) (1976). Forestry for local community development. FAO Forestry Paper No. 7. Food and Agricultural Organization (FAO), Rome, Italy, 114 pp.

Jeffries, M. (2005). Biodiversity and conservation, 2nd ed. Routledge, New York, NY, USA.

King, K.F.S. (1987). The history of agroforestry. In: Steppler, H.A. and Nair, P.K.R. (eds.) Agroforestry: a decade of development. International Council for Research in Agroforestry, Nairobi, Kenya, pp. 3-12.

Macqueen, D.J. and Bila, A. (2004). Gleanings on governance: learning from a two year process of forest policy support to ProAgri. IIED, London, UK, 31 pp.

Rhodes, R.A.W. (1996). The new governance. governing without government. Political Studies 44(4): 652-667.

Ros-Tonen, M.A.F and Kusters, K. (2011). Pro-poor governance of non-timber forest products: the need for secure tenure, the rule of law, market access and partnerships. In: Shackleton, S., Shackleton, C. and Shanley, P. (eds.) Non-timber forest products in the global context. Tropical Forest Series, Springer Publishers, Heidelberg, Germany, pp. 189-207.

Ros-Tonen, M.A.F., Zaal, F. and Dietz, T. (2005). Reconciling conservation goals and livelihood needs: new forest management perspectives in 21st century. In: Ros-Tonen, M.A.F. and Dietz. T. (eds.) African forests between nature and livelihood resources. Interdisciplinary studies in conservation and forest management. African Studies No. 81, The Edwin Mellen Press, Lewinston/Queenston/Lampeter, USA/UK, pp. 3-30.

Western, D. and Wright, M. (1994). The background to community-based conservation. In: Western, D., Wright, M. and Strum, S. (eds.) Natural connections. Perspectives in community-based conservation. Island Press, Washington, DC, USA, pp. 1-12.

Wiersum, K.F. (1997a). Indigenous exploitation and management of tropical forest resources: An evolutionary continuum in forest-people interactions. Agriculture, Ecosystems and Environment 63: 1-16.

Wiersum, K.F. (1997b). From natural forests to tree crops, co-domestication of forests and tree species, an overview. Netherlands Journal of Agricultural Science 45: 425-438.

Wiersum, K.F. (1999). Social forestry: changing perspectives in forestry science or practice? PhD Thesis, Wageningen Agricultural University, Wageningen, the Netherlands, 211 pp.

Wilshusen, P.R., Brechin, S.R., Fortwangler, C.L. and West, P.C. (2002). Reinventing a square wheel: critique of a resurgent 'protection paradigm' in international biodiversity conservation. Society and Natural Resources 15: 17-40.

Wilson, E., 2006. Nature revealed – selected writings 1949-2006. The Johns Hopkins University Press, Baltimore, MD, USA, 719 pp.

Part I. Non-timber forest products and agroforestry

2. Non-timber forest product extraction as a productive bricolage process

Mirjam Ros-Tonen
University of Amsterdam, Amsterdam Institute of Social Science Research (AISSR), Nieuwe Prinsengracht 130, 1018 VZ Amsterdam, the Netherlands; m.a.f.ros-tonen@uva.nl

Abstract

This chapter explores the usefulness of the 'productive bricolage' concept, coined by Croll and Parkin (1992) and further elaborated by Batterbury (2001), in understanding the role of non-timber forest products (NTFPs) in people's livelihoods and the forested landscape. I argue that NTFP extraction as part of a productive bricolage process – defined as 'the flexible and dynamic crafting together of various livelihood options and its associated impacts on the landscape' – holds limited potential for poverty alleviation as it is mostly a sign of economic precariousness. Regarding the impact of NTFP production on the landscape, I demonstrate that the productive bricolage concept is useful for reinterpreting Wiersum's writings on the evolutionary continuum of forest-people interactions and the co-domestication of forests and trees. However, a more encompassing approach is needed considering the decreasing autonomy of community forestry and the growing integration of NTFP production into commercial networks and multilevel governance regimes. I propose political ecology as the overall perspective to deal with such multiscalar influences.

Keywords: non-timber forest products, productive bricolage, political ecology, livelihood diversification, scale

2.1 Introduction

Since the end of the 1980s, it has become widely acknowledged that forests not only provide valuable timber, but also a large variety of non-timber forest products (NTFPs), such as food, fodder, medicines, construction materials and tools (for reviews, see Belcher *et al.*, 2005; Neumann and Hirsch, 2000; Ros-Tonen and Wiersum 2005). These products comprise plant and animal products for subsistence and trade from forested landscapes, including human-modified ones (Ros-Tonen, 2000; Ros-Tonen and Wiersum, 2005). Firewood is usually excluded from the definition of NTFPs as it does not fit the conservation paradigm very well, although several developing countries include fuelwood in the definition of NTFPs in national legislation (Ingram, pers. comm.; see Chapter 5 by Jolien Schure in this volume for a more extensive discussion on the inclusion of fuelwood in the definition of NTFPs). It has become increasingly clear that forests provide multiple livelihood assets, not only productive, but also religious, spiritual and cultural (Cocks and

Wiersum, 2003; Cocks *et al.*, 2011). It is now common wisdom that forest-based livelihood activities usually form part of complex livelihood strategies consisting of multiple components. Due to seasonality and the low densities in which non-timber forest products generally occur, forest-based livelihoods rarely involve specialised full-time activities, but are usually combined with other economic activities (Ros-Tonen and Wiersum, 2005; Wiersum, 1999).

In this chapter, I explore whether the 'productive bricolage' concept, coined by Croll and Parkin (1992) and further elaborated by Batterbury (2001), is useful for understanding the role of NTFPs in people's livelihoods and the forested landscape. Here this concept is understood as 'the flexible and dynamic crafting together of various livelihood options and its associated impacts on the landscape'. I argue that NTFP extraction as part of a productive bricolage process holds limited potential for poverty alleviation because productive bricolage, as Ellis (2000) has stated for livelihood diversification, is a sign of economic precariousness. Regarding the impact of NTFP production on the landscape, I will demonstrate that the productive bricolage concept is useful for reinterpreting Wiersum's writings on the evolutionary continuum of forest-people interactions (Wiersum, 1997a) and the co-domestication of forests and trees (Wiersum, 1997b), but that the decreasing autonomy of community forest management (Wiersum, 2009) and the growing integration of NTFP production into commercial networks and multilevel governance regimes (Ros-Tonen and Wiersum, 2005; Wiersum 2009) require a more encompassing approach. To take such multiscalar influences into account I propose political ecology – labelled 'environmental political geography' by Dietz (1999) – as the overall perspective. The idea of using the productive bricolage concept as part of a political ecological analysis of livelihood and landscape dynamics is not new and was proposed by Simon Batterbury in his article on livelihood diversification in southwestern Niger (Batterbury, 2001). However, as I will show, the concept has hardly been adopted in the NTFP literature.

First, I outline some general characteristics of NTFP production and its contribution to livelihoods, before introducing the concept of productive bricolage and how it relates to, but also differs from, the concept of livelihood diversification. I then elaborate on how cross-scale and multilevel interactions impact on people's livelihoods and consequently on the landscape. I conclude by integrating these ideas into an analytical framework for the study of NTFP production systems.

2.2 Non-timber forest products: general characteristics and their role in peoples' livelihoods

A large number of studies and reviews carried out during the past two decades (e.g. Belcher *et al.*, 2005; Kusters *et al.*, 2006; Neumann and Hirsch, 2000; Ros-Tonen and Wiersum 2005; Vedeld *et al.*, 2007) provide insight into how NTFPs are used worldwide (Table 2.1). The use patterns are remarkably similar across the world.

On all continents, people in remote areas use a considerable number of plant species for subsistence, with over five hundred different species being no exception. With over 50% of the species being used for medicinal purposes (Van Andel, 2000; Van Dijk, 1999), wild medicines rank first in subsistence use, but lower in terms of their share in cash income (6.9% in the study by Vedeld *et al.*, 2007). Animal species – fish and game – are important sources of protein, while hunting for and eating bushmeat are also culturally important activities (Davies and Brown, 2007; Demmer and Overman, 2001; Van Vliet *et al.*, 2011). Even when people migrate to cities, bushmeat continues to be a favourite food item, particularly in Africa. Comparative research indicates wild foods, fuelwood and fodder to be the main sources of income from forests (Vedeld *et al.*, 2007). Whether it is classified as an NTFP or not, fuelwood is the most important forest product worldwide According to a CIFOR study encompassing 8,000 households from 60 sites in 24 countries, fuelwood contributes 20% of forest income on average (press release, CIFOR Poverty and Environment Network (PEN), September 2011). The general tendency is that the poorest consume relatively more forest products than people who are

Table 2.1. Non-timber forest products and their uses (adapted from FAO, 1991, excluding fuelwood).

Products category	Examples
A. Plant products	
Food	Edible plants and plant parts (seeds, roots, tubers, stems, leaves, shoots, flowers, fruits, nuts) providing vegetables, snacks, beverages, edible fats and oils, spices, flavourings, etc.
Forage	Plants used as food for livestock and wildlife
Medicinal products	Medicinal herbs, plants, and plant parts (leaves, bark, etc.)
Construction materials	Bamboo, rattan, smallwood, fibbers, cork, leaves for roofing (roof thatch)
Utensils	Smallwood for handicrafts and tools, leaves for wrapping food, fibbers for basketry and cloths
Biochemicals	Inedible fats and oils, waxes, gums, latex, dyes, tannins, biochemicals for plastics and coatings, paints and varnish, toxins for hunting, hallucinogens
Aromatics	Essential oils for cosmetics, perfumes and incense
Ornamentals	Aesthetically pleasing plants, cut and dried flowers
B. Animal products	
Food	Meat and protein from mammals, birds, fishes, reptiles and insects, eggs, edible nests, honey
Forage	Fish oil, bones
Medicinal products	Pharmaceuticals extracted from mammals, fish and reptiles
Utensils	Horn, feathers, bones
Biochemicals	Wax, silk, propolis, guano, toxins
Ornamentals	Live animals and animal products like feathers, hides, skins, shells and horn

better off. These products can be considered 'gap fillers' that complement farm produce, both qualitatively (complementary food items) and quantitatively (a fall-back during the slack season) (Sunderlin *et al.*, 2005).

Gathering NTFPs tends to be replaced by other activities once alternative options become available (e.g. Dove, 1993; Godoy and Bawa, 1993). Exceptions to this are some specialised and culturally important activities, such as hunting, handicraft making and the extraction of products for stable external markets, including those for specialty food like edible birds' nests, certain ginseng roots and exclusive mushrooms (Kusters and Belcher, 2004) and products in high-value trade, such as Brazil nuts, damar resin and *Prunus africana* (Kusters *et al.*, 2006). This results in a diversified picture of the importance of NTFPs for people's livelihoods, with access to markets being the main determining factor. Belcher *et al.* (2005) grouped them as follows:

- *Subsistence.* Households extract low-value products like palm fibres, canes and grasses for tools and thatching as well as medicinal plants from forests and fallows, and process and use these mainly within the household for their own use. Income from NTFPs in this group is less than 50% of the household's total income (cash and subsistence), but it is their main or only source of cash income.
- *Supplementary*. Households extract fruits and medicinal plants from the wild for local processing, consumption or the regional market. Households in this group receive more than half of their income in cash, but less than 50% from NTFPs.
- *Integrated.* Households cultivate NTFPs or manage them intensively in a limited area and process them locally to sell on local and domestic markets. Examples of products in this group are bamboo, high-value wood carvings, fruits and resin. As is the case in the supplementary group, these households earn a large proportion of their income in cash from a combination of farming, off-farm activities and NTFP production, with the last providing less than 50% of their household income.
- *Specialised-natural.* Households exploit high-value food items and medicinal plants from the wild for national and international markets. This generates more than 50% of their income, which is at an intermediate level.
- *Specialised-cultivated.* Households cultivate and intensively manage specialty food products, resins and dyes for international markets, which provide them with incomes higher than the local average, with more than 50% derived from NTFPs.

The processing technologies employed in NTFP production are generally perceived to be simple. This is true for such products as Brazil nuts, basketry and rattan furniture, but Belcher and Schreckenberg (2007) note that some NTFPs used in medicines or vegetable oil require processing in very sophisticated laboratories.

A lively debate has evolved around the potential of NTFPs to contribute to conservation and development aims since Peters *et al.* (1989) and De Beer and McDermott (1989) put the term on the research agenda. The underlying reasoning of such a strategy is that local authorities and forest resource managers will have an interest in preventing indiscriminate forest use or the conversion of forest to other land uses when NTFP extraction contributes to the gross national product (GNP) and export earnings. As far as local communities are concerned, increased income from the trade in NTFPs was thought to provide a stimulus for protecting their forests and managing them sustainably. Since it was assumed that many NTFPs can be harvested without significantly altering the forest structure, their exploitation was expected to maintain the forest's environmental services and biological diversity (Peters *et al.*, 1989). All these factors have led to the notion that the commercial extraction of NTFPs would be a potentially sound conservation strategy, while also contributing to local development (Ros-Tonen, 2000; Ros-Tonen *et al.*, 1995). In Brazil, in particular, NTFPs like rubber and Brazil nuts also became a symbol of rubber tappers' struggle for the demarcation of extractive reserves to continue a sustainable forest-based way of life, protected from claims by cattle ranchers and other actors to forest land (Allegretti, 1990).

Three major changes can be observed in the NTFP debate since then. First, the discussion has extended beyond natural tropical rain forests to include products from semi-arid areas and human-modified vegetation types like forest gardens and agroforestry systems (Ros-Tonen and Wiersum, 2005; Wiersum and Shackleton, 2005). One of the reasons is that NTFP exploitation is economically more feasible in anthropogenic (human-made) forests where product density is higher (Wiersum, 1999), as a result of which livelihood outcomes of NTFP production in these systems are better (Kusters *et al.*, 2006). Second, recent literature refutes the claims that NTFP extraction is environmentally benign, as even local and regional trade can lead to overexploitation and extinction (Clark and Sunderland, 2004). Third, recent insights also make it clear that NTFPs are indeed an important safety net for poor forest-dwelling people, but have little potential to lift people out of poverty (Belcher *et al.*, 2005; Kusters *et al.*, 2006; Vedeld *et al.*, 2007). A lack of storage and processing facilities, bad roads and high transportation costs hinder the trading of NTFPs, especially in isolated areas. Consequently, perishable products in particular have to be consumed locally.

Due to the seasonality and unpredictability of production cycles, the low densities of products in the wild and their wide distribution mean volumes are low and supplies irregular. Markets are also fragmented because NTFPs provide input to a wide range of industries, including food and beverages, pharmaceuticals, cosmetics and botanical medicines. Producers often face the difficulty of meeting international quality standards prevalent on these markets, and these are also subject to trends and therefore changeable. All this, combined with a lack of organisation among harvesters, often makes NTFP producers dependent on intermediaries and subject to exploitative production and trading relations (Ros-Tonen, 2000). As a result, the contribution of commercial NTFPs to poverty alleviation is generally limited. Even the products in the 'specialised' group (e.g. Brazil nuts (*Bertolletia excelsa*) harvested from the wild in Bolivia, Brazil and Peru, or damar resin, harvested from complex agroforests in Indonesia), which generate incomes that are relatively higher than local averages, are still not high in absolute terms (Belcher *et al.*, 2005). NTFP extraction alone is rarely sufficient to build a secure livelihood. The productive bricolage concept that I introduce in the next section can help with understanding the role of NTFPs in people's efforts to maintain their livelihoods.

2.3 NTFP production as productive bricolage

The term 'bricolage' is primarily known in studies on theories of natural resource governance thanks to Frances Cleaver (2002), who proposed the concept to understand the dynamic, ad hoc and flexible nature of building and reconstructing institutions. Drawing on Douglas (1973), who extended Levi-Stauss' concept of 'intellectual bricolage' to institutional thinking, and Ostrom's (1992) notion of 'institutional design and crafting' as a deliberate process of selecting institutional mechanisms for a particular purpose, Cleaver (2002: 15) perceives bricolage as an evolutionary process in which institutions of cooperation are embedded in everyday relations, networks of reciprocity and the negotiation of cultural norms. Unlike Ostrom, she does not perceive this process as one of 'conscious selection of mechanisms fit for the collective action task', but as 'a messier process of piecing together shaped by individuals acting within the bounds of circumstantial constraint'. De Koning (2011) applied this concept to NTFP contexts but, like Cleaver, also with a focus on institutions. To my knowledge, the concept has not been applied in the NTFP literature to study the interplay between livelihoods, natural resource use and forested landscapes in an integrated way as proposed by Batterbury (2001).

The bricolage concept first appeared in association with livelihoods in the work of Croll and Parkin (cited in Batterbury, 2001), who coined the concept of 'productive bricolage' to refer to 'tasks over which agents see themselves as having some control, as distinct from work controlled by others outside the home' (*ibid.*: 12). One of the first authors (and as far as I was able to identify, the only one) who employed this concept in relation to NTFPs was Clare Madge (1994) in her study of collected food and domestic knowledge in The Gambia. The publication went largely unnoticed in the NTFP literature, as she does not use the term NTFP and theoretically linked her study to indigenous knowledge and gender studies of the 1980s and early 1990s. Madge showed that rural households in The Gambia 'follow a seasonal and adaptable survival plan' in which 'various activities are combined to produce an integrated, flexible strategy' (Madge, 1994: 282). She showed that the use of collected food items had increased to cope with reduced rainfall and associated crop failures, and that some collected food items tended to replace cultivated ones in times of labour shortage because they required less cooking time. Women who had experienced severe drought periods had learned to innovate with collected food items, which in that context was therefore referred to as 'famine food' (*ibid.*: 287).

Similarly, Batterbury (2001) draws on the productive bricolage concept to analyse how rural people in Niger have constructed livelihoods in a 'fluid, non-linear, and dynamic process' (*ibid.*: 437) in response to local constraints and opportunities, and income-generating possibilities elsewhere. Scoones (2009), although not elaborating on it to a great extent, uses the bricolage concept as a synonym for 'portfolio' to understand the dynamics of economic diversification from local perspectives.

Based on these definitions, I define productive bricolage as 'the flexible and dynamic crafting together of various livelihood options and its associated impacts on the landscape'. This definition is narrowly related to, but also different from, the livelihood diversification concept put forward by Ellis (1998, 2000), which is much more common in the NTFP literature. Ellis (1998: 4) defined livelihood diversification as 'The process by which rural families construct a diverse portfolio of activities and social support capabilities in their struggle for survival and in order to improve their standard of living'. This comprises cash and non-cash income as well as the social institutions (kin, family, compound, village, etc.), gender relations and property rights required to build and sustain a livelihood. Access to social services such as healthcare, water supply and education is also part of this process. Stressing that livelihood diversification is a sign of the 'precariousness' of rural livelihoods, Ellis (2000) elaborates on six factors that help to explain households' engagement in multiple livelihoods. These are seasonality, risk mitigation, labour markets, thin credit markets, asset strategies (investments in future livelihood security) and coping behaviour. The last is related to 'diversification by necessity', whereas responding to labour market opportunities by migration and investing in assets for the future (e.g. social networks or trees that can be harvested in the future) are examples of 'diversification by choice'. Overall, livelihood diversification is directed towards reducing income variability within and between years, to prevent overall income failure and improve livelihood security.

This conceptualisation of livelihood diversification is frequently echoed in the NTFP literature in notions like 'safety net' and 'gap filling', which indicate the role of NTFPs in coping strategies (mainly in the subsistence group distinguished above). Also, the livelihoods in 'supplementary' and 'diversified' groups reflect the strategies adopted by families that do not 'put all their eggs into one basket' (Ellis, 2000: 294). In the words of Belcher *et al.* (2005: 1444-1445):

> NTFPs must be considered in terms of systems – people use them in various ways, but most often in combination with other economic activities ... Using various NTFPs in an economic portfolio ... allows households to spread risk and to modulate the timing of income ... Diversification also allows households to balance seasonal labor requirements.

Why then use the 'productive bricolage' concept to analyse the role of NTFPs in livelihoods? In my view, the attractiveness of the productive bricolage concept lies in its consideration of the interaction between livelihood activities and changes in the landscape. This is an important issue in NTFP studies, which are still strongly driven by the wish to combine development and conservation goals. Ellis (1998) acknowledges the importance of the physical environment in determining the options for diversification, but mainly in terms of constraints (e.g. access to land or availability of resources). Batterbury (2001), in contrast, argues that livelihood activities themselves are important drivers of change in the landscape. In other words:

> The point here is that the intentional mix, or bricolage of agriculture, migration, livestock, and market trading are imprinted in the local landscape, and particularly in land cover (Batterbury, 2001: 452).

Erosion caused by the intensification of farming activities is one example, but also the extension of woodland cover and creation of peri-village forest patches in Western Africa (Fairhead and Leach, 1996, 1998). Productive bricolage thus plays a major role in the forms and dynamics of landscapes. According to Batterbury, landscapes can therefore not be understood without an analysis of livelihood dynamics and decisions about everyday resource use at the local level.

This is also acknowledged by Giannecchini *et al.* (2007), who analysed historical patterns in land-cover change in three villages in a former Bantustan region in South Africa between 1974 and 1997. In general, land-cover change in the region was the combined outcome of socioeconomic dynamics (e.g. the weakening control by traditional authorities over natural resource use since the dismantling of Apartheid, the influx of migrants and resulting expansion of cropland, and increasing exploitation of natural resources such as fuelwood by outsiders) and environmental dynamics (such as rainfall fluctuations and periodic droughts). Interestingly, despite their proximity and comparable biophysical settings, and the similar national and international contexts, the three villages showed dissimilar trends in land-cover change due to site-specific and intra-household differences in livelihood diversification. Productive bricolage thus results in heterogeneous landscape mosaics (*ibid.*: 27). Examples are the depletion of woodlands and certain shrubs and trees around the village as a result of a growing number of households engaging in the trade in fuelwood (Batterbury, 2001), and increasing soil erosion in plots further away from the village when temporary labour migration forces households to allocate scarce labour and organic fertilisation primarily to the plots located closest to the village (Giannecchini *et al.*, 2007).

Madge (1994) is less explicit about the interaction between bricolage and the landscape as her focus is more on women's knowledge of food collection, storage and preservation techniques. Yet she points to the fact that the vegetation types involved in food collection (e.g. mangrove swamps, uncultivated swamps, fallow areas, closed woodland and public spaces in villages), as well as rivers and the sea, form a 'bricolage' in themselves, with some 'wild' food items being 'collected from cultivated fields, while crops are often grown in an agroforestry complex' (*ibid.*: 292). She thus challenges the dichotomy between 'wild' and 'cultivated' food items, as Wiersum did when he coined the concepts of 'evolutionary continuum in forest-people interactions' (Wiersum, 1997a), the 'co-domestication of forests and trees' (Wiersum, 1997b) and 'intermediate land-use systems and multifunctional landscapes' (Wiersum, 2004).

Although Wiersum does not use the productive bricolage concept, his work can be used to shed further light on how the NTFP component in productive bricolage processes impacts on the landscape. In his article on indigenous exploitation and management, Wiersum (1997a) stressed that people living in and near forests are active forest managers who purposively modify the landscape to safeguard the availability of valuable forest resources. He says of this process (Wiersum, 1997a: 3):

> The protection and purposeful regeneration of useful species does not necessarily involve the transformation into an agricultural landscape, but may result ... in a mosaic of managed forests and agricultural systems.

Based on this statement, he presents a continuum adapted from Harris (1989) that goes from uncontrolled procurement of wild tree products via protection and controlled utilisation to purposeful regeneration of wild trees and, ultimately, the cultivation of domesticated trees in crop plantations. This is reflected in the landscape as a 'gradual transformation of the natural forest into an agro-ecosystem' (*ibid.*: 11) – not as a sequential and unidirectional process, but one in which different phases can coexist. The domestication of trees can also occur in the forest by changing a 'wild' forest into a managed one in which the growth and productivity of preferred trees are enhanced (Wiersum, 1997b). The resulting modalities include: (1) protected natural forests (sacred groves, watershed protection forests, clan/village forests); (2) resource-enriched native forests (including enriched fallows); (3) reconstructed forests (forest gardens, planted temple forests, fortification fallows); and (4) mixed arboriculture (homegardens and mixed smallholder plantations). The place of NTFPs in these forest/tree management systems is illustrated in Table 2.2.

All the NTFP production systems listed in Table 2.2 are part of productive bricolage processes that result in diversified livelihoods in which people combine the collection of NTFPs with the cultivation of food crops, livestock rearing and/or other economic activities. The classification also illustrates how the NTFP component of productive bricolage processes may impact on the landscape. These characteristics have been integrated in Figure 2.1, which shows that NTFP extraction and/or cultivation is an integral part of an agroecological system that is based on the crafting together of different economic activities, including extraction, farming and livestock rearing. In addition, the livelihood system includes cash

Table 2.2. NTFP production systems as reflected in the landscape (Wiersum, 1999).

NTFP production system	Forest type	Example
A. (Modified) forests with predominantly tolerant[1] forest management practices		
1. Gathering of non-timber products in natural forests in which NTFPs are protected	Specific areas or specific tree species in natural forests that are favoured and protected because of their value for providing useful materials	Individually claimed trees
2. Resource-enriched natural forests	Natural forests, either old-growth or fallow vegetation, whose composition has been altered by selective protection and incidental or purposeful propagule dispersion of food and/or commercial species	Enriched natural forests; enriched fallows
B. Transformed forests with predominantly intrusive[1] forest management practices		
1. Reconstructed natural forests	Cultivated and semi-cultivated forest stands with several planted useful species, tolerated or encouraged wild species of lesser value, and non-tree plants (herbs, lianas) composed mainly of wild species	Forest gardens
2. Mixed arboriculture	Cultivated mixed stands, almost exclusively of planted, and often domesticated, tree species	Homegardens; smallholder plantations.
3. Interstitial trees on croplands	Either naturally regenerated or protected trees, or planted and sometimes domesticated trees scattered over agricultural fields	Scattered fruit trees cultivation on/ along crop fields
4. Commercial plantation s with associated agroforestry practices	Plantations of domesticated tree crops which are (temporarily) intercropped with food plants or grazed by livestock	(Mixed) tree crop plantations

[1] The distinction between forest-tolerant management practices (by which the native vegetation is largely conserved) and forest-intrusive management practices (by which the native vegetation is replaced by (mixed) tree plantations) comes from Anderson (1990).

income largely acquired outside the agroecological system, through off-farm and non-farm income (locally or through migration), social provisions and remittances, and – for those who manage to invest in land and physical capital – rent. The NTFP production system manifests itself in the landscape as a bricolage process as well, with NTFPs being collected from various vegetation systems as distinguished by Wiersum (1997a, 1999), ranging from natural forest via modified forests to domesticated NTFPs in agroforestry and farming systems (see also Michon, 2007 for a rich collection of illustrative cases in Southeast Asia). Similar evolutionary domestication processes have been described for animals (e.g. Peterson *et al.*, 2005; Posey, 1985). Here too, the domestication of species occurs as a gradual process

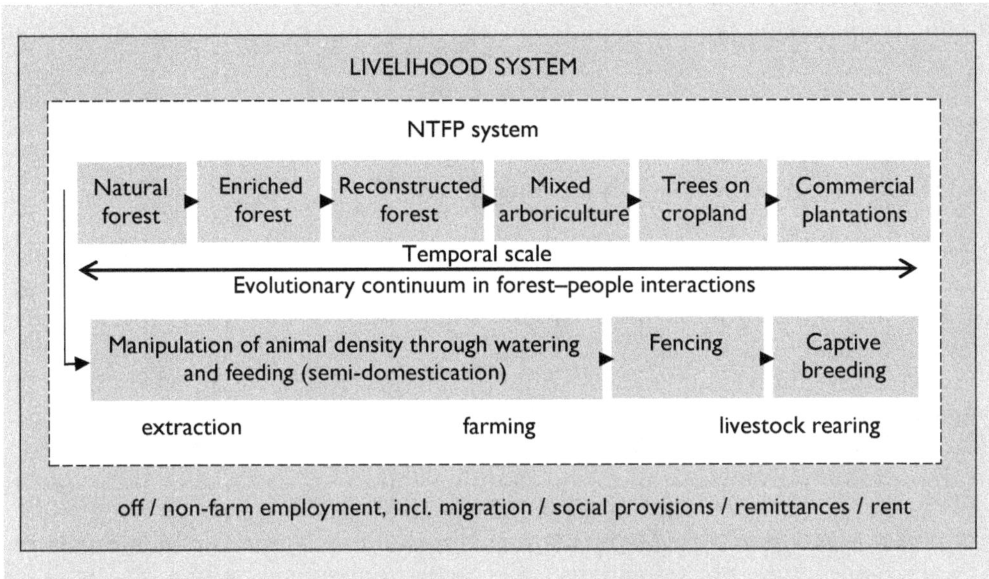

Figure 2.1. The NTFP system as part of a productive bricolage process.

that starts with manipulating wildlife and fish densities through watering (in the case of wildlife) and feeding (directly and/or by planting species that attract game and fish) and evolves via fencing to captive breeding of livestock and smallstock, and aquaculture.

Despite the similarities between Table 2.2 and Figure 2.1, Wiersum's approach differs from the productive bricolage literature in stressing human creativity in the reconstruction of landscapes, rather than the need to adapt to the dynamics of climate change, deforestation and/or changes in the socioeconomic and political sphere. Wiersum's analysis thus corresponds with what Ellis (1998) labels 'diversification by choice', whereas productive bricolage is the combined effect of 'diversification by choice' and 'diversification by necessity'. Therefore, a more encompassing approach is needed for a better understanding of the drivers behind decisions about livelihood strategies and natural resource use, also considering the growing complexity of the entrenchment of natural resource management in multiscalar and multilevel frameworks. This issue will be addressed in the next section.

2.4 The need to consider levels of scale: adding a political ecological perspective

Considering global environmental change and the increasing integration of local resource use systems into broader economic and governance networks (Ros-Tonen and Wiersum, 2005; Wiersum, 2009), a local-level analysis of resource use and productive bricolage fails to adequately explain landscape dynamics. Multiscalar influences of different kinds (e.g. geographical, temporal and/or institutional) should be taken into account. Batterbury (2001) acknowledges this and proposes such an analysis from a political ecological perspective as a bridging framework to explain landscape features as the result of various processes operating at different scales and with different underlying forces.

Political ecology became popular, particularly among human geographers, as an approach to analysing the politics of human interaction with the environment. Inspired by neo-Marxism, the early political ecological literature (Blakie, 1985; Watts, 1983) focused mainly on uneven access to natural resources to explain the socioeconomic vulnerability of rural people and the resulting environmental impacts. But already in this early phase of the discipline, explanations for land degradation and human adaptation to changing environments were sought in the combined influence of local features of the biophysical and institutional environment, and ecological and political processes at higher scales (cf. Watts, 2000). Blakie and Brookfield (1987: 17), for example, emphasised that the complexity of human-environment interactions can only be understood by investigating 'the contribution of different geographical scales and hierarchies of socioeconomic organisations (e.g. person, household, village, region, state, world)'. This statement was further elaborated in Blakie's (1995) 'chain of explanation', which explains land degradation by combining local physical changes in soil and vegetation and associated local economic symptoms and land-use practices with the resources, skills, assets, time horizons and technologies of land users, the (institutional) nature of agrarian society, the state and, ultimately, the international economy. Neumann (2009) notes that at that time scale was seen in terms of both spatial extent ('geographical scale') and organisational levels ('hierarchy of socioeconomic organisations'). According to this distinction, examples of geographical scale are individual farm plots, villages, regions, countries and the world, while households, village councils, national governments and multilateral organisations are examples of organisational levels. The more recent literature follows Cash *et al.* (2006), who define scale as an analytical dimension (e.g. geographical, temporal or institutional scales) and levels as the unit of analysis within these scales. In this view, farm plots, villages, etc. are levels of geographical scale and households, village councils, etc. are levels within the jurisdictional scale (Figure 2.2).

Political ecology, as it developed in the 1990s, is also about power imbalances and social action, gender's place in the political ecological landscape, the role of the politics of knowledge and discourses in natural resource management, and institutional politics (see Peet and Watts, 1996 for an overview of the prevailing themes). Since the turn of the century, the political ecological literature has been more concerned with cross-scale interactions (e.g. Adger *et al.*, 2006; Neumann, 2009) and resilience and adaptation to global change (e.g. Adger, 2000; Batterbury and Mortimore, 2011). Figure 2.2 takes these themes into account and situates the NTFP production system in an overall livelihood strategy that is influenced by ecological,

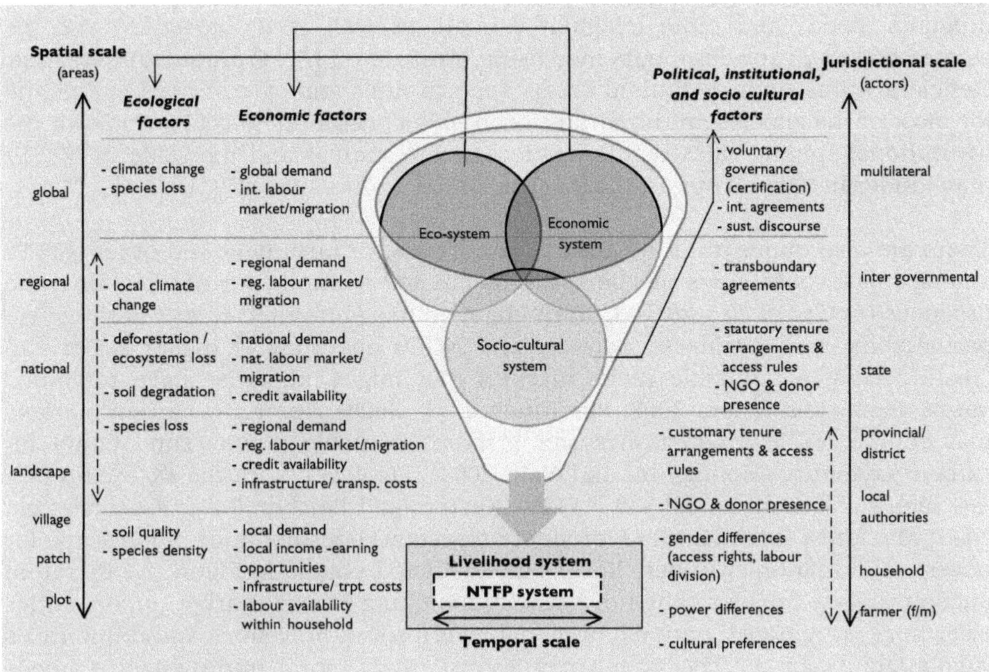

Figure 2.2. A political ecological approach to understanding productive bricolage in NTFP systems.

economic, sociocultural and political factors at different scales. The components that form part of the overall bricolage process depend on ecological, economic and political, institutional and sociocultural factors at different scales. The scales included in the figure are the most commonly researched: spatial, jurisdictional (here also including institutional aspects) and temporal. This is not to deny the importance of other scales identified in Cash *et al.* (2006), such as management, networks and knowledge. Interactions between levels of knowledge (with local/ traditional knowledge at one end of the scale and knowledge of international research institutions at the other, and bridging organisations mediating between them) may result in joint learning and problem solving (Berkes, 2009; see also Buizer *et al.*, 2011). Networks (producer associations at the one end and for instance global NGO networks at the other) also have a significant impact on decisions regarding NTFP use (Ros-Tonen *et al.*, 2008). Some of these elements have been integrated into the figure, such as the presence of donors and NGOs promoting NTFP trade or engagement in global voluntary governance mechanisms, such as certification. Similarly, discourse use, which is usually part of the knowledge scale, is included here as the impact of sustainability and empowerment discourses are important drivers promoting NTFP trade (see e.g. Banjade and Paudel, 2008).

The different factors included in the scheme pose constraints as well as opportunities. Global environmental change – climate change, with associated droughts, floods and more irregular rainfall, as well as deforestation and the associated loss of species – puts increasing pressure on NTFP production systems. Deficient infrastructure, limited credit opportunities and lack of market demand for most of the species are important economic constraints to NTFP trade. In the institutional sphere, rules regulating access to resources and the trade in NTFPs may hinder livelihood opportunities (Ros-Tonen and Kusters, 2011).

There are also opportunities. Cultural preference for bushmeat and other NTFPs among migrants to cities or overseas has created a significant market for these products (Cocks *et al.*, 2011; Cunningham, 2011; Van Vliet *et al.*, 2011). Also, partnerships with businesses, donors and NGOs operating at higher scales can create access to profitable niche markets that might otherwise have remained out of reach (Ros-Tonen, 2007; Ros-Tonen *et al.*, 2008). Emerging carbon markets may render NTFP production systems more profitable if they can qualify for carbon payments (Montagnini and Nair, 2004). However, with the exception of a few high-value commercialised NTFPs (Kusters and Belcher, 2004; Kusters *et al.*, 2006), the forest (and the NTFP production system) does not offer much scope for poverty elimination (Wunder, 2001). The livelihood system in Figure 2.1 therefore encompasses a broader range of options, including labour market opportunities elsewhere (at regional, national or global level), social provisions and remittances from migrants abroad. The latter can be a substantial factor in improving livelihoods in rural communities (see e.g. Kabki *et al.*, 2004).

The results in terms of landscape change are as complicated as the productive bricolage process itself. Gray (1999) illustrated this nicely for southwestern Burkina Faso, as Batterbury (2001) did for Niger and Woldeamanuel (Chapter 3 in this volume) for the lowlands in Ethiopia. Gray's case is exemplary for the importance of scale. Seen from the regional level, aerial photographs revealed land degradation in the form of increased deforestation, expansion of farmland and increases in areas characterised as degraded. This was due to technological change (animal traction, cotton production) and the influx of migrants. However at the micro level, this had not resulted in a decline of soil fertility, as farmers responded to increasing land pressure with erosion prevention, organic fertilisation, nurturing trees and other forms of intensification. This reminds us of the famous 'Machakos miracle' in Kenya described by Tiffen *et al.* (1994): despite a population growth from 240,000 in 1930 to 1.4 million in 1990, Machakos district experienced a threefold increase in agricultural output per capita in the same period thanks to farmers' investments in terracing and soil conservation. Although they did not use the term, the work of Burgers and Zaal (2009) illustrates that such investments in terracing and soil conservation (which impact on the landscape) are related to productive bricolage and the complex of scale factors brought together in Figure 2.2. They did so by showing the importance to farmers' livelihood portfolios of (1) location (to nearby labour markets, allowing farmers to earn non-farm income on a seasonal basis, which they can then use for investments in their farm plots), (2) infrastructure (a road that gives access to markets), and (3) developments on the world market (e.g.

favourable coffee prices). These examples illustrate once again the importance of considering multiple scales in the analysis of livelihood dynamics and their impacts on the landscape.

Where NTFP production plays an important role in the productive bricolage process (as among the supplementary and integrated households distinguished by Belcher *et al.* (2005)), Wiersum's evolutionary continuum in forest-people interactions and co-domestication of forest and trees may apply and result in corresponding changes in the landscape. In such cases, vegetation change is to a large extent directly related to NTFP production. This also applies to households that specialise in profitable NTFPs, such as damar farmers in Sumatra, Indonesia, who establish and maintain biodiverse agroforests (Kusters, 2009).

In most cases, however, people's decisions regarding resource and land use are influenced by a more complex set of factors than their desire for NTFPs. The very same productive bricolage process can also undermine the NTFP system. Several studies demonstrate that increased income, from both NTFP trading and non-farm activities, is often invested in the expansion of farming land (Demmer and Overman, 2001; Escobal and Aldana, 2003), and that (with the exception of some culturally important species) overall NTFP consumption tends to decline with increasing incomes (Dove, 1993; Godoy and Bawa, 1993). This suggests that what is an opportunity in terms of livelihoods may lead to a decline in the NTFP system, which then gives way to farmland and pasture, with corresponding impacts on the landscape.

2.5 Conclusion

In this chapter I have explored the relevance of the productive bricolage concept for NTFP research. This concept is more encompassing than livelihood diversification, and offers an approach by which to examine livelihoods, everyday decisions on natural resource use and landscape change in a holistic way. This fits nicely into Wiersum's conceptualisation of human-environment interactions, such as the co-domestication of forests and trees. Studying NTFP production systems as part of productive bricolage processes helps us to understand the role of NTFPs in people's struggles to build their livelihoods and how they modify the landscape to that end. Whereas Wiersum stresses the human creativity in this process, reliance on NTFPs in a process of productive bricolage is mostly a sign of precarious livelihoods in which livelihood dynamics are not exclusively driven by choice, but also by necessity. In most of the cases it holds little promise for lifting people out of poverty, as several authors have claimed before (Belcher *et al.*, 2005; Sunderlin *et al.*, 2005).

However, the productive bricolage concept in itself does not provide a sufficient basis to fully understand the dynamics of livelihood portfolios and land-cover change. Increasing integration of local natural resource management in multilevel and cross-scale networks requires a broader approach. Political ecology is such an approach, with a rich tradition in studying the interaction between ecological, economic, institutional, political and sociocultural phenomena at various scales. Such a perspective enables us to gain a deeper insight into the drivers of land-cover change and the constraints and opportunities that determine people's choices regarding land and natural resource use. In combination, productive bricolage and political ecology provide the ingredients for a more holistic analysis of the NTFP production system as a component of both livelihoods and landscapes.

Acknowledgements

I would like to thank my colleagues at the Natural Resources and Wellbeing subgroup within the Governance and Inclusive Development (GID) programme group at AISSR for their useful comments on an earlier version of this paper.

References

Adger, W.N. (2000). Social and ecological resilience: are they related? Progress in Human Geography 24(3): 347-364.

Adger, W.N., Brown, K. and Tompkins., E.L. (2006). The political economy of cross-scale networks in resource co-management. Ecology and Society 10(2): 9. Available at: http://www.ecologyandsociety.org/vol10/iss2/art9/.

Allegretti, M.H. (1990). Extractive reserves: an alternative for reconciling development and environmental conservation in Amazonia. In: Anderson, A.B. (ed.) Alternatives to deforestation: steps towards sustainable use of the Amazon rain forest. Columbia University Press, New York, USA, pp. 252-264.

Anderson, A.B. (1990). Extraction and forest management by rural inhabitants in the Amazon estuary. In: Anderson, A.B. (ed.) Alternatives to deforestation: steps towards sustainable use of the Amazon rain forest. Columbia University Press, New York, USA, pp. 65-85.

Banjade, M.R. and Paudel, N.S. (2008). Economic potential of NTFPs in Nepal: myth or reality? Journal of Forest and Livelihood 7(1): 36-48.

Batterbury, S. (2001). Landscapes of diversity: a local political ecology of livelihood diversification in South-Western Niger. Cultural Geographies 8(4): 437-464.

Batterbury, S.P.J. and Mortimore, M.J. (2011). Adapting to drought in the West African Sahel. In: Palutikof, J. and Karoly, D. (eds.) Natural disasters and climate change adaptation. Cambridge University Press, Cambridge, UK, pp. 134-143.

Belcher, B. and Schreckenberg, K. (2007). Commercialisation of non-timber forest products: a reality check. Development Policy Review 25(3): 355-377.

Belcher, B., Ruiz-Pérez, M. and Achdiawan, R. (2005). Global patterns and trends in the use and management of commercial NTFPs: implications for livelihoods and conservation. World Development 33(9): 1435-1452.

Berkes, F. (2009). Evolution of co-management: role of knowledge generation, bridging organizations and social learning. Journal of Environmental Management 90: 1692-1702.
Blakie, P. (1985). The political economy of soil erosion. Wiley, New York, NY, USA.
Blakie, P. (1995). Understanding environmental issues. In: Morse, S. and Stocking, M. (eds.) People and the environment. UCL Press, London, UK, pp. 1-30.
Blakie, P. and Brookfield, H. (1987). Land degradation and society. Methuen, New York, NY, USA, 296 pp.
Buizer, M., Arts, B. and Kok, K. (2011). Governance, scale and the environment: the importance of recognizing knowledge claims in transdisciplinary arenas. Ecology and Society (in press). Available at: http://www.ecologyandsociety.org/vol16/iss1/art21/ES-2010-3908.pdf.
Burgers, C.P.J. and Zaal, F. (eds.) (2009). Sustainable land management in the tropics: explaining the miracle. International Land Management Series. Ashgate, UK, 217 pp.
Cash, D.W., Adger, W.N., Berkes, F., Garden, P., Lebel, L., Olsson, P., Pritchard, L. and Young, O. (2006). Scale and cross-scale dynamics: governance and information in a multilevel world. Ecology and Society 11(2): 8. Available at: http://www.ecologyandsociety.org/vol11/iss2/art8/.
Clark, L.E. and Sunderland, T.C.H. (eds.) (2004). The key non-timber forest products of Central Africa: state of the knowledge. Technical Paper No. 122, US Agency for International Development, Washington, DC, USA, 186 pp.
Cleaver, F. (2002). Reinventing institutions: bricolage and the social embeddedness of natural resource management. The European Journal of Development Research 14(2): 11-30.
Cocks, M., Lópes, C. and Dold, T. (2011). Cultural importance of non-timber forest products: Opportunities they pose for bio-cultural diversity in dynamic societies. In: Shackleton, S., Shackleton, C. and Shanley, P. (eds.) Non-timber forest products in the global context. Tropical Forest Series. Springer-Verlag, Heidelberg, Germany, pp. 107-128.
Cocks, M.L. and Wiersum, K.F. (2003). The significance of biodiversity to rural households in Eastern Cape Province of South Africa. Forests, Trees and Livelihoods 13: 39-58.
Croll, E. and Parkin, D. (1992). Cultural understandings of the environment'. In: Croll, E. and Parkin, D. (eds.) Bush base, forest farm: culture, environment and development. Routledge, London, UK, pp. 11-36.
Cunningham, A.B. (2011). Non-timber forest products and markets: lessons for export-oriented enterprise development from Africa. In: Shackleton, S., Shackleton, C. and Shanley, P. (eds.) Non-timber forest products in the global context. Tropical Forest Series. Springer-Verlag, Heidelberg, Germany, pp. 83-106.
Davies, G. and Brown, D. (2007). Bushmeat and livelihoods: wildlife management and poverty reduction. Blackwell Publishing, London, UK.
De Beer, J.H. and McDermott, M. (1989). The economic value of non-timber forest products in Southeast Asia with emphasis on Indonesia, Malaysia and Thailand. IUCN Amsterdam, the Netherlands, 197 pp.
De Koning, J. (2011). Reshaping institutions. Bricolage processes in smallholder forestry in the Amazon. PhD thesis Wageningen University, Wageningen, the Netherlands, 268 pp.
Demmer, J. and Overman, H. (2001). Indigenous people conserving the rain forest? The effect of wealth and markets on the economic behaviour of Tawahka Amerindians in Honduras. Tropenbos Series 19. Tropenbos International, Wageningen, the Netherlands, 382 pp.
Dietz, T. (1999). Political environmental geography of the tropics. Development 42(2): 13-19.
Douglas, M. (1973). Rules and meanings. Penguin, Harmondsworth, UK, 319 pp.

Dove M.R. (1993). A revisionist view of tropical deforestation and development. Environmental Conservation 20(1): 17-24/56.

Ellis, F. (1998). Household strategies and rural livelihood diversification, Journal of Development Studies 35(1): 1-38.

Ellis, F. (2000). The determinants of rural livelihood diversification in developing countries. Journal of Agricultural Economics 51(2): 289-302.

Escobal, J. and Aldana, U. (2003). Are nontimber forest products the antidote to rainforest degradation? Brazil nut extraction in Madre De Dios, Peru. World Development 31: 1873-1887.

Fairhead, J. and Leach, M. (1996). Misreading the African landscape: society and ecology in a forest-savannah mosaic. Cambridge University Press, Cambridge, UK, 354 pp.

Fairhead, J. and Leach, M. (1998). Reframing deforestation: global analyses and local realities – studies in West Africa. Routledge, London, UK, 238 pp.

Food and Agriculture Organization of the United Nations (FAO) (1991). Non-wood forest products: The way ahead (FAO Forestry Paper 48). FAO, Rome, Italy, 37 pp.

Giannecchini, M., Twine, W. and Vogel, C. (2007). Land-cover change and human-environment interactions in a rural cultural landscape in South Africa. Geographical Journal 173(1): 26-42.

Godoy, R. and Bawa, K.S. (1993). The economic value and sustainable harvest of plants and animals from the tropical forest: assumptions, hypotheses and methods. Economic Botany 47(3): 215-219.

Gray, L.C. (1999). Is land being degraded? A multi-scale investigation of landscape change in southwestern Burkina Faso. Land Degradation and Development 10(4): 329-343.

Harris, D.R. (1989). An evolutionary continuum of people-plant interaction. In: Harris, D.R. and Hillman, G.C. (eds.) Foraging and farming. The evolution of plant exploitation. Unwin Hyman, London, UK, pp. 11-26.

Kabki, M., Mazzucato, V. and Appiah, E. (2004). 'Wo benane a eye bebree': the economic impact of remittances of Netherlands-based Ghanaian migrants on rural Ashanti. Population, Space and Place 10: 85-97.

Kusters, K. (2009). Non-timber forest product trade. A trade-off between conservation and development. Assessing the outcomes of non-timber forest product trade on livelihoods and the environment, with special emphasis on the damar agroforests in Sumatra, Indonesia. Rozenberg Publishers, Amsterdam, the Netherlands, 240 pp.

Kusters, K., Achdiawan, R., Belcher, B. and Ruíz Pérez, M. (2006). Balancing development and conservation? An assessment of livelihood and environmental outcomes of nontimber forest product trade in Asia, Africa, and Latin America. Ecology and Society 11(2): 20. Available at: http://www.ecologyandsociety.org/vol11/iss2/art20/.

Kusters, K. and Belcher, B. (eds.) (2004). Forest products, livelihoods and conservation. Case studies of non-timber forest product systems. Volume 1 – Asia. CIFOR, Bogor, Indonesia, 365 pp.

Madge, C. (1994). Collected food and domestic knowledge in The Gambia, West Africa. The Geographical Journal 160(3): 280-294.

Michon, G. (2007). Domesticating forests. How farmers manage forest resources. CIFOR & ICRAF, Bogor, Indonesia, 187 pp.

Montagnini, F. and Nair, P.K.R. (2004). Carbon sequestration: an undervalued environmental benefit of agroforestry systems. Agroforestry Systems 61: 281-295.

Neumann, R.P. (2009). Political ecology: theorizing scale. Progress in Human Geography 33(3): 398-406.

Neumann, R.P. and Hirsch, E. (2000). Commercialisation of non-timber forest products: Review and analysis of research. Center for International Forestry Research, Bogor, Indonesia, 176 pp.

Ostrom, E. (1992). Crafting institutions for self-governing irrigation systems. ICS Press, San Francisco, CA, USA, 111 pp.

Peet, R. and Watts, M. (1996). Liberation ecologies. Routledge, London, UK, 273 pp.

Peters, C.M., Gentry, A.H. and Mendelsohn, R.O. (1989). Valuation of an Amazonian rainforest. Nature 339: 655-656.

Peterson, M.N., Lopez, R.R., Laurent, E.J., Frank, P.A., Silvy, N.J. and Liu, J. (2005). Wildlife loss through domestication: the case of endangered key deer. Conservation Biology 3: 939-944.

Posey, D.A. (1985). Indigenous management of tropical forest ecosystems: the case of the Kayapó Indians of the Brazilian Amazon. Agroforestry Systems 3(2): 139-158.

Ros-Tonen, M.A.F. (2000). The role of non-timber forest products in sustainable tropical forest management. Holz als Roh- und Werkstoff 58: 196-201.

Ros-Tonen, M.A.F. (ed.) (in collaboration with H. Van den Hombergh and A. Zoomers) (2007). Partnerships in sustainable forest resource management: learning from Latin America. CLAS Series. Brill Publishers, Leiden, the Netherlands, 329 pp.

Ros-Tonen, M.A.F., Van Andel, T., Morsello, C., Otsuki, K., Rosendo, S. and Scholz, I. (2008). Forest-related partnerships in Brazilian Amazonia: there is more to sustainable forest management than reduced impact logging'. Forest Ecology and Management 256: 1482-1497.

Ros-Tonen, M.A.F., Dijkman, W. and Lammerts van Bueren, E. (1995). Commercial and sustainable extraction of non-timber forest products. Towards a policy and management oriented research strategy. The Tropenbos Foundation, Wageningen, the Netherlands, 32 pp.

Ros-Tonen, M.A.F and Kusters, K. (2011). Pro-poor governance of non-timber forest products: the need for secure tenure, the rule of law, market access and partnerships. In: Shackleton, S., Shackleton, C. and Shanley, P. (eds.) Non-timber forest products in the global context. Tropical Forest Series. Springer-Verlag, Heidelberg, Germany, pp. 189-207.

Ros-Tonen, M.A.F. and Wiersum, K.F. (2005). The scope of improving rural livelihoods through non-timber forest products: an evolving research agenda'. Forests, Trees and Livelihoods 15(2): 129-148.

Scoones, I. (2009). Livelihoods perspectives and rural development. Journal of Peasant Studies 36(1): 171-196.

Sunderlin, W.D., Angelsen, A., Belcher, B., Burgers, P., Nasi, R., Santoso, L. and Wunder, S. (2005). Livelihoods, forests, and conservation in developing countries: an overview. World Development 33(9): 1383-1402.

Tiffen, M., Mortimore, M.J. and Gichugi, F. (1994). More people, less erosion: environmental recovery in Kenya. Wiley, Chichester, NH, USA, 311 pp.

Van Andel, T.R. (2000). Non-timber forest products of the North-West District of Guyana. Part I and Part II. (PhD thesis Utrecht University). Tropenbos-Guyana Series 8a/b. Tropenbos International, Wageningen, the Netherlands, 684 pp.

Van Dijk, J.F.W. (1999). Non-timber forest products in the Bipindi-Akom II region, Cameroon. A socio-economic and ecological assessment. Tropenbos-Cameroon Series 1. The Tropenbos-Cameroon Programme, Kribi, Cameroon, 197 pp.

Van Vliet, N., Nasi, R. and Taber, A. (2011). From the forest to the stomach: bushmeat consumption from rural to urban settings in Central Africa. In: S. Shackleton, C. Shackleton and Shanley, P. (eds.) Non-timber forest products in the global context. Tropical Forest Series. Springer-Verlag, Heidelberg, Germany, pp. 129-145.

Vedeld, P., Angelsen, A., Bojö, J., Sjaastad, E. and Berg, G.K. (2007). Forest environmental incomes and the rural poor. Forest Policy and Economics 9: 869-879.

Watts, M. (1983). Drought, environment and food security. In: Glantz, M. (ed.) Drought and hunger in Africa. Cambridge University Press, Cambridge, UK, pp. 171-212.

Watts, M. (2000). Political ecology. In: Sheppard, E. and Barnes, T.J. (eds.) A companion to economic geography. Blackwell Publishers, Malden, MA, USA, pp. 257-274.

Wiersum, K.F. (1997a). Indigenous exploitation and management of tropical forest resources: An evolutionary continuum in forest-people interactions. Agriculture, Ecosystems and Environment 63: 1-16.

Wiersum, K.F. (1997b). From natural forests to tree crops, co-domestication of forests and tree species, an overview. Netherlands Journal of Agricultural Science 45: 425-438.

Wiersum, K.F. (1999). Understanding diversity in NTFP management: a neglected issue in NTFP research. In: Ros-Tonen, M.A.F. (ed.) NTFP Research in the Tropenbos programme. Results and Perspectives. Proceedings of a seminar held on 29 January 1999, The Tropenbos Foundation, Wageningen, the Netherlands, pp. 161-165.

Wiersum, K.F. (2004). Forest gardens as an 'intermediate' land-use system in the nature-culture continuum: characteristics and future potential. Agroforestry Systems 61: 123-134.

Wiersum, K.F. (2009). Community forestry between local autonomy and global encapsulation: quo vadis with environmental and climate change payments? Paper presented at First Community Forestry International Workshop 'Thinking globally — acting locally: community forestry in the international arena', Pokhara, Nepal, September 15-18, 2009.

Wiersum, K.F. and Shackleton, C. (2005). Rural dynamics and biodiversity conservation in Southern Africa. In: Ros-Tonen. M.A.F. and Dietz, T. (eds.) African forests between nature and livelihood resources. Interdisciplinary studies in conservation and forest management. African Studies 81. Edwin Mellen Press, Lampeter, UK, pp. 67-91.

Wunder, S. (2001). Poverty alleviation and tropical forests – what scope for synergies? World Development 29(11): 1817-1833.

3. Gum and resin-producing species in the drylands of Ethiopia: productive bricolage footprints on the landscape

Teshale Woldeamanuel
Wageningen University, Forest and Nature Conservation Policy Group, P.O. Box 47, 6700 AA Wageningen, the Netherlands; Wondo Genet College of Forestry and Natural Resources, Hawassa University, P.O. Box 128, Shashemene, Ethiopia; teshale.woldeamanuelhabebo@wur.nl; twoldeamanuel@yahoo.com

Abstract

In different parts of the world humans have for centuries maintained highly structured landscape mosaics. These have recently been recognised as important components of the global cultural heritage. Based on a study of the use and management of gum and resin-producing trees in three areas in the drylands of Ethiopia, this chapter clarifies what aspects of the livelihood portfolio or productive bricolage process determine landscape dynamics and diversity. The livelihood strategies of the households in Ethiopia's drylands used to be based on just a few economic activities, such as pastoralism in Borana and mixed farming in Abergelle. Nowadays, households pursue a mix of livelihood strategies, all based on natural-resource-based activities. The practice of switching from livestock production or crop cultivation to mixed farming and gum and resin production is becoming more widespread. Although gum and resin have become important components of the productive bricolage process, in none of the study areas are the woodlands that host these resources intensively managed. This means that the contribution of gum and resin to the total income of the household is not reflected in the management of the landscapes that host the gum and resin-producing species of *Acacia*, *Boswellia* and *Commiphora* ('ABC landscapes'). Individual households play a limited role in the use and management of ABC landscapes, implying that sustainable use of natural resources and biodiversity conservation is not only an outcome of the productive bricolage process, but also depends on how formal and informal institutions are arranged. Thus, the productive bricolage concept alone is not sufficient to explain landscape diversity and domestication processes; the rights of access and tree property should also be examined.

Keywords: gum and resin production, *Acacia, Boswellia, Commiphora,* rural livelihoods, productive bricolage, institutions

3.1 Introduction

Cultural landscapes are 'geographical areas in which the relationships between human activity and the environment have created ecological, socioeconomic, and cultural patterns and feedback mechanisms that govern the presence, distribution, and abundance of species assemblages' (Farina, 2000). These landscapes are not necessarily sterile monocultures, but can provide a mix of production and environmental functions (Sayer and Elliott, 2003). Biodiversity is often richer in cultural landscapes than in remnants of natural landscapes (Farina, 2000). Rather than endangering species by ruthless overexploitation, local communities may consciously preserve and even cultivate endemic and endangered plant species in anthropogenic landscapes with a view to fulfilling household and conservation needs. For instance, agroforests, which include a range of human-modified and created forests, form an integral part of local land-use systems and contribute to both production and biodiversity conservation. These may include a variety of indigenously-evolved forest resource management types ranging from protected forest patches to tree cultivation systems. Some of these systems are still based on the natural vegetation, while others hold an intermediate position between natural vegetation and agricultural/tree crop cultivation systems that mimic the natural ecosystems. Indeed, human-induced changes do not always involve deforestation, which shows that the dichotomy between 'pristine' nature and cultivated fields as a cultural artefact is wrong when considering biodiversity conservation (Asbjornsen *et al.*, 2004; Wiersum, 2004). Despite the changes in the distribution of humans and the use of resources, people in different parts of the world have for centuries maintained highly structured landscape mosaics, which have recently been recognised as important components of our global cultural heritage (Farina, 2000).

Landscape approaches to natural resource management are receiving increasing attention, because inherent synergies can be delivered by addressing rural livelihoods, food security and biodiversity conservation at a landscape scale in an integrated and coordinated manner (Scherr and Rhodes, 2005). In the assessment of potentials for the use and conservation of natural resources, it is important to consider the overall mosaic of forested landscapes rather than only natural vegetation (Laurance and Bierregaard, 1997; Zuidema, 2003). In this vein, Ros-Tonen and Wiersum (2005) argue for a 'resource-in-context' approach that combines insights into community-level creativity and livelihood dynamics with those of macroeconomic and spatial processes and which allows for a more realistic assessment of the development and conservation potential of non-timber forest products (NTFPs).

A more specific explanation regarding landscape diversity is associated with household livelihood strategies. Most rural poor people maintain diversified rural livelihood strategies, as they cannot obtain sufficient income from any single strategy (Ellis, 2000). Agricultural diversification has long been known and the common forms include enterprise diversification (increasing the number of marketable activities), geographical diversification (increasing the number of locations), income diversification (increasing the number of income sources through off-farm work or investments) and crop diversification (increasing the number of varieties) (Wandel and Smith, 2000). In many rural areas, becoming less dependent on agriculture is part of the process of becoming better off, and occupational multiplicity is becoming more common and more pronounced. This shifts the household income from farm to non-farm sources (Deshingkar, 2005; Lopez and Vales, 2000). The ways in which rural people construct a livelihood is a response to local constraints and opportunities and to broader patterns of income-generating possibilities. Batterbury (2001) labelled the resulting livelihood portfolio as 'productive bricolage' (see Chapter 2 by Ros-Tonen in this volume). The productive bricolage households choose to pursue can be manifested in the fields and farms as well as in the economic and locational choices of households, and ultimately determines the diversity of landscapes.

Based on a study carried out in the drylands of Ethiopia, this chapter aims to make clear how landscape diversity and dynamics are related to livelihood portfolios, institutional arrangements at different levels and the characteristics of the production and use of gum and resin. Specifically, it describes the management of the landscapes dominated by the gum and resin-producing *Acacia*, *Boswellia* and *Commiphora* (ABC) species, focusing on the embeddedness and interaction of the livelihood strategies, the institutional factors and the nature of the production system. The following questions are addressed:

1. What are the livelihood strategies of the households and how do these strategies interact?
2. What is the management status of the ABC species in the drylands of Ethiopia?
3. What factors determine the management status of the woodlands that produce gum and resin?

The data presented in this chapter was collected between July 2007 and June 2010 within the framework of the FRAME (FRAnkincense, Myrrh and gum arabic: sustainable use of dry woodland resources in Ethiopia) integrated research project coordinated by Wageningen University and various institutions in Ethiopia. This project aimed to improve gum and resin production and sustainable woodland management in the drylands of Ethiopia. The methods included participatory research methods and a household survey. Open-ended questionnaire interviews were held with community groups and key informants (elders, office representatives, cooperative representatives, NGO delegates and traditional historians). Groups and key informants were selected from various sites in the three areas, Abergelle, Metema and Borana, in order to obtain a broad representation of the areas. The key informants were selected on the recommendation of field guides. A total of 19 focus group discussions and 37 key informant interviews were held. The household survey was held among a total of 326 respondents: 99 from Abergelle, 97 from Metema and 130 from Borana (Lemenih *et al.*, 2011; Woldeamanuel *et al.*, 2011). The data was complemented with information from other studies carried out in the study area (notably Lemenih *et al.*, 2007; Leminih *et al.*, 2011; and also Dejene, 2008; Eshete *et al.*, 2005; Worku, 2006).

3.2 Dry forests in Ethiopia

Most forest resources in Ethiopia can be found in the dry woodlands in the lowlands. Ethiopia's dry forests are rich in ABC species (Chikamai and Tchatat, 2004; Chikamai *et al.*, 2009; Lemenih and Teketay, 2004; Lemenih *et al.*, 2003), which produce hardened resinous plant exudates (Lemenih *et al.*, 2003; Wickens, 1995). At present about 35 ABC species have been identified as potential producers of commercial gums and resins. Nevertheless, gums and resins are currently collected from only a few species (Lemenih and Kassa, 2011). The gum and resin products contribute significantly to rural livelihoods, the national economy and ecosystem stability. They contribute to local livelihoods as a subsistence crop and as a source of cash income. The products are used as herbal medicines, insecticides, and hygienic and sanitation detergents, and have also been traded for centuries on international and domestic markets. Myrrh and frankincense production and trade in Ethiopia are claimed to go back to the Aksumite Empire around the first millennium AD (Bard *et al.*, 2000; Butzer, 1981; Gebremedhin, 1997), although the earliest recorded accounts of commercial gum and resin production date from the 1940s (Taib, 1982). Currently, about 10,000 metric tonnes are consumed locally each year, with an estimated value of USD 12 million, 60% of which is used in religious ceremonies (Lemenih and Kassa, 2011). Some of the produce is exported. The total export volume and foreign currency earnings from gums and resins have increased during the past decade (Lemenih and Kassa, 2011). In the period between 1998 and 2007, Ethiopia exported annually an average of 2,500 tonnes of natural gum and resin and earned USD 34.1 million from the trade (Lemenih and Kassa, 2010). During the 2007/2008 fiscal year, Ethiopia exported 4,612 tonnes of gums and resins and earned about USD 7.7 million (Lemenih and Kassa, 2011).

The drylands of southern and northern Ethiopia represent areas with contrasting ABC production and marketing conditions. This chapter examines the conditions in the Abergelle district in the north, the Metema district in the northwest and the Borana zone in the south (Figure 3.1). All the study sites host woodland that are rich in ABC species (Kuchar, 1995; Lemenih and Teketay, 2004; Vollesen, 1989). However, there are several differences in woodland types, ABC species compositions, socioeconomic and livelihood conditions, and the organisation of the gum and resin production. *Boswellia papyrifera*, which is the source of the most widely marketed type of incense – the Tigray type (white) frankincense – is the important commercial species in Abergelle and Metema (Lemenih, 2005; Lemenih *et al.*, 2007) and Assossa. In Borana there are a variety of gum and resin-producing species, including *Acacia senegal* and *A. seyal* for gum arabic production, *B. neglecta* and *B. rivae* for frankincense and *Commiphora myrrha* and *C. guidotti* for myrrh and opoponax (Lemenih, 2005).

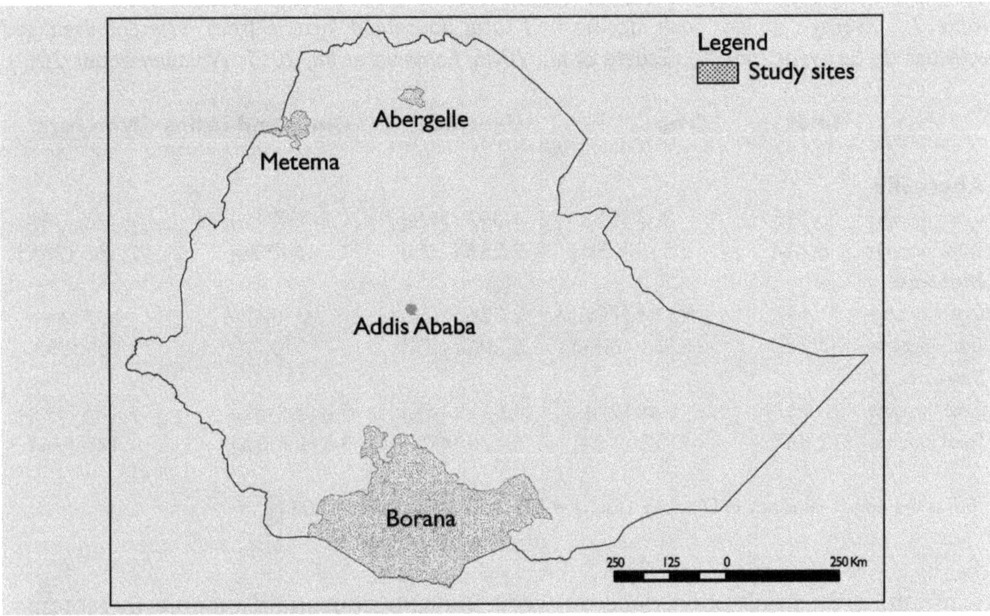

Figure 3.1. Map showing the location of the study areas in the gum and resin-producing drylands of Ethiopia. The Federal Republic of Ethiopia consists of regions (or kililoch*) based on ethnolinguistic communities, which are subdivided into zones that consist of various districts (*woredas*). Abergelle and Metema are districts; Borana is a zone.*

3.3 Gum and resin in the dryland livelihoods of Ethiopia: livelihood strategies and dynamics

In the drylands of Ethiopia the household livelihood strategies generally comprise a combination of crop cultivation, livestock husbandry, gum and resin production and trade, and non-farm activities. However, the degree of attachment of households to various livelihood activities differs between the study areas. Income from remittances is negligible in all three areas. The contribution made by the different livelihood activities to total income and cash income varies considerably between the three areas (Table 3.1).

In Abergelle, mixed farming supports the economy of the households. Crops form the main source of income (28%), while other important sources of household income are livestock production (26%), non-farm income (26%) and gum and resin (20%). Cash income forms 64% of total household income. The crops are mostly directly consumed and provide only 5% of the cash income. The main cash income is derived from non-farm activities (38%), gum and resin (31%), and livestock (26%) (Woldeamanuel, 2011).

Table 3.1. Average annual cash income and total household income from different livelihood activities (in Birr/year[1] and %) (Eshete et al.*, 2005; Lemenih* et al.*, 2007; Woldeamanuel, 2011).*

	Total	Crops	Livestock	Gums and resins	Non-farm
Abergelle					
Cash income	5,612	304 (5%)	1,397 (26%)	1,746 (31%)	2,165 (38%)
Total income	8,714	2,518 (28%)	2,285 (26%)	1,746 (20%)	2,165 (26%)
Metema					
Cash income	11,610	9,181 (79%)	2,326 (20%)	0 (0%)	103 (1%)
Total income	42,323	36,819 (87%)	5,395 (13%)	0 (0%)	109 (0%)
Borana					
Cash income	10,457	948 (9%)	4,552 (44%)	3,812 (36%)	1,145 (11%)
Total income	12,847	1,770 (14%)	6,120 (47%)	3,812 (30%)	1,145 (9%)

[1] Birr is the unit of currency in Ethiopia. USD 1 = ETB 16.81 (November 2011).

Gum and resin production does not contribute to household income in Metema. Woldeamanuel (2011) has shown that in this district crops form the major share of total household income (87%), with livestock accounting for the rest (13%). These two activities are also the main sources of cash income (79% and 20% respectively). The woodland resource dominated by *Boswellia papyrifera* provides the households in Metema with a wide diversity of goods (wood for construction, firewood, charcoal, farm tools, household furniture and utensils, fodder and shade for their animals, wild fruit) and services (wild meat from hunting, herbal medicines, recreation, bee forage for honey production, ecological services such as soil and water conservation, climate amelioration and habitat for wildlife), but is not valued for its gum and resin products (Lemenih *et al.*, 2007). Hence, gum and resin resources do not play any role in the livelihood strategies of the households in Metema. Most of the inhabitants originate from the highlands and not only lack the expertise required for frankincense tapping, but also regard it to be insufferable, arduous and unrewarding work compared with farming, as well as being a shameful and culturally unacceptable activity (Leminih *et al.*, 2007). Indeed, none of the respondents are engaged in frankincense production and therefore gain no income from it (Dejene, 2008; Lemenih *et al.*, 2007; Woldeamanuel, 2011). In this region, gum and resin extraction is carried out by external labourers rather than local households. Eshete (2002) also indicated that in the North Gondor zone, which includes Metema, on average 1,400 individuals are employed each year in frankincense production, but all of them come from outside Metema, notably from the neighbouring Tigray region. In general, the inhabitants of Metema district utilise the *B. papyrifera* dominated woodland in a way that does not differ from non-incense bearing woodland resources found elsewhere.

In Borana, a mixed economy prevails, but here livestock husbandry forms the main household activity (47% of total income). Other important livelihoods activities are gum and resin (30%), crops (14%) and non-farm activities (9%). No remittances are received. The cash income forms 81% of the total household income, derived from livestock (44%), followed by gum and/or resin (36%), non-farm income (11%), and crops (9%). The important role of gum and resin is demonstrated by the fact that they contribute 30% to total household income and 36% to cash income (Woldeamanuel, 2011; Worku, 2006).

Where gum and resin contribute to incomes, they also play an important role in times of emergencies, such as droughts. In Borana, 94% of the respondents depend on gums and resins to cope with emergencies, such as droughts and other livelihood crises. During times of emergency they collect and sell the products to purchase grains. In rare cases, the gums and resins are consumed as chewing gums. Other coping strategies consist of the receipt of emergency aid (92% of respondents), livestock sales (37% of respondents), support from relatives or neighbours (26% of respondents) and consumption of savings (6% of respondents). In Abergelle, during periods of livelihood stress households used to migrate to the highlands, where food crop production is relatively abundant. Recently, alternative coping strategies have been developed in the form of emergency aid by government organisations and NGOs (66% of respondents). Moreover, 62% of the respondents sell livestock when they face livelihood stress and purchase grain from areas as far as six hours travel away. Other coping strategies consist of support from relatives (20% of respondents) and consumption of savings (15% of respondents). Only 20% of the respondents use gums and resins as a coping strategy (Woldeamanuel, 2011).

The livelihood strategies of households in the study areas have considerably changed in recent decades. According to Woldeamanuel (2011), mixed farming involving a combination of crop and livestock production has increased, whereas sole cropping (in Abergelle and Metema) and sole livestock keeping (in Borana) have decreased. The production of gums and resins has gradually increased. In Abergelle the number of households engaged in gum and resin production doubled from 28% in the years before 1991 to 59% in 2010. However, the change in livelihood activities in Abergelle was relatively small compared with Metema and Borana. In Metema the change in livelihood activities was greatly influenced by migration. Before migrants started to settle in the district in the 1960s, 75% of the local population were engaged in shifting cultivation and only 25% in mixed farming. From the 1960s until 1991, 75% of the new settlers were solely engaged in crop cultivation, but the proportion of the total population solely engaged in crop cultivation has since fallen to 32%. The wealth accumulated from sesame and cotton production has helped them to diversify into animal husbandry.

In Borana pure pastoralism was the sole livelihood activity until the 1960s, but this decreased to 85% by the early 1990s. Crop cultivation started in the 1960s and has rapidly grown since the early 1990s. More than three-quarters (78%) of the respondents began cultivating crops in response to the frequent droughts, which affected livestock production and productivity. Since 1991 a massive government agricultural extension programme in the area has contributed to the further expansion of crop cultivation, and currently over 80% of the respondents cultivate crops, although crop income is still low. Recently, households have further diversified their livelihoods by increasing production of gums and resins as well as activities not based on natural resources. The engagement in gum and resin production increased from less than 50% of the respondents prior to 1991 to 90% at present.

To sum up, the livelihood systems in the three regions are characterised by a diversified household economy in Abergelle, crop cultivation in Metema and a traditional pastoralist system gradually changing towards mixed farming in Borana.

3.4 Status of the woodlands dominated by *Boswellia* species in the drylands of Ethiopia

Lemenih *et al.* (2011) show that the woodlands that produce gum and resin in the drylands of Ethiopia are not well conserved or developed. Three major issues characterise the status of the ABC woodlands in Ethiopia.

First, little effort has yet been made to develop and conserve the ABC resources in these areas. The *Acacia*, *Boswellia* and *Commiphora* species that produce the gum and resin are still in their early phases of domestication in terms of enrichment planting and cultivation. Management techniques that enhance production do exist, but show regional variation. In Abergelle and Metema, the extraction methods have been intensified by employing tapping techniques that increase production levels. Especially in Abergelle, the long-established use of this production technology has resulted in over-tapping and loss of production capacity. Although conscious efforts are now being undertaken to allow the regeneration of trees by establishing enclosures, cultivation of selected wild trees in plantations is not common in the drylands. Cultivation trials have recently been started in Abergelle, but cultivation of genetically selected or improved gum or resin-producing tree crops in intensively managed plantations does not yet exist in Ethiopia. Neither production enhancement through the employment of certain tapping techniques nor efforts to conserve the resource base have been observed in Borana, where gum and resin production is carried out by collecting natural exudates.

Second, the gum and resin-producing woodlands have been subject to clearance for alternative uses, such as crop cultivation and commercial fuelwood, charcoal and scent wood production. In Abergelle, the *Boswellia* woodlands have been converted into agricultural woodlands as a result of frequent land redistributions. In Metema, agricultural expansion is the major cause of ABC woodland depletion. So far, 19,271 people have been settled in the Metema area, with an average landholding of about

8 ha per household. About 23,247 ha of woodland have been granted to private companies that produce cash crops such as cotton and sesame (Anonymous, 2011). As will be seen below, customary institutions stimulate such conversions. In Borana, charcoal, firewood and scent wood utilisation and, more recently, also the cultivation and establishment of private grazing enclosures have provoked the clearance of ABC woodlands (Woldeamanuel, 2011). Indeed, utilisation of the ABC trees for gum and resin production has relatively little impact on the resource base. In some of the areas the impact is even very minimal, as the only gum and resin collected is that which has oozed naturally from the bark. The ABC resource base has been affected mainly by non-gum and non-resin uses. In combination with the limited efforts to conserve and develop the resource base, the ABC trees species are gradually disappearing from Ethiopia's drylands.

Third, although control of access to the gum and resin resource base and the market has restricted individual access to ABC resources, lack of implementation and enforcement by the state authorities has prejudiced the resource base. In Abergelle and Metema, the state authorities have formalised and implemented procurement control, but state control has not always been effective. The result has been forest loss (due to the transformation of woodlands into a financially more rewarding cash-cropping system) and/or degradation and loss of production capacity (due to low-level extraction management). In order to counter such processes of ecosystem and resource degradation, new organisational forms of control are being stimulated. The Abergelle cooperative case described in the next section shows that these developments may involve the transfer of at least some control from central government to local community organisations. In Borana, the access rights and harvesting intensities are still predominantly controlled under the indigenous *gada* system for rangeland management. The cooperative arrangements have been established only recently (Woldeamanuel, 2011).

3.5 Understanding the management status of gum and resin-producing woodlands

Despite diversification of livelihoods into gum and resin production, in none of the areas are the ABC resources intensively managed. Where the ABC species are used as a diversification strategy (e.g. Abergelle), the domestication status of the ABC species is not as advanced as would be expected (Woldeamanuel, 2011). The contribution made by gum and resin to the total household income is not reflected in the management of the ABC species in any of the three areas. To fully understand the management status of the ABC woodlands, we need to consider several factors, addressed below: the embeddedness of ABC exploitation in the productive bricolage process, institutional and policy factors, and production and use characteristics.

Embeddedness of ABC exploitation in the productive bricolage strategies of local communities

The unfavourable farming conditions in Abergelle are reflected in the relatively low incomes from crop cultivation, but the contribution from gum and resin production to the cash income of households is high. The *Boswellia* trees in the farmlands perform important functions. Farmers appreciate the *Boswellia* trees for having leaves that augment soil fertility. The trees are also used to provide shade and as fodder for the cattle and goats. However, *Boswellia* trees are not fully integrated into crop cultivation because their presence in farmland requires careful management. Their shallow root system makes them vulnerable to uprooting and drying when the land around them is cultivated. However, due to the long tradition of gum and resin production in Abergelle and the substantial contribution made by gums and resin to livelihoods, gum and resin resource utilisation is more controlled in Abergelle than in the other areas. Local people in this region have recognised the commercial value of these products and have gradually started to integrate gum and resin-producing species into their mixed farming systems. Initial efforts have also been undertaken to protect the trees on croplands and in homegardens. *Boswellia* trees are also conserved as part of the land-use system because the law prohibits felling them wherever they grow or ploughing within two to five metres of them. Farmers respect the ban on cutting down *Boswellia* trees even on their own land, although they do not always adhere precisely to the ploughing distance stipulated in the law (Woldeamanuel, 2011). As a result, the ABC resources are better conserved in Abergelle than in the Metema area. The majority of the households that use gum and resin in Abergelle are from the relatively rich and middle classes, for whom gum and resin production is part of a specialising livelihood diversification strategy (see also Chapter 2 by Ros-Tonen and Chapter 5 by Schure in this volume).

In Metema, crop production is the dominant livelihood activity, with livestock production also being an important component of household livelihood strategies. The lowland of Metema provides the settlers with fertile soils for cash crop production and hence with a good agriculture-based livelihood opportunity. The average land holding per household in Metema district for crop farming alone was found to be 8 ha, compared with the average holdings of only 0.5-1.0 ha in the highlands (e.g. Bekele, 2003). Of these 8 ha, the largest share (3.5 ha on average) is allocated to sesame, which is an expensive cash crop. The remaining land is divided between sorghum (2.6 ha), cotton (the second cash crop, 1.5 ha) and other cereals and homegardens (0.4 ha) (Dejene, 2008). The gum and resin resources do not play a role in the livelihood strategies of the households. As a result, the ABC woodlands in this region are more degraded than the other dryland areas and the stock of ABC trees is progressively declining. The *Boswellia* species are outcompeted by sesame and cotton production, which flourish well in the well-drained soils. The Metema case shows that where biophysical conditions favour woodland conversion to financially more rewarding land uses, woodland degradation is more severe than in regions where woodlands form an essential land-use component in ecologically fragile environments, such as in Borana (discussed below). This demonstrates how options for alternative land uses greatly influence decisions on whether to domesticate or convert forests.

The biophysical conditions in Borana are not conducive to crop cultivation, which therefore presents little competition as an economic activity to the use and management of ABC trees. As mentioned above, in the predominantly pastoral livelihood system in Borana, livestock accounts for the lion's share of total household income. The people of Borana value livestock-based livelihood activities over any other activity. Livestock husbandry is carried out not only as a livelihood source, but also because the Borana attach considerable social, cultural and ritual value to livestock. They believe livestock are for the people and the people are for the livestock. Due to this traditional strong reliance on livestock, their livelihoods have shown little diversification and they have only recently begun adopting alternative livelihood activities, such as crop cultivation and increased use of natural resources. As a result, hardly any efforts have yet been made to domesticate gum and resin-producing trees in this region. Although the gums and resins contribute substantially to the household income in Borana, they are mostly utilised by poor households in times of emergency. Gum and resin production acts as a safety net in times of emergency or as a strategy to obtain supplementary income. These households mostly collect the gum and resin from communal lands and so the local people are not engaged in their conservation and development. Rather, the ABC species are open access resources for both traditional and non-traditional uses. In Borana, livestock production and gum and resin production are interconnected to a certain extent, but in critical circumstances livestock production is favoured and the *Boswellia* trees are cleared for pasture development. The ABC species are even considered as invaders on the rangelands (Woldeamanuel, 2011).

Institutional and policy factors

Although the analysis above illustrates how the interaction between biophysical and socioeconomic factors may impact on the landscape, the relative importance of gum and resin in the productive bricolage process and the interconnectedness with other livelihood components is insufficient to explain how landscapes are managed. For a more comprehensive understanding, a closer look is needed into the institutions that regulate access to resources and markets. In all the areas, livelihoods and landscape management are strongly governed by both formal (statutory) and informal (customary) institutions.

In Abergelle, formal rules and regulations strongly influence household livelihood strategies and, ultimately, landscape diversity. In the first place, government interventions in the form of safety net programmes and emergency aid supplement the livelihoods of households that do not produce sufficient food or income for their annual consumption needs. Secondly, formal rules restrict access to the gum and resin production systems and their markets, even more so since the emergence of cooperative associations in each *kebele* (neighbourhood). To produce gum and resin from *Boswellia* trees on public lands, membership of general purpose cooperatives is a must. Moreover, tapping skill is a necessary requirement for obtaining use rights to *Boswellia* trees on public lands, but because of the limited resource base, cooperative members receive use rights, on a rotational basis, only once every two or three years (Lemenih *et al.*, 2011). None of the gum and resin producers (whether individuals or members of the cooperatives) are allowed to sell the gum and resin they produce in markets of their preference, but are obliged to sell to the cooperative shops in their respective *kebele*. Furthermore, producers are not allowed to sell when they want to; the cooperative announces the date it wants to buy the gum and resin from them. As a result of these arrangements, gum and resin do not play a role as a safety net in the livelihood strategies of the households in Abergelle. Thirdly, the landscapes that are home to the gum and resin trees have been subject to frequent land redistributions and are characterised by land and tree insecurity. Although gum and resin are important components of household livelihoods, access to the ABC trees and the market for gum and resin is liable to strict formal regulations. In Abergelle, private and state companies were the sole gum and resin exploiters for more than six decades and smallholders still have relatively little access to the resources and market.

In Metema, gum and resin production is undertaken solely by external labourers employed by commercial enterprises (Woldeamanuel, 2011). Here, the people-woodland interaction is influenced predominantly by state-sponsored resettlement schemes that have altered demographic conditions and land-use practices. Most of the residents in the Metema district are recent settlers from the highlands, where agriculture is the main subsistence strategy. According to the respondents, the most culturally acknowledged and respected activity is agriculture that involves crop cultivation and livestock rearing. Since the woodland of Metema offers them vast agricultural resources, fertile land and good pasture, they continue to cultivate crops and herd livestock after they settle in the Metema district. This is facilitated by customary institutions that enable easy entry into agriculture, including those related to share cropping, land sharing and oxen renting. Rather than exploiting gum and resin-producing tree resources, the local people convert the ABC woodlands into crop fields. One of the customary institutions that stimulates such conversion is the local *Mofer Zemet* arrangement that allows non-local people who are not registered as a resident in a *kebele* to clear a plot of woodland and temporarily cultivate it during the rainy season. Another custom is the widespread *Mate Qedem* arrangement that enables de facto private cultivation of plots in public forestlands without the formal approval of local authorities. Large numbers of households have been and still are settling illegally in the woodlands, and it is very likely that they will eventually obtain legal rights to become established there. In Metema, the actual conversion of forestland and existing land-use practices are therefore the combined result of demographic pressures caused by immigration and informal rules that stimulate forest destruction. Formal interventions have very minimal impact on the livelihoods and the management of the gum and resin-producing landscapes (Woldeamanuel, 2011).

In Borana, gum and resins have always been produced by individual pastoralist households and are well-recognised emergency resources. They are collected freely from communal or public woodlands without any access restriction. Market access is still not yet fully formalised and so the poor can produce and sell these products on a daily basis in any market of their choice. The main livelihood component is livestock, which is reflected in a strong traditional institutional arrangement that regulates access to rangeland and water resources, called the *gada* system. Within this setting there is also a long-established tradition of helping each other during periods of livelihood stress, for example by giving livestock to a fellow Borana who has lost his own cattle in a time of crisis. However, local regulatory systems are gradually weakening as a result of multiple biophysical stresses (e.g. climate change) and socioeconomic stresses (e.g. political interference). The introduction of formal rules and regulations as well as modern administrative units have suppressed the traditional administrative units and associated institutions. As a consequence, the traditional natural resource management system, which has been resilient for centuries, has come under stress (Lemenih *et al.*, 2011; Woldeamanuel, 2011).

The cooperatives have become the dominant arrangement for gum and resin utilisation in all the drylands of Ethiopia, except for Metema, where production is still predominantly carried out by concessionaires (Leminih *et al.*, 2011). The combination of concession and cooperative arrangements for gum and resin exploitation has resulted in the exclusion of individual gum and resin producers in favour of the cooperatives and companies. This exclusion has prompted smallholders to clear the ABC resources on or around their farms to defend their land claims and has restricted investments in ABC tree domestication. Moreover, in several cases gum and resin exploitation has become a source of competition and conflict between cooperatives and private companies, further limiting opportunities for individuals to participate in the production and management of the ABC resources and restricting endogenous development of domestication (Woldeamanuel, 2011).

In summary, in Abergelle the people-woodland interaction is shaped by strong state control over access to and use of the woodlands, such as the rotational allocation of exploitation rights. In Metema, cultural preferences and customary institutions have resulted in the expansion of crop cultivation. In Borana, the interaction is mainly shaped by customary institutions and the prevalence of pastoralism. In regions where either strong statutory regulations (Abergelle) or customary rules (Borana) apply, the state of the woodlands is generally good, even though the stage of domestication is further advanced in Abergelle than in Borana (Woldeamanuel, 2011).

Production and use characteristics

The production system and the landscape are not only affected by household livelihood strategies and the institutional context, but also by the harvesting techniques. Gum and resin products can be extracted in a variety of production systems that use different techniques. For instance, the collection of gums and incense does not involve any tapping in the Borana zone in southern Ethiopia, where the people collect exudates that ooze naturally from the bark (Lemenih *et al.*, 2003; Woldeamanuel, 2011; Worku, 2006). However, tapping the *Boswellia* trees is the major technique for producing frankincense in northern Ethiopia (Eshete *et al.*, 2005; Gebremedihn, 1997). The impact of tapping on the ABC species can already be seen. It damages the trees physically and also exposes them to insect attack. Rijkers *et al.* (2006) indicated that tapped trees produce fewer flowers, fruits and seeds than untapped trees, and smaller fruits with seeds of lower weight and reduced vitality.

3.6 Discussion: productive bricolage and landscape management

The livelihood strategies in the gum and resin-producing drylands of Ethiopia used to depend on a single livelihood strategy, such as pastoralism (e.g. Borana) and crop cultivation (e.g. Abergelle and Metema). Currently, however, the livelihood strategies are based on a mix of livelihood activities. Unlike the productive bricolage process in Niger described by Batterbury (2001), where households diversified their livelihood strategies towards off-farm activities not based on natural resources, in the drylands of Ethiopia livelihood diversification occurred through natural resource-based activities only. As a result, the livelihood components in the productive bricolage processes in the drylands of Ethiopia dominated by gum and resin-producing trees are strongly related to each other and to the landscape.

However, the extent to which the household livelihood portfolio depends on gum and resin is not entirely reflected in the conservation and management of the woodlands that host the gum and resin-producing resources. Even in areas where gum and resin are a major part of the productive bricolage process, such as Abergelle, conservation and development of gum and resin-producing trees is not as advanced as that of *Coffea arabica* in the Ethiopian highlands (Wiersum, 2010) or of *Acacia senegal* in Sudan, which plays an important role in the livelihood mix of the households (Ballal *et al.*, 2005; Rahim, 2006). Considering the long history of gum and resin production in people's productive bricolage in Abergelle, one would expect a relatively advanced stage of gum and resin-producing tree management via conservation and development (e.g. planting). In reality, however, the management of gum and resin-producing tree species is still not advanced in this region, but has been largely limited to the introduction of social control mechanisms, such as controlled access to the ABC resources and the market for gum and resin products. This is what Wiersum (1997) has termed the 'acculturalisation phase'. Although the institutionalisation of gum and resin production in Ethiopia has a long history and is a dynamic process, the ecological and biological management of the resource base in the drylands of Ethiopia cannot be fully explained by their relative contribution to the cash and total income of the households. This means that productive bricolage alone is not enough to explain the relation between livelihood strategies and landscape conditions. Rather, three major factors stand out regarding the relation between productive bricolage and NTFP conservation and management in Ethiopia.

First, the degree of management of the landscapes that host the resources is determined by the way NTFPs fit into the livelihood strategies of households and how they interact with other livelihood components. It was found that management (conservation and development) of NTFPs becomes progressively more intensive as their role in livelihood strategies evolves from a coping strategy (e.g. Borana) to a diversification strategy (e.g. Abergelle). It means that situations in which NTFPs form an important component in a diversified livelihood strategy offer a good starting point for a gradual intensification of their management. Such processes have been described by various authors (e.g. Belcher *et al.*, 2005; Homma, 1992, Ros-Tonen and Wiersum, 2005). The spatial interaction among the various livelihood components is also important as the extent to which a livelihood activity competes with another for physical space affects the landscape. The case of Metema shows that gum and resin resources are not compatible with crop cultivation and that the cultivation of the new cash crop sesame in particular results in the clearing of *Boswellia* trees. However, in Abergelle, despite competition between crop cultivation and woodlands that host gum and resin-producing trees, these woodlands are relatively protected on both farmland and public lands owing to strong formal regulations that require the conservation and development of such trees. In Borana, pastoralism as a major livelihood source complements gum and resin production and marketing, but in some cases the gum and resin-producing trees are cleared to enhance pasture growth.

Second, the institutional environment that governs access to both the resources and the market determines how productive bricolage is manifested in NTFP conservation and development. Gum and resin utilisation has been dominated by exploitation arrangements that disfavour local people's rights of ownership and their access to resources and the market. Moreover, the various institutional arrangements for gum and resin utilisation (e.g. cooperatives, companies) compete with private exploitation arrangements. This has limited the possibility for individual households to participate in the production and management of ABC resources, and ultimately restricts the development and conservation of the ABC woodlands. Exclusion of smallholders by the concessionaries has forced the smallholders to clear the ABC resources on or around their farms, limiting their ability to invest in and intensively manage NTFPs (Belcher *et al.*, 2005). A similar case has been reported from India, where the nationalisation of NTFPs has resulted in a decline of their production and management (see Chambers *et al.*, 1990).

Third, regulations and policies formulated in other sectors often work against the provisions of macro-level institutions that aim to stimulate forest management. The case of gum and resin-producing trees shows that although their value is acknowledged by specific regulations that advocate their use and management, the woodlands that host them have become targets for other uses, such as resettlement and the development of mechanised agriculture, under rural development and investment policies. This observation is in line with the claim that the root causes of forest degradation are often found outside the forestry sector (De Montalembert, 1992) and that forest use, development and conditions are the fundamental consequences of the wider configuration of national policy and economic development (Kowero *et al.*, 2003). Moreover, the results show that it is not only productive bricolage at the community level, but also the mix of livelihood strategies pursued outside the community that affects the landscape. This means that the livelihood mix pursued in other areas may have an impact on the landscape beyond its territory.

Fourth, customary institutions play crucial roles in shaping productive bricolage and ultimately the landscape. Bassi and Tache (2011) show how the 'law of grasses and water', *seera marraa bisaanii,* governs forest management in Borana. Under this law, the rangeland is managed by norms of inclusion and exclusion designed for pastoral activity. Although no family can be directly denied access to the rangeland, the law differentiates between dry season pastures (with permanent water points) and wet-season pastures (with good grass, but only accessible during and immediately after rains). It imposes the maximum use of wet-season pasture whenever possible, thus minimising the pressure on the most intensely utilised rangelands served by permanent water points. There are also provisions to restrict access to certain areas, which are to be kept as reserve for certain stock categories during the dry season. These norms and practices have direct impacts on the ecology of rangelands, particularly on the composition of grass species. Additional practices contribute to the control of the composition of the bushes and trees, such as controlled fires, selective cutting of bush for firewood and the periodical movement of villages to avoid depletion of trees. Woldeamanuel (2011) shows that certain tree species in Borana are protected due to the cultural values and the beliefs associated with them, rather than their being used in the portfolios of livelihood strategies.

3.7 Conclusion

This chapter has shown that economic and locational choices of individual households play a role in the use and management of landscapes, but that landscape diversity also depends on how formal and informal institutions are arranged. The management of the woodlands dominated by gum and resin-producing tree species in the drylands of Ethiopia depends on a mixture of formal statutory rules designed at different levels and informal rules and norms established by local people. This means that local strategies for resource enrichment and conservation are not only related to the relative role of NTFPs in livelihood strategies, but also depend on the presence of proper institutional arrangements. This and other studies (Wiersum, 1997, 2008) have shown that domestication processes are narrowly related to the nature of rights of access and tree property, and that this relationship should be carefully examined. This aligns with the call from Ros-Tonen (in this volume) that productive bricolage in NTFP systems can only be properly understood if due attention is paid to political, institutional and sociocultural factors that determine how the NTFP production system is shaped and impacts on the landscape. This further substantiates the resource-in-context approach (Ros-Tonen and Wiersum, 2005), which combines insights into community level creativity and livelihood dynamics and into macro and spatial processes to produce a more realistic assessment of the development and conservation potential of NTFPs.

References

Anonymous (2011). Socioeconomic survey of Metema District. Report of the Office of Agriculture of Metema district, Ethiopia.

Asbjornsen, H., Angelsen, A., Belcher, B., Michon, G., Ruiz-Perez, M. and Wijesekara, V.P.R. (2004). Cultivating (in) tropical forests? The evolution and sustainability of intermediate systems between extractivism and plantations. In: Proceedings International Workshop, 28 June to 1 July 2000, Kraemmervika, Lofoten, Norway. Available at: http://www.etfrn.org/etfrn/workshop/lofoten/etfrn_series_3.pdf.

Ballal, M.E., Siddig, A., Elfadl, M. and Luukkanen, O. (2005). Gum Arabic yield in differently managed *Acacia senegal* stands in western Sudan. Agroforestry Forum 63: 237-245.

Bard, A.K., Coltorti, M., DiBlasi, C.M., Dramis, F. and Fattovich F. (2000). The environmental history of Tigray (Northern Ethiopia) in the middle and late Holocene: a preliminary outline. African Archaeological Review 17: 65-86.

Bassi, M. and Tache, B. (2011). The community conserved landscape of the Borana Oromo, Ethiopia: opportunities and problems. Management of Environmental Quality 22(2): 174-186.

Batterbury, S. (2001). Landscapes of diversity: a local political ecology of livelihood diversification in south-western Niger. Ecumene 8(4): 437-464.

Bekele, W. (2003). Economics of soil and water conservation. Theory and empirical application to subsistence farming in the Eastern Ethiopian Highlands. Doctoral Thesis (AGRERIA 404), Swedish University of Agricultural Sciences, Department of Economics, Uppsala, Sweden.

Belcher, B., Michon, G., Angelsen, A., Ruiz-Perez, M. and Asbjørnsen, H. (2005). The socioeconomic conditions determining the development, persistence, and decline of forest garden systems. Economic Botany 59(3): 245-253.

Butzer, K.W. (1981). Rise and fall of Axum, Ethiopia: a geo-archaeological interpretation. American Antiquity 46: 471-495.

Chambers, R., Saxena, N.C. and Tushaar, S. (1990). To the hands of the poor: water and trees. Westview Press, Boulder, CO, USA, 273 pp.

Chikamai, B. and Tchatat, M. (2004). Forest management for non-wood products and services in Africa. A report prepared for lessons learnt on sustainable forest management in Africa. AFORNET/FAO/KSLA, Nairobi, Kenya.

Chikamai, B., Tchatat, M., Tieguhong, J. and Ndoye, O. (2009). Forest management for non-wood forest products and services in sub-Saharan Africa. Discovery and Innovation 21(1): 50-59.

De Montalembert, M.R. (1992). Intersectoral policy linkages affecting the forestry sector. In: Gregerson, H., Oram, P. and Spears, J. (eds.) Priorities for forestry and agroforestry research. IFPRI, Washington, DC, USA, pp. 79-92.

Dejene, T. (2008). Economic valuation of alternative land uses in *Boswellia papyrifera* dominated woodland area of Metema, North Gondar Ethiopia. MSc thesis, Hawassa University, Hawassa, Ethiopia.

Deshingkar, P. (2005). Maximizing the benefits of internal migration for development. Paper presented at Regional Conference on Migration and Development in Asia, Lanzhou, China, 14-16 March 2005. Available at: http://www.odi.org.uk/resources/download/1606.pdf.

Ellis, F. (2000). Rural livelihoods and diversity in developing countries. Oxford University Press, Oxford, UK, 273 pp.

Eshete, A. (2002). Regeneration status, soil seed banks and socio-economic importance of *Boswellia papyrifera* (Del.) Hochst. in two woredas of North Gonder zone, Northern Ethiopia. MSc thesis, Swedish University of Agricultural Sciences, Skinnskatteberg, Sweden.

Eshete, A., Teketay, D. and Hulten, H. (2005). The socio-economic importance and status of populations of *Boswellia papyrifera* (Del.) Hochst. in northern Ethiopia: the case of North Gonder Zone. Forests, Trees and Livelihoods 15: 55-74.

Farina, A. (2000). The cultural landscape as a model for the integration of ecology and economics. Bioscience 50: 313-320.

Gebremedihin, T. (1997). *Boswellia papyrifera* from the Western Tigray: Opportunities, constraints, and seed germination responses. MSc thesis, Swedish University of Agricultural Sciences, Skinnskatteberg, Sweden.

Homma, A.K.O. (1992). The dynamics of extraction in Amazonia: a historical perspective. In: Nepstad D.C. and Schwartzman, S. (eds.) Non-timber forest products from tropical forests: evaluation of a conservation and development strategy. New York Botanical Garden, New York, NY, USA, pp. 23-32.

Kowero, G., Campbell, B. and Sumaila, U.R. (2003). Policies and governance structures in woodlands of southern Africa. Center for International Forestry Research, Bogor, Indonesia, 438 pp.

Kuchar P. (1995). Identification and characterization of *Boresraceae* in the Southeastern Ethiopia. South-eastern rangelands project technical paper. Addis Ababa, Ethiopia, 57 pp.

Laurence, S.G. and Bierregaard, R.O. (eds.) (1997). Tropical forest remnants, ecology, management, and conservation of fragmented communities. University of Chicago Press, Chicago, IL, USA, 616 pp.

Lemenih M. (2005). Production and marketing of gums and gum resins in Ethiopia. In: Ben, C. and Enrico, C. (eds.) Production and marketing of gum resins: frankincense, myrrh and opoponax. FAO/NGARA, Nairobi, Kenya, pp. 55-70.

Lemenih, M. and Kassa, H. (2010). Socio-economic and environmental significance of dry land resources of Ethiopia and their development challenges. Journal of Agriculture and Development 1: 71-91.

Lemenih, M. and Kassa, H. (eds.) (2011). Opportunities and challenges for sustainable production and marketing of gums and resins in Ethiopia. CIFOR, Bogor, Indonesia.

Lemenih, M. and Teketay D. (2004). Natural gum and resin resources: opportunity to integrate production with conservation of biodiversity, control of desertification and adapt to climate change in the drylands of Ethiopia. In: Proceedings of a Workshop on Conservation of Genetic Resources of Non Timber Forest Products (NTFPs) in Ethiopia, 5-6 April 2004, Addis Ababa, pp. 37-49.

Lemenih, M., Abebe, T. and Mats, O. (2003). Gum and resin resources from some *Acacia, Boswellia,* and *Commiphora* species and their economic contributions in Liban, South-East Ethiopia. Journal of Arid Environments 55: 465-482.

Lemenih, M., Feleke, S. and Tadesse, W. (2007). Constraints to smallholders production of frankincense in Metema district, North-western Ethiopia. Journal of Arid Environments 71: 393-403.

Lemenih, M., Wiersum, F., Woldeamanuel, T. and Bongers, F. (2011). Diversity and dynamics of management of gum and resin resources in Ethiopia: a trade-off between degradation and domestication. Land Degradation and Development, in press. DOI: http://dx.doi.org/10.1002/ldr.115.

López, R. and Valdés, A. (2000). Fighting rural poverty in Latin America: new evidence of the effects of education, demographics and access to land. Economic Development and Cultural Change 49(1): 197-211.

Rahim, A.H.A. (2006). Economic analysis of deforestation. The case of the gum Arabic belt in Sudan. PhD Thesis, Wageningen University, Wageningen, the Netherlands.

Rijkers, T., Ogbazghi, W., Wessel, M. and Bongers, F. (2006). The effect of tapping for frankincense on sexual reproduction in *Boswellia papyrifera.* Journal of Applied Ecology 43: 1188-1195.

Ros-Tonen, M.A.F. and Wiersum, K.F. (2005). The scope for improving rural livelihoods through nontimber forest products: An evolving research agenda. Forests, Trees and Livelihoods 15: 129-148.

Sayer, J. and Elliott, C. (2003). The role of commercial plantations in forest landscape restoration. In: Mansourian, S., Vallauri, S. and Dudley, N. (eds.) Forest restoration in landscapes: beyond planting trees. Springer, New York, NY, USA, pp. 379-383.

Scherr, S. and Rhodes, C. (2005). Building up on synergies to deliver the millennium development goals: landscape-scale approaches to enhancing rural livelihoods, food security and environmental sustainability. PLEC News and Views New Series 8: 11-15.

Taib, A. (1982). The Swiss, Italian and Finnish markets for Ethiopian *Gum Olibannum.* Programme for Development Cooperation, Market Research Report No. 5, The Helsinki School of Economics, Helsinki, Finland.

Vollesen, K. (1989). *Burseraceae.* In: Hedberg, I. and Edwards, S. (eds.) Flora of Ethiopia, Volume 3. National Herbarium, Addis Ababa University and Uppsala University, Uppsala, Sweden, pp. 442-478.

Wandel, J. and Smit, B. (2000). Agricultural risk management in light of climate variability and change. In: Millward, H., Beesley, K., Ilbery, B. and Harrington, L. (eds.) Agricultural and environmental sustainability in the new countryside. Rural Research Centre, Nova Scotia Agricultural College, Truro, Canada, pp. 30-39.

Wickens, G.E. (1995). Role of *Acacia* species in the rural economy of dry Africa and the Near East. FAO Conservation Guide No. 27, FAO, Rome, Italy, 56 pp.

Wiersum, K.F. (1997). From natural forest to tree crops, co-domestication of forests and tree species, an overview. Netherlands Journal Agricultural Science 45: 425-438.

Wiersum, K.F. (2004). Forest gardens as an 'intermediate' forest system in the nature – culture continuum: Characteristics and future potential. Agroforestry Systems 61: 123-134.

Wiersum, K.F. (2008). Domestication of trees or forests: Development pathways for fruit tree production in South-east Asia. In: Akinnifesi, F.K., Leakey, R.R.B., Ajayi, O.C., Sileshi, G., Tchoundjeu, Z., Matakala, P. and Kwesiga, F.R. (eds.) Indigenous fruit trees in the tropics: domestication, utilization and commercialization. CABI, Wallingford, UK, pp. 70-83.

Wiersum, K.F. (2010). Forest dynamics in Southwest Ethiopia: interfaces between ecological degradation and resource enrichment. In: Bongers, F. and Tennigkeit, T. (eds.) Degraded forests in Eastern Africa: management and restoration. Earthscan, London, UK, pp. 323-342.

Woldeamanuel, T. (2011). Dryland resources, livelihoods and governance: diversity and dynamics in use and management of gum/resin trees in Ethiopia. PhD Thesis, Wageningen University, Wageningen, the Netherlands.

Worku A. (2006). Population status and socio-economic importance of gum and resin bearing species in Borana Lowlands, southern Ethiopia. MSc. Thesis, Addis Ababa University, Department of Biology, Addis Ababa, Ethiopia.

Zuidema, P.A. (2003). Tropical forests in multi-functional landscapes. In: Proceedings of two seminars on Issues in International Nature Conservation. Prince Bernhard Centre for International Nature Conservation, Utrecht University, Utrecht, the Netherlands.

4. Land-use dynamics in enset-based agroforestry homegardens in Ethiopia

Tesfaye Abebe[1] *and Frans Bongers*[2]
[1]*Hawassa University, College of Agriculture, P.O. Box 5, Hawassa, Ethiopia;* [2]*Wageningen University and Research Centre, Centre for Ecosystem Studies, P.O. Box 47, 6700 AA Wageningen, the Netherlands; tesfayea@hu.edu.et; frans.bongers@wur.nl*

Abstract

In many tropical countries homegardens sustain large numbers of people. Households depend for their livelihoods on the variety and continued production of food and other products that are provided by their own gardens. Such homegardens combine production with the maintenance of biodiversity. Long-term sustainability of the system is crucial for the long-term wellbeing of the population, including food security. The enset-coffee system of Southern Ethiopia is an example of such agroforestry homegardens. Enset is the main food crop and coffee is the main cash crop. We analysed the homegardens of 144 households covering a range of geographical conditions. Household wealth status was the main determinant of homegarden size. In total, 198 species of cultivated crops (78) and trees (120) were found. Enset and crops were represented by 42 and 24 cultivars respectively. The homegardens were covered by around one-third of enset and one-third of coffee. Enset provides the backbone of the system as it provides food security. Our results show that the stability of the system depends both on its diversity and on the specific characteristics of the two main crops, enset and coffee. The enset-coffee homegardens are changing as farmers increasingly make individual choices and respond to emerging challenges and opportunities. The dynamics of the agroforestry homegarden of Southern Ethiopia can be considered as a productive bricolage process on the micro scale.

Keywords: coffee, *Enset ventricosum*, Ethiopia, livelihood, agricultural intensification

4.1 Introduction

The production of a regular supply of sufficient food will continue to be the primary goal of most human populations living in the rural tropics (Godfray *et al.*, 2010; Trenbath, 1999). The long-term sustainability of agricultural production and the environmental consequences of the current intensification of agricultural systems are crucial elements in the debate on how to reconcile the need to protect and conserve decreasing biodiversity with the need to feed, clothe and shelter a growing human population (Padoch and Pinedo-Vasquez, 2010). Many studies have shown that intensification and simplification of food-producing systems renders them liable to instability and collapse (Altieri, 1995; Cox and Atkins, 1979; Perfecto *et al.*, 2009). Therefore, attention needs to be given to achieving stability in the use of land in the long term, while at the same time meeting the needs of the local population (Godfray *et al.*, 2010; Matson *et al.*, 2002; Swift and Ingram, 1996; Tilman *et al.*, 2002).

Modern technologies are often not the best option for improving agricultural production in smallholder farming systems in the tropics. In many remote areas a better option for achieving sustainable agriculture is likely to be the adoption of integrated forms of land use that make better use of local resources and natural processes (Abebe *et al.*, 2006; Godfray *et al.*, 2010; Nair, 1993; Perfecto *et al.*, 2009; Wiersum and Gonzalez, 2000). One such integrated land-use system is the tropical homegarden agroforestry system (Kumar and Nair, 2006).

The diversity of crop and tree species in agroforestry homegardens fosters recycling of nutrients, increases efficiency in the use of moisture, nutrients and sunlight, and reduces the incidence of weeds, pests and diseases (Abebe *et al.*, 2010; Altieri, 1995; Trenbath, 1999), while the decomposition of litter and manuring (Hoogerbrugge and Fresco, 1993) and the low export of harvested products (Nair, 2001) maintain soil fertility. The diverse and complex nature of these systems contributes to their productivity and sustainability (Perfecto *et al.*, 2009; Wojtkowski, 1993). Multispecies homegarden agroforestry systems in the tropics have been producing sustained yields for centuries, with a relatively low energy input for their establishment and maintenance (Mergen, 1987), and they are generally considered to be economically efficient, ecologically sound and biologically sustainable agroforestry systems (Kumar and Nair, 2006).

Homegardens conserve biodiversity, including agrobiodiversity (Albuquerque *et al.* 2005; Casas *et al.* 1996; Engels, 2002; Goddard *et al.*, 2009). We can distinguish between planned diversity, to produce goods and services, and associated diversity (Vandermeer *et al.* 1998). Planned diversity includes tolerated and favoured spontaneous useful plants (Blankaert *et al.* 2004); associated diversity refers to the diversity that is brought in, for example by animals attracted by the planned diversity in the homegardens.

An example of an old homegarden system is the enset-coffee homegarden in Southern Ethiopia (Abebe, 2005; Abebe *et al.*, 2006, 2010). This system is characterised by a unique combination of two native perennial crops, enset and coffee. Enset (*Enset ventricosum* Welw. Cheesman) is the staple food in the region and coffee (*Coffea arabica* L.) the main cash crop. These crops are grown together with a multitude of crop and tree species and livestock in multistorey agroforestry systems. The homegardens are located at altitudes of 1,500-2,300 m, where moisture and temperature conditions are favourable for agriculture. The sites are characterised by a moist to sub-humid warm subtropical climate with average annual rainfall of

1000-1,800 mm and a mean temperature of 15-20 °C. The dominant soils are eutric nitrosols. Unlike most homegardens in the tropics, which are supplementary food production systems in addition to farm fields located elsewhere, the enset-coffee homegardens are 'integral' or complete farm systems from which farmers derive almost all their subsistence and cash needs (Abebe *et al.*, 2006).

The total area of enset-coffee homegardens in Southern Ethiopia is estimated at 576,000 ha (BODEP, 1996). This region is one of the most densely populated areas of the country, with an average population density of about 500 people/km^2. This chapter addresses the question of how this increasing population pressure is affecting the dynamics and characteristics of enset-coffee agroforests and what can be done to ensure the ecological, socioeconomic and agricultural sustainability of this land-use system. We first assess the diversity and composition of these enset-coffee homegardens in the Sidama region (see Figure 4.1) between 1999 and 2002 (Abebe, 2005) and then evaluate the effects of socioeconomic factors on the diversity and composition of crops and trees. Detailed data were collected from 144 homegardens located in 12 different sites from four districts. The sites were selected to systematically cover the various geographic conditions in the study area. Within each administrative unit the homegardens were selected randomly (for further details see Abebe, 2005).

In the sections below we characterise the composition and structure of the enset-coffee agroforestry homegardens, evaluate the central role of enset in the

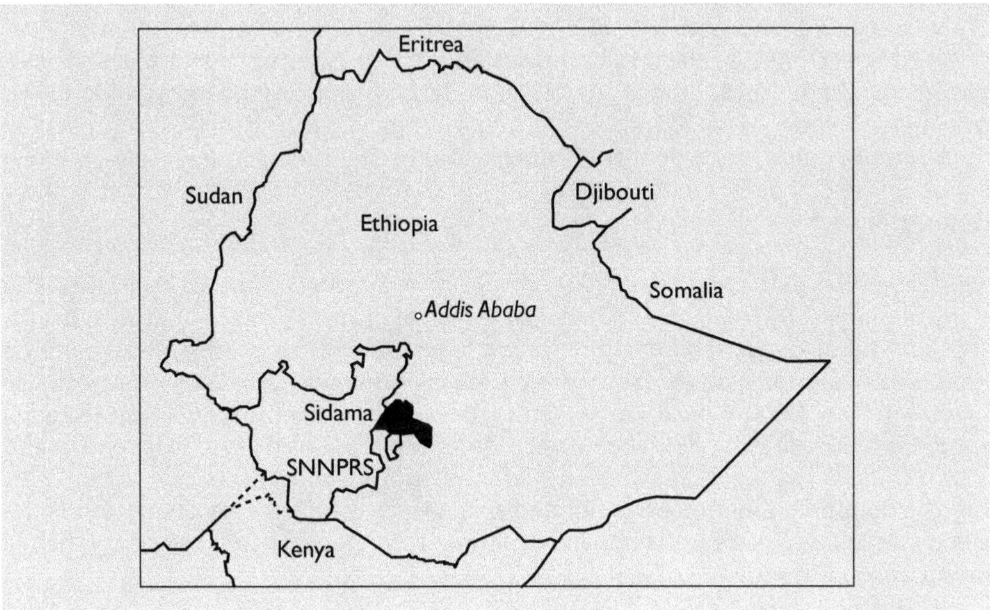

Figure 4.1. Location of Sidama in Ethiopia.

system, discuss the recent trends in land-use changes and the factors that impact on these changes, and discuss the implications of these dynamics for agricultural sustainability.

4.2 Structure and composition of enset-coffee agroforestry homegardens

Most of the homegardens in Sidama have evolved from forests. Farmers maintain the upper storey trees and clear the undergrowth to open up space for planting enset, coffee and other crops. More species and varieties of crops and trees are gradually introduced. Partial harvesting of the upper storey trees takes place to obtain wood and to create favourable growing conditions for the undergrowth. In recent years, farmers have been observed to convert their plot of grazing land into enset-coffee gardens.

The homegardens were on average 0.70 ha in area (range: 0.18-7.46 ha). The land holdings included residential and grazing areas (14%), cultivated lands (82%) and sometimes some specialised woodlots (4%). The wealth status of the households was the major variable influencing farm size, with average sizes of 0.55, 1.46 and 2.75 ha for the poor, middle income and rich farmers respectively. Coffee and enset dominate in over 50% of the homegarden area, while the share of other crops is much smaller.

The homegardens contain three main components of agricultural biodiversity: crops, trees and livestock. A total of 198 species of cultivated crops (78) and trees (120) were recorded in the 144 homegardens. The mean number of plant species per homegarden was 37 (range: 15-78). In addition to species diversity, a high level of genetic diversity was found in two major crops, enset and coffee, represented by 42 and 24 cultivars respectively (Abebe, 2005). The homegardens also included seven livestock species: cattle, goats, sheep, donkeys, horses, mules and poultry (mainly chickens).

The diversity and density of the crops vary spatially and temporally. The homegardens show four distinct vertical layers of crops. Vegetables, spices, beans, root and tuber crops occupy the lowest strata up to 1.5 m. Coffee, enset, maize, khat, sugarcane and some fruit trees, such as banana and papaya, occupy the layer between 1.5 and 5 m. Larger fruit trees, such as avocado and white sapote, some shrubs and pollarded shade trees dominate the third layer between 5 and 12 m. The fourth layer, above 12 m, which could sometimes extend up to 35 m high, is dominated by timber-producing shade trees. The lowermost stratum is the richest in species (64%); the second stratum is the densest because of the heavy dominance of enset and coffee.

The key components of these homegardens are the perennial crops enset and coffee, with an area share of 26.4 and 36.6% respectively (Abebe *et al.*, 2010). These crops also have a relatively long life span: 5-8 years for enset and 24-30 years for coffee. Other perennials, such as khat (*Catha edulis* Vahl. Forssk, ex. Endl.) and fruit trees are also widely grown. While perennials are present throughout the year, herbaceous crops, such as cereals and vegetables, are grown only during the rainy period, which lasts from six to nine months. Hence, the diversity and composition of crops increases during the rainy season and declines when the herbaceous annuals are harvested.

The high diversity of species, which combines crops, trees and livestock with different uses and production cycles, is considered an essential component of sustainable agriculture because of the wide socioeconomic and ecological roles it plays in these systems. The roles include, among others, year-round production of food and wood products, decreased risks of production failure, increased productivity over time (Netting and Stone, 1996), maintenance of genetic diversity and landscape protection (Perfecto *et al.*, 2009; Trenbath, 1999).

The number of crop species grown in a farm is an important indicator of diversity. However, from the utility point of view, it is not only the number that matters, but also the varied functions of the crops. In order to fulfil the dietary and cash requirements of the households, the systems should include both food crops – supplying carbohydrates, proteins, fat and vitamins – and cash crops. The Sidama homegardens are diverse in functional crop types. In addition to tree species, ten functional groups of plants were identified: fruit crops (24%), root and tuber crops (16%), vegetables (15%), stimulant crops (10%), cereals (9%), pulses (6%), spices and condiments (5%), oil crops (3%), medicinal crops (3%) and miscellaneous crops (9%). Each functional crop type was represented by 3 to 15 species (Abebe *et al.*, 2006).

Food species with high energy content are dominant crop species in the homegardens, and enset is the most important staple food crop (Table 4.1). The supply of proteins comes from pulses, but poor farmers with small landholdings produce very low quantities of pulses (e.g. beans). Rich and middle-income farmers produce relatively large quantities of pulses and they also obtain protein from animal sources. Most homegardens have a combination of different functional crop types. This combination, coupled with the presence of livestock in the homegardens, illustrates the integrated nature of the farming system.

4.3 The central role of enset in homegardens and livelihoods

Enset (Figure 4.2) is the staple food crop for more than 15 million people in the Southern region of Ethiopia. Food is extracted from its pseudostem and corm, and its by-products and other parts provide different products, such as fibre,

Table 4.1. Production area and nutritive value of common food crops in the enset-coffee agroforestry homegardens of Sidama, Southern Ethiopia.

Functional groups of crop	Main food crops	Mean area coverage (%)	Nutrient value per 100 g of edible portions[1]	
			Calories (kcal)	% protein
Root and tuber crops	enset	26.4	190	1.5
	sweet potato	2.4	114	1.5
	yam	0.8	41	1.0
Cereals	maize	17.0	363	10.0
Vegetables	cabbage	1.6	28	2.0
Fruits	avocado	0.8	165	1.5
	banana	0.6	116	1.0
Pulses	haricot bean	1.1	339	24.0

[1] Nutrient value as given in World Food Programme (1991).

*Figure 4.2. Matured enset plant (*Enset ventricosum *Welw. Cheesman).*

wrapping material and fodder. In addition to these products, enset delivers vital environmental services. It protects the soil from erosion and runoff, it serves as shade and improves the microclimate for the undergrowth, while the litter from the leaves and pseudostems improves soil fertility. Unlike annual plants, only a small portion of the biomass is taken out of the system when it is harvested, while the largest portion is returned directly as litter or indirectly through animal manure. Enset guarantees food security and stability in the household economy in that processed produce can be stored for a long time and live plants can be maintained on farm and harvested any time when the need arises (Negash, 2001; Tsegaye, 2002), which is why enset areas are not prone to famine (Rahmato, 1995). In addition to providing food and cash, enset also plays an important social role. People in Southern Ethiopia consider it as part of their cultural heritage because of its diverse uses, for example in medicine, ritual offerings and compensation for payment (Murdock, 1959; Negash, 2001; Shack, 1966; Yentiso, 1996).

Finally, enset is an ornamental crop. According to Sidama elders, 'a cottage without enset in its surroundings is like a bird without feathers' (meaning 'unattractive or without grace'), and giant enset plants around houses are a source of pride for the owners. Enset production and management practices in Southern Ethiopia are extensively documented (Negash, 2001; Tesfaye, 2002; Tsegaye, 2002). In general, enset has ideal attributes for low-input sustainable agricultural production systems: it is high yielding, it can be harvested any time once it is about four years old, it does not require external inputs, and it has multiple functions, including protecting and/or enhancing the environment. It is, therefore, no wonder that it has supported a very dense population for a long time.

Women are more involved in managing the enset crop than men. Land preparation, planting and transplanting are usually done by men, while women are active in transporting planting materials (suckers), manuring, harvesting, processing and marketing (Negash, 2001; Woldetensaye, 1997). Women are also more knowledgeable about the attributes of the different landraces of enset (Negash, 2001) and can select suitable ones for different end uses.

Enset covers a high proportion of the homegarden areas. Out of the total area of the farms allotted for crop production, enset takes up an average of 26.4%, the area varying across sites and between households (Table 4.2). This variability is mainly related to farm size and access to markets, as described in detail in the following section.

Surprisingly, the proportion of land allotted to enset is more or less similar across the different wealth groups of farmers. In contrast, the share of coffee increases with growing wealth, either directly or indirectly through an increase in the area of land holdings. Large farm holders can produce sufficient enset for consumption and allot an increasingly larger area to grow coffee as a cash crop. Poor farmers have a slightly higher share of land under maize because, with a lower farm size,

Table 4.2. Mean area share of the major crops in the homegardens of Sidama region, Southern Ethiopia.

Wealth status of farmers	Area share of the major crops (%)[1]		
	Enset	Coffee	Maize
Poor (n=48)	26.9±15.9	31.8±16.1	19.9±17.3
Medium (n=48)	28.1±14.7	36.2±19.8	14.8±13.5
Rich (n=48)	24.3±10.5	41.9±17.8	14.5±9.3
Mean (n=144)	26.4±13.9	36.6±18.3	16.4±13.4

[1] ± shows standard deviation of the mean.

farmers cannot depend solely on enset, which takes at least four years to mature, and also need early maturing crops such as maize and sweet potato. This illustrates the fact that enset, as the main staple crop of the household, is a strategic crop and determines cropping plans, land use, use of technology and consumption and marketing decisions (Rahmato, 1995). If a household has sufficient enset plants for consumption, it will plant high value crops for the market. If there is no market, cash crops will be curtailed in favour of food crops. If the number of mature, ready-to-harvest enset plants on the farm is insufficient to feed the family during the year, more land will be used to grow annual food crops such as maize and sweet potato. These additional annual crops are planted in the part of the homegarden usually allocated for transplanting enset, or in areas freed up by thinning the young enset plantations. The nature of enset cultivation lends itself well to such flexible decisions. For the first two years enset is often grown in very high densities of about 10,000 or more plants per ha. If there is ample space, the thinnings are transplanted into the new plots; if there is not enough space, the thinnings are used as livestock feed or incorporated in the soil.

The crucial decisions made by the households are the area to be planted with enset and the distribution of enset plants over age classes, as households aim to obtain a sustainable yearly harvest. For instance, if enset is harvested after a rotation period of six years, dividing the enset fields into six equal plots of different age classes (1-6) could give a more or less uniform and sustainable harvest every year. This is applicable to enset seedlings planted in their final density of about 2,000 plants per ha. If a two-year old sapling is transplanted, it is expected to stay in the field for four years and hence the enset area should be divided into four equal plots of different age classes. In this region, enset is normally harvested after five to eight years. If we take the average of six years, as stated in the example above, the ideal proportional area of each age class of enset plantation should be 16.7%. In this example, one would expect mature enset plants to be harvested from a 16.7% share of the enset field each year, giving a more or less equal and sustainable yield. This is not the case, however. In a study of age structure of enset in the agroforestry homegardens of Sidama, Abebe (unpublished data) found that enset fields are dominated by immature enset plants: about 90% of the plants were less than four years old, while only 10% were five to eight years old (Figure 4.3). Poor farmers in particular have a smaller area under harvestable enset. The ratio of mature enset (age older than five years) increased with wealth: 4.7% (poor), 10% (medium) and 17.3% (rich). This confirms that farmers with a smaller area of farmland cannot afford to depend on enset only, as it takes at least four years to mature.

A poor farmer with a land holding of about 0.5 ha allots a quarter of it to enset, of which 4.7% (60 m^2) is mature enset. With a common spacing of 2×2.5 m for mature enset plants, the farmer will have only 12 plants to harvest during the whole year, which – at 31 kg dry matter per plant (Makiso, 1975; Shank and Ertiro, 1996) – corresponds to a total of 372 kg of edible enset (*Qocho*) per year. The average consumption of enset as a staple food is high, at 0.55 kg per person per day in the Gurage area (Pijls *et al.*, 1994) and 0.80 kg in the Sidama area (Abebe, 2005). Even with the lower consumption rate of 0.55 kg/person/day, the poor farmer with a family of five requires about 1000 kg of *Qocho* for the whole year. Thus, the potentially harvestable mature enset from these small farms satisfies only about 40% of the yearly consumption. These results are confirmed by Ashagari (1984), who found that a family of five using enset as a staple food consumes 60 mature plants per year.

Poor farmers are therefore forced to fill the consumption gap with annual food crops, such as maize and sweet potato – crops that are often grown as monocultures because of their high light requirement. As the dry matter yield of enset per unit

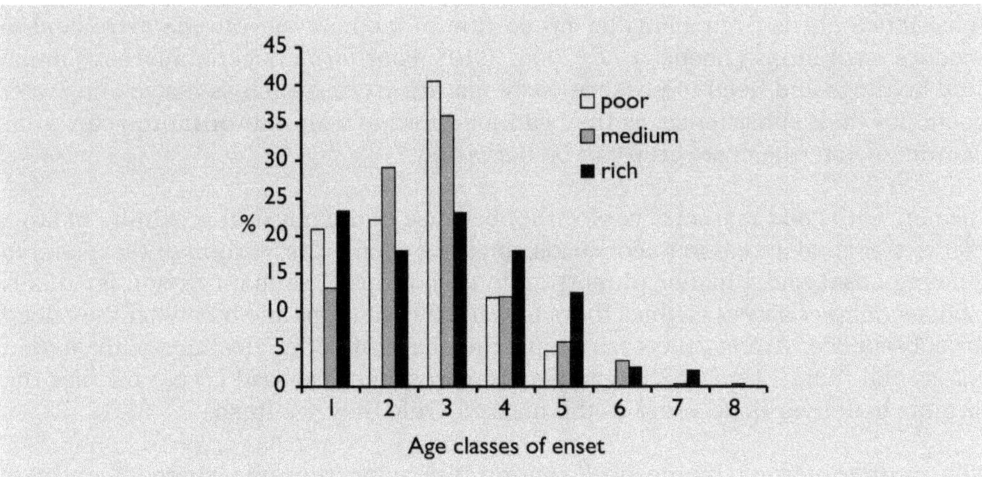

Figure 4.3. Average proportion of different age classes of enset plantations in Sidama, Ethiopia. Results are given for rich (n=16), medium (n=16) and poor (n=16) households.

area and time is much higher than other crops (Tsegaye, 2002), research and extension efforts should focus on reversing the declining trend of enset plantations on small farms through a systematic allotment of enset plantations. To achieve this, the number of enset plants sufficient for annual consumption of a household should be estimated, and that number should be planted every year during the whole rotation period. For instance, if 60 mature enset plants are required for the household's annual consumption and the rotation length (harvesting age) is six years, then a total of 360 enset plants divided into six age categories (1-6) of 60 enset plants each should be grown. At the end of the sixth year the mature 60 plants can be harvested and replaced by new seedlings. This rotation would help to obtain a sufficient and regular yield from enset and could contribute to the maintenance of the complexity and integrated nature of the homegardens. During the establishment phases of the different age classes of enset plantations, poor farmers might face food shortage because of the lack of space to accommodate seasonal crops. In such cases, safety-net or other relief support programmes could fill the gap.

4.4 Dynamics in enset area composition

Although households on average have roughly one quarter of their homegarden area devoted to enset, there are large variations across sites and between households within a site. Various socioeconomic factors (e.g. farm size, access to markets) and biophysical factors (e.g. altitude) influence how much land farmers use for enset production.

Farmers with large farms have a lower proportion of their land under enset and a higher proportion under the cash crops coffee and khat. They meet their subsistence needs from a smaller proportion of land and devote the extra land to produce cash crops (Abebe *et al.*, 2006, 2010). Poor farmers generally have small land holdings and need to produce early maturing crops such as maize and sweet potato for their subsistence, as they cannot afford to wait four or more years after planting before the enset crop can be harvested.

Farmers with good market access, either because of the physical proximity of large markets or good access to major roads, devote a smaller proportion of their land to growing enset and a higher proportion to cash crops. The main reason for this is obvious: market access enables them to sell the cash crops and buy what they need for subsistence. Among the cash crops, the share of coffee declines with market access and that of *khat* and pineapple increases, because road access enables the farmers to deliver these crops to the market while they are fresh.

The most important biophysical factor influencing the area share of enset is altitude. In this intermediate highland (*Woyna dega*) agroecological zone where altitude ranges from 1,500 to 2,200 m, the share of enset increases with altitude.

In the warmer and relatively drier low altitude sites (1,500-1,750 m) farmers grow more of other food crops, mainly maize and sweet potato (Abebe, 2005). Pineapple is also grown on these sites.

The variation among farmers and the impacts of such drivers show that the land-use systems in the Sidama region are dynamic: farmers make individual choices and respond to emerging challenges and opportunities. When per capita land holding declines, mostly as a result of increased population pressure, farmers tend to produce more staple food crops than cash crops. Among the food crops, they give priority to annuals rather than perennials (such as enset), because of immediate subsistence needs. Farmers also assess their comparative economic advantage and respond to increasing market opportunities, as was manifested in the expansion of khat and pineapple in the study areas. Such responses by farmers to market opportunities have also been reported in earlier studies (Abebe *et al.*, 2010; Abdoellah *et al.*, 2006; Kaya *et al.*, 2002; Wiersum, 1982, 2006).

The dynamics of the agroforestry homegarden system of Southern Ethopia can be considered as a productive bricolage process on the micro scale – with productive bricolage being defined as 'the flexible and dynamic crafting together of various livelihood options and its associated impacts on the landscape' (Ros-Tonen, Chapter 2 in this volume). These homegardens already carry a very dense population, which is still growing fast. The high population growth (2.5%) is likely to result in increased fragmentation of farmlands. Based on the results of this study, we expect that this will lead to a further expansion of annual crops and a reduction in the perennial components (crops and trees) and livestock, which are vital for the sustenance of the system. Likewise, the increasing commercialisation of crops such as khat and pineapple, which is associated with an increased share of annual food crops (maize and sweet potato), is leading to the reduction in the key native perennial crops, enset and coffee. The share of native and ecologically friendly multipurpose trees declines with better road access, while that of eucalyptus tends to increase. This will lead to a further uniformisaton of the landscape.

4.5 Implications of the dynamics in enset-coffee agroforestry homegardens for agricultural sustainability

A high species diversity which combines crops, trees and animals having different uses and production cycles, and the perennial nature of major components that help in the recycling and efficient use of nutrients, are generally considered to be essential components of sustainable agriculture (Perfecto *et al.*, 2009; Tilman *et al.*, 2002; Trenbath, 1999). At the same time, there are weighty arguments in favour of agricultural intensification, which state that the food demand of the growing population can be met by maximising yield per unit area of land by using different inputs. Obviously, such intensification could increase production in the short term, but it might affect the long-term sustainability of the systems by reducing the production potential of the resource base. Of course, this would depend on the quantity and quality of the inputs into the system. Many studies propose the use of 'sustainable intensification' (Godfray *et al.*, 2010; Pretty *et al.*, 2003; Royal Society of London, 2009; Tilman *et al.*, 2002), which seeks to combine increased production of food from the same area of land with a reduction in environmental impacts (Royal Society of London, 2009). While the enset-coffee homegardens of Southern Ethiopia display the characteristics of sustainable agriculture, there is room for increasing productivity through intensification. If intensification involves a reduction in plant diversity and the perennial nature of the systems, leading to the development of monoculture fields, this could disrupt the ecosystem services provided by the integrated multistorey agroforestry systems. However, integrating high value crops into the systems without significantly affecting the composition and diversity of components could lead to sustainable intensification.

We argue that maintaining the existing components of the enset-coffee agroforestry homegardens would contribute to its ecological and socioeconomic sustainability. Ecological sustainability indicates the extent to which natural resources are conserved so that farming can be continued, while economic sustainability shows its suitability and adaptability to local farming conditions and its economic viability (Holden and Linnerud, 2007; Neher, 1992; Ojiem *et al.*, 2006; Peyre *et al.*, 2006; Pretty *et al.*, 2003). Below, we evaluate the ecological and socioeconomic sustainability aspects of the enset-coffee homegardens and justify the need for their maintenance.

Ecological sustainability aspects

Several features of the enset-coffee homegardens are important for the ecological sustainability of the enset-coffee homegarden, including (1) maintenance of species diversity, which is important for risk spreading and minimisation, genetic conservation of native species, efficient resource use and biological pest control; (2) reduced use or elimination of soluble or synthetic fertilisers, increased or improved use of manure and other organic materials as soil ameliorants, and soil conservation; (3) reduced use or elimination of chemical pesticides, replacing these with integrated pest management practices and system diversity; and (4) self-sufficiency, by using on-farm or locally available 'internal' resources and a minimum or conditional use of purchased 'external' resources, which contributes to the long-term conservation of the resource base and environmental resilience of the systems (Abebe *et al.*, 2006). Considering the importance of these elements (Holden and Linnerud, 2007; Pretty *et al.*, 2003; Trenbath, 1999), we strongly believe that enset-coffee homegardens are ecologically sustainable, at least compared with many other agricultural systems. Of course, the recent land-use changes currently threaten this sustainability.

The results also show that the stability of the system depends both on its diversity and on the specific characteristics of the two main components, enset and coffee.

As an evergreen perennial crop, enset gives permanent shade to understorey crops, including coffee. Soil management is facilitated by the use of enset residues as a mulching material. Coffee is an ideal complimentary crop to enset: not only is it architecturally and ecologically compatible with enset, but the harvests of both enset and coffee involve only selected plant parts and do not involve major exports of soil nutrients (Abebe *et al.*, 2006). Thus, enset and coffee can also be considered as keystone species contributing to the ecological sustainability of the system.

Socioeconomic sustainability aspects

The maintenance of high species diversity in the enset-coffee homegardens also contributes to socioeconomic stability. As in other homegarden systems, the diversity of crop, tree and livestock species with different uses and production cycles enables the year-round production of different products, reduces the risk of production failure, allows flexible use of labour and enables efficient cycling of locally available resources, which means that production does not depend on external inputs (Abebe, 2005; Kumar and Nair, 2004). Moreover, the enset-coffee homegardens also possess several specific features that promote socioeconomic sustainability. They not only have a high species diversity, but also a high diversity in functional crop types, notably staple food crops and cash crops, in addition to the usual supplementary homegarden crops (Abebe *et al.*, 2006). The carbohydrate-rich basic food crops enset and maize are supplemented by pulses, vegetables, fruits and animal products that provide proteins, fats and vitamins, and by trees that provide resources for construction and household energy. This is of crucial importance for households, especially poor families. The cash crops coffee, khat and pineapple are increasingly being incorporated into the homegarden, giving a more balanced household income. Farmers clearly adjust the proportion of subsistence and cash crops to their household requirements. Both the diversity of crops and the inclusion of the perennial enset spread the risk to households of individual crop failures. The possibility of flexibly harvesting enset as a staple food is one of the main reasons why the Southern highlands are relatively free from hunger (Brandt *et al.*, 1997; Rahmato, 1995).

Like the ecological sustainability of the agroforestry homegarden system, its socioeconomic sustainability cannot be explained by its species diversity alone, but also by the specific features of the two key species enset and coffee. Enset is both a food crop and a provider of different products, such as fibre and fodder. It is therefore ideally suited to low-external input agricultural production systems (Negash, 2001; Tesfaye, 2002), while its high productivity and multiple functions provide sustenance for a very dense population, which is often two to three times higher than in the cereal-based systems found in other parts of Ethiopia. Coffee serves as a main cash crop supplementing the mainly subsistence-oriented enset production. The combined production provides a good safety net in times of crop or market failure (Abebe *et al.*, 2006). Also, processing and marketing of coffee creates employment for many people. Consequently, not only from an ecological point of view, but also from a socioeconomic point of view, coffee and enset can be considered as keystone species.

4.6 Conclusions

The traditional enset-coffee agroforestry homegardens are characterised as a sustainable land-use system, but this does not mean that they are not subject to change. Decreasing farm size and increased commercialisation is affecting the system. The shift from the traditional enset-coffee systems towards inclusion of other food and cash crops has diversified the diet and increased household income. The expansion of open-field food crops (maize, sweet potato) and of monocultural cash crops (khat, pineapple), causes not only a gradual loss of species diversity and tree biomass, but also a decrease in the perennial crops and native tree species, to the detriment of the dominance of the two key species enset and coffee. As these are considered to play a significant role in the stability and resilience of the homegarden system (Montagnini, 2006; Trenbath, 1999), we expect this to have negative impacts on the landscape. It results in a gradual reduction of the ecological benefits derived from these integrated and complex systems, which threatens their long-term sustainability. We therefore argue for the maintenance and use of the perennial component in homegardens and the integration of new crops into the existing multistorey system, without affecting the biodiverse nature of enset-coffee agroforestry homegardens and without losing their essential keystone species, enset and coffee. This can be achieved through directed extension services by government institutions and local organisations.

Additional research should focus on the integration of expanding cash crops (such as khat and pineapple) into the existing systems without changing the multistorey structure of the homegardens. As enset produces the highest volume of food per unit area and time (Tsegaye, 2002), and because of its different end uses and diverse ecological roles, the future of these homegardens depends on the maintenance of enset-based staple food production. Thus, strategies should be developed to reverse the increasing dependence on maize and enhance the systematic production of enset.

References

Abdoellah, O.S., Hadikusumah, H.Y., Takeuchi, K., Okubo, S. and Parikesit, G.B. (2006). Commercialization of homegardens in an Indonesian village: vegetation composition and functional changes. In: Kumar B.M. and Nair P.K.R. (eds.) Tropical homegardens: a time-tested example of sustainable agroforestry. Springer Science, Dordrecht, the Netherlands, pp. 233-250.

Abebe, T. (2005). Diversity in homegarden agroforestry systems of southern Ethiopia. Tropical Resource Management Paper No. 59, Wageningen University, Wageningen, the Netherlands, 143 pp.

Abebe, T., Wiersum, K.F. and Bongers, F. (2010). Spatial and temporal variation in crop diversity in agroforestry homegardens of Southern Ethiopia. Agroforestry Systems 78: 309-322.

Abebe, T., Wiersum, K.F., Bongers, F. and Sterck, F. (2006). Diversity and dynamics in homegardens of Southern Ethiopia. In: Kumar B.M. and Nair P.K.R. (eds.) Tropical homegardens: a time-tested example of sustainable agroforestry. Springer Science, Dordrecht, the Netherlands, pp. 87-103.

Albuquerque, U.P., Andrade, L.H.C. and Caballero, J. (2005). Structure and floristics of homegardens in Northeastern Brazil. Journal of Arid Environments 62: 491-506.

Altieri, M.A. (1995). Agroecology: the science of sustainable agriculture (2nd ed). Westview Press, Boulder, CO, USA.

Ashagari, D. (1984). Studies on the bacterial wilt of enset and prospects for its control. Ethiopian Agricultural research Conference, Addis Ababa, Ethiopia.

Blanckaert, I., Swennen, R.L., Paredes-Flores, M., Rosas-López, R. and Lira-Saade, R. (2004). Floristic composition, plant uses and management practices in home gardens of San Rafael Coxcotlaán, Valle of Tehuacán-Cuicatlán, Mexico. Journal of Arid Environments 57: 179-201.

Bureau of Development and Economic Planning of Southern Nations, Nationalities and Peoples' Regional State (BODEP) (1996). Regional Conservation Strategy, Volume 1. BODEP, Awassa, Ethiopia.

Brandt, S.A., Spring, A., Hiebsch, C., McCabe, J.T., Tabogie, E., Diro, M., Wolde-Michael, G., Yntiso, G., Shigeta, M. and Tesfaye, S. (1997). The tree against hunger. American Association for the Advancement of Science, Washington D.C., 56 pp.

Casas, A., Vázquez, M., Viveros, J. and Caballero, J. (1996). Plant management among the Nahua and the Mixtec in the Balsas River Basin, Mexico: an ethnobotanical approach to the study of plant domestication. Human Ecology 24: 455-478.

Cox, G.W. and Atkins, M.D. (1979). Agricultural ecology: an analysis of world food production systems. WH Freeman and Co., San Francisco, 721 pp.

Engels, J.M.M. (2002). Home gardens – a genetic resource perspective. In: Watson, J.W. and Eyzaguirre P.B. (eds.) Proceedings of the second international home garden workshop. Bioversity International, Rome, Italy, pp. 3-9.

Goddard, M.A., Dougill, A.J. and Benton, T.G. (2009). Scaling up from gardens: biodiversity conservation in urban environments. Trends in Ecology and Evolution 25: 90-98.

Godfray, H.C.J., Beddington, J.R., Crute, I.R., Haddad, L., Lawrence, D., Muir, J.F., Pretty, J., Robinson, S., Thomas, S.M. and Toulmin, C. (2010). Food security: the challenge of feeding 9 billion people. Science 327: 812-818.

Holden, E. and Linnerud, K. (2007). The sustainable development area: Satisfying basic needs and safeguarding ecological sustainability. Sustainable Development 15: 174-187.

Hoogerbrugge, I.D. and Fresco, L.O. (1993). Homegarden systems: agricultural characteristics and challenges. Gatekeeper Series No. 39. International Institute for Environment and Development, London, UK.

Kaya, M., Kammesheidt, L. and Weidelt, H.J. (2002). The forest garden system of Saparua Island, central Maluku, Indonesia, and its role in maintaining tree species diversity. Agroforestry Systems 54: 225-234.

Kumar, B.M. and Nair, P.K.R. (2004). The enigma of tropical homegardens. Agroforestry Systems 61: 135-152.

Kumar, B.M. and Nair P.K.R. (eds.) (2006). Tropical homegardens. A time-tested example of sustainable agroforestry. Springer, Dordrecht, the Netherlands.

Makiso, T. (1975). A study on the yield of Enset. Proceedings of the 6th IAR Research Seminar. Addis Ababa, Ethiopia.

Matson, P.A., Parton, W.J., Power, A.G. and Swift, M.J. (2002). Agricultural intensification and ecosystem properties. Science 277: 504.

Mergen, F. (1987). Research opportunities to improve the production of homegardens. Agroforestry Systems 5: 57-67.

Montagnini, F. (2006). Homegardens of MesoAmerica: biodiversity, food security, and nutrient management. In: Kumar B.M. and Nair P.K.R. (eds.) Tropical homegardens: a time-tested example of sustainable agroforestry, Springer Science, Dordrecht, the Netherlands, pp. 233-250.

Murdock, G.P. (1959). Africa: its peoples, their cultures and history. McGraw Hill, New York.

Nair, P.K.R. (1993). An introduction to agroforestry. Kluwer Academic Publishers, Dordrecht, the Netherlands.

Nair, P.K.R. (2001). Do tropical homegardens elude science, or is it the other way around? Agroforestry Systems 53: 239-245.

Negash, A. (2001). Diversity and conservation of enset (*Ensete ventricosum* Welw. Cheesman) and its relation to household food and livelihood security in South-western Ethiopia. PhD dissertation, Wageningen University, the Netherlands.

Neher, D. (1992). Ecological sustainability in agricultural systems: Definition and measurement. Journal of Sustainable Agriculture 2: 51-61.

Netting, R.M.C. and Stone, M.P. (1996). Agrodiversity on a farming frontier: Kofyar smallholders on the Benue plains of Central Nigeria. Africa 66: 52-69.

Ojiem, J.O., De Ridder, N., Vanlauwe, B. and Giller, K.E. (2006). Socio-ecological niche: a conceptual framework for integration of legumes in smallholder farming systems. International Journal of Agricultural Sustainability 4: 79-93.

Padoch, C. and Pinedo-Vasquez, M. (2010). Saving swidden to save biodiversity. Biotropica 42: 550-552.

Perfecto, I., Vandermeer, J. and Wright, E.L. (2009). Nature's matrix: linking agriculture, conservation and food sovereignty. Earthscan Ltd., London, UK.

Peyre, A., Guidal, A., Wiersum, K.F. and Bongers, F. (2006). Homegarden dynamics in Kerala, India. In: Kumar B.M. and Nair P.K.R. (eds.) Tropical homegardens: a time-tested example of sustainable agroforestry. Springer Science, Dordrecht, the Netherlands, pp. 87-103.

Pijls, L.T.J., Timmer, A.A.M., Wolde-Gebriel Z. and West C.E. (1995). Cultivation, preparation and consumption of ensete (*Ensete ventricosum*) in Ethiopia. Journal of the Science and Food of Agriculture 67: 1-11.

Pretty, J.N., Morison, J.I.L. and Hine, R.E. (2003). Reducing food poverty by increasing agricultural sustainability in developing countries. Agriculture, Ecosystems and Environment 95: 217-234.

Rahmato, D. (1995). Resilience and vulnerability: enset agriculture in Southern Ethiopia. Journal of Ethiopian studies 28: 23-51.

Royal Society of London (2009). Reaping the benefits: science and sustainable intensification of global agriculture. Royal Society of London, London, UK.

Shack, W.A. (1966). The Gurage: a people of enset culture. International African Institute, Oxford University Press, London, UK.

Shank, R. and Ertiro, C. (1996). A linear model for predicting enset plant yield and assessment of *Qocho* production in Ethiopia. World Food Programme/SNNPRS/UNDP units for Ethiopia. Addis Ababa, Ethiopia.

Swift, M.J. and Ingram, J.S.I. (eds.) (1996). Effects of global change on multi-species agroecosystems. Global Change and Terrestrial Ecosystems, Report no. 13, GCTE Activity 3.4, GCTE Focus 3 Office, Wallingford, UK.

Tesfaye, B. (2002). Studies on landrace diversity, *in vivo* and *in vitro* regeneration of enset (*Enset ventricosum* Welw.). PhD dissertation, Humboldt University, Berlin, Germany.

Tilman, D., Cassman, K.G., Matson, P.A., Naylor, R. and Polasky, S. (2002). Agricultural sustainability and intensive production practices. Nature 418: 671-677.

Trenbath, B.R. (1999). Multispecies cropping systems in India: Predictions of their productivity, stability, resilience and ecological sustainability. Agroforestry Systems 45: 61-107.

Tsegaye, A. (2002). On indigenous production, genetic diversity and crop ecology of enset (*Enset ventricosum* (Welw.) Cheesman). PhD dissertation, Wageningen University, Wageningen, the Netherlands.

Vandermeer, J., Van Noordwijk, M., Anderson, J., Ong, C. and Perfecto, I. (1998). Global change and multispecies agroecosystems: concepts and issues. Agriculture, Ecosystems and Environment 67: 1-22.

Wiersum, K.F. (1982). Tree gardening and Taungya on Java: examples of agroforestry techniques in the humid tropics. Agroforestry Systems 1: 53-70.

Wiersum, K.F. (2006). Diversity and change in homegarden cultivation in Indonesia. In: Kumar B.M. and Nair P.K.R. (eds.) Tropical homegardens: a time-tested example of sustainable agroforestry. Springer Science, Dordrecht, the Netherlands, pp. 233-250.

Wiersum, K.F. and Gonzalez, I.C.G. (2000). Intermediate forest types as nature – human systems: characteristics and future potential. Paper presented at the international workshop 'Cultivating in tropical forests, the evolution and sustainability of intermediate systems between extractivism and plantations', June 28-July 1, 2000, Lofoten, Norway.

Wojtkowski, P.A. (1993). Toward an understanding of tropical homegardens. Agroforestry Systems 24: 215-222.

Woldetensaye, A. (1997). The ecology and production of *Ensete ventricosum*. Doctoral thesis, Swedish University of Agricultural Sciences, Uppsala, Sweden.

World Food Programme (1991). Manual on food nutritional status. World Food Programme, Rome, Italy.

Yentiso, G. (1996). Economic and socio-cultural significance of enset among the Ari of Southwest Ethiopia. In: Abate, T., Hiebsch, C., Brandt, S.A. and Gebremariam, S. (eds.) Enset-based sustainable agriculture in Ethiopia. Institute of Agricultural Research, Addis Ababa, Ethiopia.

5. Woodfuel and producers' livelihoods in the Congo Basin[1]

Jolien Schure
Centre for International Forestry Research (CIFOR), Wageningen University, Forest and Nature Conservation Policy Group (FNP), P.O. Box 47, 6700 AA Wageningen, the Netherlands; jolien.schure@wur.nl

Abstract

Fuelwood and charcoal are often among the first forest products to be mentioned that provide income to the poor. However, there is little quantitative data about the exact contribution made by these products in terms of household income and how they may help to reduce poverty. The objective of this study was to gain a better understanding of the contribution made by the commercial woodfuel sector to the livelihoods of producers around Kinshasa, capital of the Democratic Republic of Congo. Surveys and interviews with fuelwood and charcoal producers in the supply zones of Kinshasa indicate that (1) profits from fuelwood (USD 288 per year) and charcoal (USD 405 per year) represent a considerable part of the average household income (47% and 75% respectively); (2) annually, the poorest households produce less than half the amount of charcoal than the richest income group, but there is no significant difference between income groups for fuelwood; and (3) woodfuel revenues contribute to both poverty mitigation (used for basic needs such as food, education and healthcare) and poverty elimination (providing capital for investment in other economic activities), and hence livelihood diversification, for 72% of the fuelwood producers and 91% of the charcoal producers. In conclusion, woodfuel revenues contribute to poverty reduction, but the overall income of producers is still low and the poorest seem to benefit least. Overharvesting takes place on a large scale, which poses a risk to the sustainability of this livelihood activity.

Keywords: fuelwood, charcoal, Democratic Republic of Congo, poverty reduction, forest products

5.1 Introduction

Woodfuel (see Box 5.1) is often put high on the list of forest products that provide monetary and non-monetary values to the poor. In the absence of an accessible and

[1] This work was first presented at the workshop 'Exploring the Forest-Poverty Link' organised by the Poverty Environment Network (PEN), 13-14 June 2011, Norwich, UK. A longer version of this chapter, in co-authorship with K. Freerk Wiersum and Patrice Levang, has been submitted to World Development.

Box 5.1. Some definitions.
Woodfuel: all types of biofuels originating directly or indirectly from woody biomass, including fuelwood and charcoal.
Fuelwood: woodfuel in which the original composition of the wood is preserved, including wood in the raw and residues from wood-processing industries.
Charcoal: the solid residue derived from carbonisation, distillation, pyrolysis and torrefaction of fuelwood.
(Food and Agriculture Organization, 2004).

affordable alternative energy source, an estimated 83% of the population in sub-Saharan Africa depend on woodfuel for cooking (Daurella and Foster, 2009). This consumption of fuelwood and charcoal represents 90% of all wood exploitation in Africa (Food and Agriculture Organization, 2011). In the Congo Basin, the second largest tropical forest reserve in the world after the Amazon rainforest, the impact of the woodfuel sector on the natural resource base is most visible around urban areas, in savannah zones and near protected areas. Urban centres in the Democratic Republic of Congo (DRC), the Republic of Congo and Cameroon show a shift from consumption of fuelwood to charcoal (Schure *et al.*, in press). Woodfuel not only serves as an energy source, but its production and trade offer income opportunities, especially with the currently increasing prices in cities, where there is little access to alternative sources of energy (Marien, 2009).

The DRC is among the poorest countries in the world, ranking 168th of 169 countries in the Human Development Index, which measures health, education and income (UNDP, 2010). The country has roughly 130 million hectares of tropical forest, amounting to 60% of the Congo Basin forests. The main question is how this abundance of resources can effectively contribute to poverty reduction. On forest products, the literature is divided into one group that believes the value of these products will give communities revenues, and therefore incentives to protect the forest (e.g. Nepstad and Schwartzman, 1992; Plotkin and Famolare, 1992), and a second group that believes open access will lead to overharvesting and elite capture of high-value products, and at best may provide a safety net function for the poor (Belcher and Schreckenberg, 2007; Wunder, 2001).

Although it is generally believed that woodfuel can provide important benefits to households, there are few quantitative studies on the exact contribution in terms of income and how it contributes to livelihoods (Arnold *et al.*, 2006). The extent to which the sector provides employment is often not adequately reflected in statistics because woodfuel production and trade take place mainly in the informal sector (Openshaw, 2010). Another aspect is the lack of understanding of the actual returns for producers, because detailed data on the cost structure are not available (Neumann and Hirsch, 2000). The increasing prices of woodfuel may suggest that there are interesting profits to be made. However, in the absence of alternative income opportunities and a relatively low entry threshold to woodfuel production, competition may be fierce and the actual profits small (Arnold *et al.*, 2006). Some recent studies offer further insights into the extent to which the woodfuel sector contributes to poverty reduction. For the whole of sub-Saharan Africa it is estimated that 13 million people could be employed in the biomass (mostly wood) energy sector (Openshaw, 2010), offering income opportunities to the rural poor (Shackleton *et al.*, 2007). Around urban centres in Uganda, charcoal revenues were found to contribute positively to household cash incomes by on average USD 122 per year (Khundi *et al.*, 2011). This seems to be of particular importance to households with low agricultural productivity. However, the poorest of the poor often do not participate in charcoal production (Khundi *et al.*, 2011). In Malawi, charcoal revenues contributed to lower income households (Zulu, 2010). This illustrates that there can be various outcomes with regard to the contribution of woodfuel revenues to people's livelihoods. Better context-specific knowledge about the driving forces and the socioeconomic aspects of woodfuel production could therefore assist in developing appropriate actions for environmental and livelihood needs.

5.2 Theoretical outlook

The concept of non-timber forest products (NTFPs), the subject of this part of the book, addresses the role of forest products in the development of rural forest communities, often linked to questions about co-benefits for environmental conservation (see Ros-Tonen in this volume). NTFPs were initially defined to encompass 'all biological materials other than timber which are extracted from natural forests for human use. These include foods, medicines, spices, essential oils, resins, gums, latexes, tannins, dyes, ornamental plants, wildlife (products and live animals), fuelwood and raw materials, notably rattan, bamboo, smallwood and fibres' (De Beer and McDermott 1989: 17). Over the past two decades, scholars have proposed many different definitions of NTFPs, often linked to certain perceptions of why the concept is important. This has led to ambiguity, both about which products are included or excluded (mainly animal products and wooden materials) and what the resource base of NTFPs is (forests, degraded forests, agroforestry, plantations) (Belcher, 2003; Ros-Tonen and Wiersum, 2005).

The status of woodfuel (charcoal and fuelwood) within the NTFP concept is often unclear. This was already noted by De Beer and McDermott (1989: 17) when they first presented the concept: 'Clearly the distinction between "timber" and "non-timber" can become blurred, as when such borderline cases as large-scale fuelwood harvesting for urban markets and the use of timber in rural house construction are considered'. In the literature about woodfuel, authors either include woodfuel as NTFPs by referring to the definition 'everything but timber', or exclude it by referring to 'non-wood forest products', meaning 'everything but wood'. Others avoid the debate entirely by discussing the product separately or by referring to 'forest products' more generally. The Food and Agricultural Organization of the United Nations (FAO) chooses to refer to 'non-wood forest products' in its definition and data collection instead of NTFPs, and placed charcoal and fuelwood into separate categories because of their overall large volumes (Killmann, 2009). In the debate on the possible benefits of NTFPs, woodfuel has been largely ignored. During the last four decades, the focus of attention has alternated between NTFPs and woodfuel.

In the 1970s, concerns arose about woodfuel shortage, or 'the other energy crisis', triggered by the oil crisis (Eckholm, 1975). Woodfuel became a priority on the policy agendas and large-scale fuelwood plantations and efficient-stove projects were implemented to assure woodfuel supplies. A decade later, in the mid 1980s, the woodfuel crisis did not materialise on a global scale because increasing urbanisation and rising incomes had slowed the demand in many regions (Arnold and Persson, 2003). It was found that woodfuel actually often came from sources other than forests (such as trees on agricultural land, village woodlots, homegardens and public land) and, in cases where local households were faced with less supply, they often coped by reducing consumption (Trossero, 2002). By the late 1980s, interest in the 'woodfuel crisis' and related intervention schemes had already waned considerably (Arnold *et al.*, 2006).

In the early 1990s, the years leading up to the 1992 UN Conference on Conservation and Development, NTFPs became more visible on the development and environment agenda following the publication of Peters *et al.* (1989) which indicated that the value of NTFPs can be higher than that of wood. The reasons why woodfuel was largely ignored in this new NTFP agenda were that it was primarily conceived of as a subsistence resource and that, as a wood product, it did not present a sustainable harvesting scenario (Sills *et al.*, 2011). The quickly increasing number of NTFP studies since the mid 1990s provided insights into the roles of NTFPs related to agriculture, conservation and poverty reduction. It became clear that forests play diverse roles in local livelihoods and the earlier high expectations about environment and development reconciliation were lowered (Arnold and Perez, 2001; Ros-Tonen and Wiersum, 2005). Moreover, there was recognition for the growing group of urban and peri-urban residents involved in NTFP use. The earlier debate on which products to include in or exclude from the NTFP concept and where the limits of the resource base are has now been replaced by a much broader perspective: 'a range of forest products (from medicinal plants to timber), a range of forest management intensities (from pure extraction to intensive forest management), and a range of markets (from international to local)' (Sills *et al.*, 2011: 43). This new understanding of NTFPs favours a 'resource-in-context' approach that focuses on the entire landscape and livelihoods (Ros-Tonen and Wiersum, 2005).

In the 2000s, it became obvious that environmental and social issues remain and that the story behind woodfuel supply is more complex than initially thought. On an international level, woodfuel gradually came to be seen as a potential modern renewable energy source (Trossero, 2002). Climate change politics concerning carbon sequestration programmes took an interest in benefits to be gained at the production and consumption level. For developing countries, it was recognised that high woodfuel demand, often related to clearing land for agriculture, could have serious impacts. This is of special concern for Africa; the only continent where household woodfuel consumption is expected to increase in the coming decades and where particular attention should be paid to the increasing urban demands of charcoal (Girard, 2002). A review of the woodfuel sector of 18 African cities by the FAO (2010) showed a wide variation in possible environmental outcomes, ranging from low pressure and weak impact to accelerated degradation of peri-urban areas. A similar variation is expected for the producers' livelihood outcomes. In this respect, woodfuel studies can learn from the experiences of NTFP studies by focusing on livelihoods and following a resource-in-context approach (Ros-Tonen and Wiersum, 2005).

Woodfuel production is often part of the livelihood strategies that people develop and use to earn a living. Livelihood is defined as 'the control an individual, family or other social group has over an income and/or a package of sources that can be used or changed to maintain a living' (Blaikie *et al.*, 1994: 9). The combination of diverse income-generating activities, often adding non-agricultural sources of income to rural livelihoods, contributes to 'livelihood diversification', which is defined as 'the process by which rural families construct a diverse portfolio of activities and social support capabilities in order to survive and to improve their standards of living' (Ellis, 1998: 1).

Shackleton *et al.* (2008: 519) classified the role of forest products in four different types of livelihood strategies (c.f. Ros-Tonen, Chapter 2 in this volume):

- Specialisation: the extraction of forest products is a primary activity that contributes over 50% of household income.
- Diversification 'by choice': the extraction of forest products is a variable activity by choice among several other income-generating activities that provides less than 50% of household income.
- Coping or 'diversification by necessity': forest products represent the most important source of income (over 50% of an overall low household income), but their extraction is mainly a response to a lack of alternatives, and as such can become a long-term strategy.
- Survival: forest products provide less than 50% of household income, generate low returns, are a last resort, and function as a safety net.

This typology of different livelihood strategies is useful for identifying the contribution made by forest products to the household portfolio and to poverty reduction. Poverty reduction can result either in poverty mitigation or poverty elimination. Poverty mitigation or avoidance can be considered as 'the use of forest resources to meet household subsistence needs, to fulfil a safety net function in times of emergency, or to serve as a "gap filler" in seasonal periods of low income'. Poverty elimination is considered as 'the use of forest resources to help lift the household out of poverty by functioning as a source of savings, investment, accumulation, asset building, and lasting increases in income and well-being' (Sunderlin *et al.*, 2005). Wealthier and poorer households are likely to gain different benefits from forest products. The wealthier households may be able to invest in cash-generating opportunities and bring these products to the urban markets where they sell for higher prices. Poorer households, on the contrary, may depend more on forest products for subsistence income and their opportunities to sell these products are likely to be restricted to local markets (Paumgarten and Shackleton, 2009).

Building upon the insights derived from NTFP and livelihood studies, this chapter assesses the role of woodfuel production and trade in local livelihoods. It takes a resource-in-context approach and highlights issues of urban woodfuel production and the role of woodfuel in the poverty reduction strategies of producer households.

5.3 Methodology

The objective of this study was to gain a better understanding of the contribution made by the commercial woodfuel sector to the livelihoods of producers around Kinshasa, the capital of the DRC. More specifically, it aimed to find answers to the following research questions:

- What is the role of woodfuel production in the livelihood strategies of producers in the supply zones of Kinshasa?
- What are the amounts and net revenues from charcoal and fuelwood production for the producers?
- How do these woodfuel revenues contribute to poverty mitigation and poverty elimination?

Answers to these questions would improve understanding of how woodfuel production interacts with other livelihood strategies of producers, how important it is to their household income and whether it has the potential to reduce poverty.

Study area

Kinshasa is the capital of the DRC and is situated on the south bank of the Congo River. With a population of 5.8 million inhabitants, Kinshasa is one of the biggest cities in sub-Saharan Africa (Division Urbaine de l'Intérieur, Décentralisation et Sécurité, 2010). The city's population is young, with half of the inhabitants under 20 year old. It is also a poor population. The poverty rate – defined by UNDP (2009) as the proportion of people living on less than USD 1.25 a day – was 41.6% in 2005 and underemployment rates are high (UNDP, 2009). The economy is dominated by the informal sector, which provides 89.5% of household income. The woodfuel supply zone of Kinshasa stretches to the southwest and northeast over an average distance of 102 km for fuelwood and 135 km for charcoal. The landscape is dominated by hilly sites and the Plateau Batéké, which is mainly savannah with pockets of forests. The provinces of Kinshasa and Bas-Congo provide most of the woodfuel (43% and 34% respectively) and most producers supplying Kinshasa are based in these two provinces (Table 5.1). Some supplies (24%) enter via the Congo River from the more distant provinces of Bandundu, Equateur and Orientale, up to 1000 km away (Schure *et al.*, in press) (Figure 5.1).

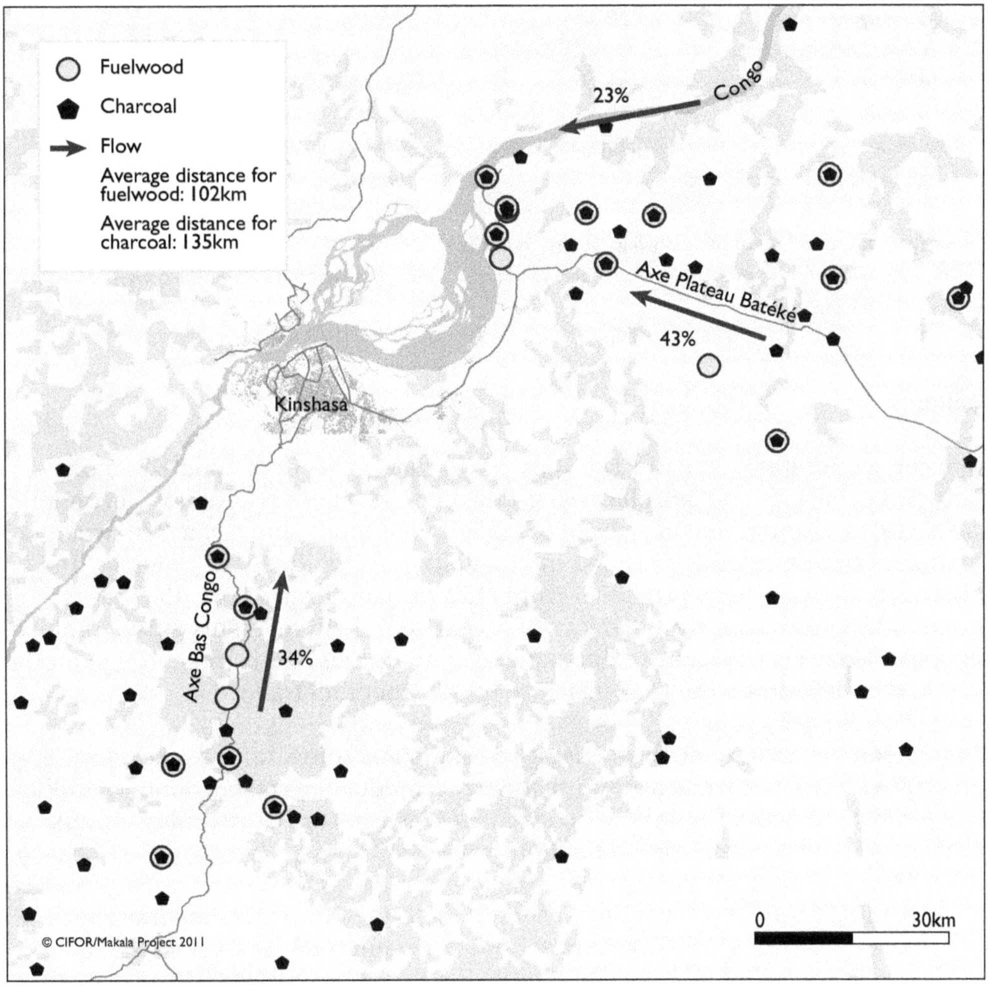

Figure 5.1. Woodfuel flows in the Kinshasa supply zone.

Table 5.1. Socioeconomic statistics for Kinshasa and Bas-Congo provinces (UNDP, 2009).

	Kinshasa province	Bas-Congo province	DRC
Area	9,965 km^2	55,920 km^2	2,344,858 km^2
Population	5.8 million	3.2 million	64.3 million
Poverty rate[1]	41.6%	69.8%	71.3%
Average household size	6.0	4.8	5.3
Primary school education rate	74.8%	71.1%	76.3%
Underemployment rate[2]	53.1%	73.4%	72.7%
Origin of household income			
Informal sector	89.5%	93.6%	94.6%
Public sector	6.1%	4.0%	3.6%
Private sector	4.4%	2.4%	1.8%
Average monthly household income	USD 84	USD 37	USD 42

[1] Poverty rate: proportion of people living on less than USD 1.25 per day.
[2] Underemployment rate: proportion of people who involuntarily work less than 35 hours a week or earn less than the minimum salary.

Methods

Between September 2009 and February 2010, surveys were conducted among 200 fuelwood collectors and 392 charcoal producers. The surveys addressed economic, social and environmental aspects of woodfuel production and trade. The woodfuel producers were randomly selected in 20 villages in the Kinshasa province and 20 villages in the Bas-Congo province. From May 2010 until April 2011, 10 villages with a total of 50 producers were subject to a year-round survey. This survey helped, among other things, to capture fluctuations of quantities and prices throughout the year and to determine the cost structure of the woodfuel trade.

The collected data were analysed using the statistical software package SPSS (16.0). Descriptive statistics were used to picture the producers' socioeconomic profiles. The producers' annual profits were calculated as the sum of revenues minus the various costs involved in the production process:

> Annual Profit = (Units sold × Average selling price) – (Transportation costs + Taxes + Labour costs + Tool and equipment costs + Costs to access the resource)

Statistical analyses used to compare differences between groups (type of product, income level) are the chi-square test for independence and one-way between-groups analysis of variance (ANOVA). For the continuous data on production quantity, the logarithm was used to transform the data into normal distribution to meet the assumptions for the parametric test (ANOVA).

5.4 Results

The role of woodfuel in overall livelihood strategies

Producers in the supply zones of Kinshasa are mostly villagers who live and produce in their village of origin. Fuelwood production, from cutting the wood to packing the bundles for sale, is often done by family members; sometimes other villagers are hired for a daily fee or compensation in kind. Urban-rural migration for production purposes has been observed to some extent, especially to meet commercial demand for fuelwood, for example by bakeries. The production cycle of charcoal, from cutting the wood to packing the bags, involves mainly local men, who are paid a daily fee (or sometimes a percentage of the production) by the landowner or the 'head producer'. Also common is the system of *ristourne*, or labour groups, by which roughly ten producers work in turns to build one member's charcoal kiln. The host producer provides drinks and food and reciprocates in labour to other members of the group. Some urban citizens have specialised in charcoal production because of limited employment opportunities in town. They mainly function as intermediaries and financiers of activities, and hire local villagers to produce the charcoal.

Producers of woodfuel are representative of average rural households in terms of working-age population and ethnic groups. The average household income of producers in the Kinshasa supply zone is less than USD 1.25 per day (poverty line standard), which is similar to other rural households in the area (Table 5.2).

For most producer households, agriculture is the primary income-generating activity besides woodfuel. Other revenues mainly come from petty trade, timber exploitation and livestock. Non-wood forest products may have important non-monetary values and provide an additional income for around 7% of all producers, but are rarely the primary source of income. Only a few producers have some paid seasonal work. This confirms the overall absence of employment opportunities in

Table 5.2. Sociodemographic information on fuelwood and charcoal producers.

	Producers Kinshasa region	
	Fuelwood producers	**Charcoal producers**
Women producers (%)	5.8	3.4
Age (years)	33.6	34.2
Working experience (years)	7.5	7.1
Household size (number of members)	6.1	5.4
Household income (USD/month)	51	45

the region and the fact that virtually all producers depend on more than one activity for their household income (Table 5.3). Overall, the income-generating activities of charcoal and fuelwood producers look quite similar, but there is a significant difference between their primary activities ($P=0.018$) [chi-square test: $\chi^2(1, n=463) = 5.6$, $P=0.018$, Cramer's $V=0.011$ (small effect size)]. Charcoal producers do relatively more farming than fuelwood producers. Timber exploitation is more common among fuelwood producers and fuelwood production by those involved in timber exploitation is likely to be a by-product of this activity, in which case the tree trunk is sold as log or sawn wood and the smaller branches and waste wood serve as fuelwood.

Woodfuel production is often related to clearing land for agriculture, either by the woodfuel producer or the landowner. Over half (52%) of the charcoal produced for the Kinshasa market comes from agricultural land, 28% from forests and 16% from other sources, such as (village) plantations. Fuelwood is mainly collected on agricultural land (62%), followed by forests (30%). Comparing these practices with the preferences voiced by producers yielded quite a difference: the preferred places for woodfuel production are forests (40%), plantations (37%) and agricultural land (20%). This suggests that producers have difficulties finding or accessing new woodfuel resources and that there is a large potential for woodfuel plantations. According to 77% of the producers, the distance to the production sites has increased over the past five years. Resource scarcity is also reflected in the proportion of producers (12.6%) who stated that they have been involved in conflicts surrounding woodfuel production. Paucity of the resource is the main cause of such conflicts.

Table 5.3. Income-earning activities mentioned by fuelwood and charcoal producers (% of producers who list this activity among their five main income-earning activities).

	Fuelwood producers	Charcoal producers
Agriculture	89.3%	94.2%
Petty trade	23.5%	20.9%
Timber exploitation	29.1%	10.8%
Livestock	10.2%	9.0%
NWFP[1]	7.7%	5.5%
Craftworker	5.6%	0.8%
Paid seasonal labour	3.1%	1.6%
Fishing	0.0%	1.1%

[1] NWFP refers to the collection of all non-wood forest products, including honey and bushmeat.

Production quantities and net revenues

Fuelwood producers produce 10.6 tonnes of wood per year (SD = 16.8) mainly meant for sale. Only 6.8% of the production is for self-consumption, in-kind payments or gifts. Of all the commercialised fuelwood, roughly half is sold in the village and the other half is brought to the urban market. The average selling price is CDF 2,236 (USD 2.7) per bundle, with a bundle weighing 43.3 kg on average.

On average, each charcoal producer produces 8.2 tonnes of charcoal per year (SD = 9.9). The charcoal is mainly sold for cash; only minor quantities (less than 4.6%) are bartered for goods or used for own consumption. A third of the producers sell the charcoal directly in the village and two-thirds of them sell the bags at the urban market. The average price per bag is CDF 7,290 (USD 8.8), with a bag corresponding to 57.0 kg on average.

The costs of production over a 12-month period are highest for transportation, followed by labour, tools, costs to access the resource base and taxes (Figure 5.2). Taxes are likely to be underrepresented because the informal taxes collected on the roads are often paid by the transporter and therefore included in the price of transportation. As may be expected, the costs of producing charcoal (73% of total revenues) are higher than the costs of fuelwood production (57% of total revenues).

Fuelwood harvest and sales provide producers with an annual net income of CDF 237,974 (USD 288) or CDF 19,831 (USD 24) per month, calculated as follows:

> Annual Profit = (Units sold × Average selling price) – (Transportation costs + Taxes + Labour costs + Tools and equipment costs + Costs of access to resource) = (228 × 2,236) – (107,952 + 17,160 + 84,696 + 17,133 + 44,893) = CDF 237,974 (USD 296) per year.

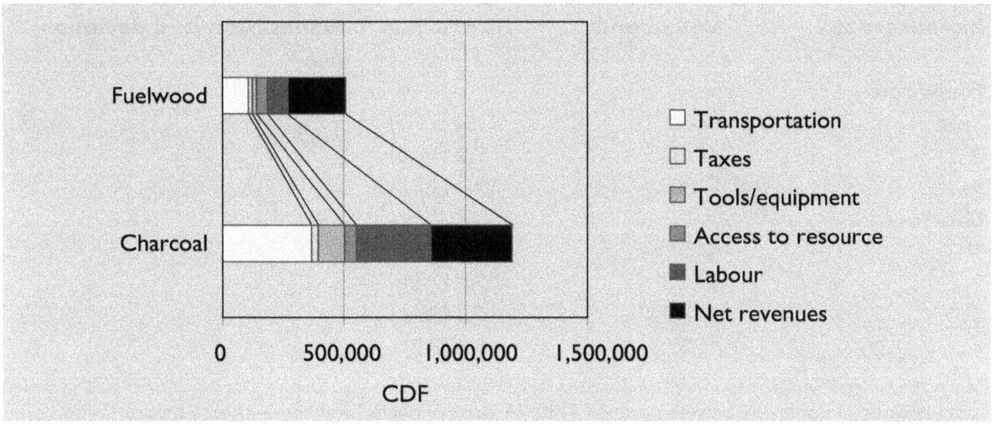

Figure 5.2. Costs and net revenues (CDF per year) to the producer of fuelwood and charcoal production.

This represents 47% of the average household income. The average charcoal producer's net revenues are CDF 334,961 (USD 405) per year or CDF 27,913 (USD 34) per month:

> Annual Profit = (Units sold × Average selling price) – (Transportation costs + Taxes + Labour costs + Tools and equipment costs + Costs of access to resource) = (163.5 × 7,290) – (368,390 + 28,490 + 304,393 + 107,103 + 48,578) = CDF 334,961 (USD 405) per year.

This represents 75% of an average charcoal producer's household income.

A one-way between-groups ANOVA was conducted to explore the impact of income level on the production quantity of woodfuel. For fuelwood production, there was no significant difference in production quantity between income groups. For charcoal production there was a statistically significant difference ($P<0.001$) in production quantities for the four income groups: $F(3, 348) = 17.7$, $P=0.000$. The mean production was lowest for the low income group and highest for the highest income groups (Table 5.4). The only non-significant difference is between groups three and four, suggesting that producer households with a monthly income over USD 50 can produce optimal quantities of charcoal.

Contribution by woodfuel revenues to poverty reduction

The benefits from woodfuel income can be divided into short-term benefits, for example the money used for daily needs, and the long-term benefits that may be obtained from investments made with the revenues. Charcoal making is not

Table 5.4. Mean production of fuelwood and charcoal (tonnes per year) per producer per income category.

Income group	USD/month	Production in tonnes (standard deviation)
Fuelwood		
1+2[a]	<50	11.4 (20.9)
3	51-100	9.2 (10.6)
4	>100	12.4 (11.4)
Charcoal		
1	<11	3.5 (3.7)
2	11-50	7.0 (7.8)
3	51-100	11.4 (13.0)
4	>100	9.6 (8.2)

[a] For fuelwood the one-way between groups ANOVA was conducted with three income groups instead of four (group one and two combined) to meet the conditions for this test.

an easy way to make money (as it involves hard physical labour), but it is quick money in times of need, also called *intervention rapide* by producers. Fuelwood collection can be more of a continuous activity and is less associated with hardship, but like charcoal it is among the very few possibilities for producers to gain cash income. Another important aspect is that the income-generating opportunity of woodfuel does not oblige the farmer to neglect agriculture, which remains the primary activity. Villagers make a clear distinction between the different types of benefits from the two activities:

> If there were only charcoal, the household would have no food! Charcoal doesn't give us *chikwange* [local staple food based on fermented cassava] or [cassava] leaves like our fields do. We make charcoal for one reason: to pay for something.

The revenues from fuelwood and charcoal production are used to pay for daily needs: mainly food (mentioned by 91% of the producers), education (88%) and healthcare (72%). The notion of 'quick intervention' by producers also illustrates that the activity serves as a 'gap filler' in times of need. Woodfuel production takes place during the slack agricultural season, mostly in the period between planting and harvesting when rural households are in need of cash income. This contribution to basic needs and the gap filler function shows a clear role for woodfuel production in poverty mitigation.

The potential contribution made by woodfuel production to poverty elimination can be seen by the fact that 72% of the fuelwood producers and 90% of the charcoal producers invest part of their woodfuel revenues in other economic activities of the household. These investments flow mainly to agriculture (90.9%), with fewer producers indicating reinvestments in petty trade (7.3%), livestock (1.1%) and fishing (0.5%). These investments can be considered as a way to accumulate assets and diversify risks and income opportunities.

5.5 Discussion

The contribution made by the commercial woodfuel sector to the livelihoods of producers in the supply zone of Kinshasa was analysed at three levels: (1) by analysing the role of woodfuel production as part of wider household strategies; (2) by quantifying the production and net revenues; and (3) by investigating how these revenues may affect poverty mitigation as well as poverty elimination.

The livelihoods of producers are characterised by a process of diversification. They combine different activities, with woodfuel adding non-agricultural income to the household. The top four listed income-generating activities besides woodfuel are agriculture, trade, timber exploitation and livestock. Other non-wood forest products (mainly honey and bushmeat) may have important non-monetary value, but are rarely the primary source of income. In terms of the fourfold classification of livelihood strategies by Shackleton *et al.* (2008) – specialisation, diversification 'by choice', coping or 'necessity', and survival – woodfuel production is for most producers a coping or 'necessity' strategy. The product provides the major part of household income and woodfuel production is a response to the lack of other cash-generating activities. Producers are generally involved in it for a long time.

Charcoal and fuelwood production in the supply zone of Kinshasa is mostly oriented towards the market and generating monetary benefits. Producers said that it is not 'easy money', but rather 'quick money', also referred to as *intervention rapide*. In terms of quantifying their financial benefits, it was found that the net benefits of charcoal are higher (USD 405 per year) than those of fuelwood (USD 288 per year). This confirms that charcoal processing is a more attractive option than fuelwood production.

Production quantities vary greatly between producers, confirming that poor producers tend to benefit least, but only for charcoal. This is probably due to the fact that the poorest households lack the financial means needed to sufficiently invest in the relatively high costs of transport, labour and equipment for charcoal production. The poorest households produce less than half of the quantity per month than the richer households. Optimal investments for these activities are made by those in the USD 51-100 per month income category. This is in line with findings on charcoal revenues in Uganda (Khundi *et al.*, 2011) and studies about other high-value forest products that indicate that the wealthier households tend to benefit more (Belcher, 2005; Paumgarten and Shackleton, 2009). In contrast, no significant difference was found in the quantities of fuelwood produced between the different income groups, probably because the initial investments and its value are lower. Further research is needed to gain more understanding about possible underlying factors.

In terms of poverty mitigation, we see that woodfuel revenues contribute to meeting subsistence needs as producers spend part of their income on daily basic needs such as food, education and healthcare. More important, however, is that woodfuel as part of diversified livelihood strategies also seems to provide capital for investments in other activities – mainly agriculture and some petty trade, livestock and fishing – which points towards poverty elimination. In this case, forests provide 'sources of rent extracted and invested in sectors with greater production potentials, allowing for a diversification of income sources' (Wunder, 2001: 1829). Khundi *et al.* (2011: 204) suggest that charcoal revenues assist in filling an income gap for those households with less agricultural production. Our study shows that woodfuel revenues contribute to both income for daily needs and investments in agriculture. The latter is likely to contribute to higher agricultural capacity and productivity.

This contribution of woodfuel revenues to accumulating capital is a key finding, but caution is warranted to avoid overoptimism, for three main reasons:

1. The overall income of producers is still low: 61.0% of producer households have an income of less than USD 50 per month, which reflects the generally high poverty level in the DRC.
2. The poorest seem to benefit less from the woodfuel trade, especially when it comes to charcoal production, where most profits can be made.
3. Environmental costs are not internalised and there is a risk of overharvesting. The latter occurs on a large scale, which poses a risk to the long-term sustainability of this livelihood activity. This may create a poverty trap, because a lack of these earnings also lessens investments in other activities and decreases these accumulated benefits.

5.6 Conclusion

This 'woodfuel in the context of Kinshasa' study highlights the integral part played by woodfuel production in household livelihood strategies that are based mainly on agriculture, and that it provides important benefits to producer households. Woodfuel sales provide households with considerable financial benefits (USD 288 per year for fuelwood and USD 405 per year for charcoal) representing a considerable part of the average household cash income (47% and 75% respectively). The poorest households produce less than half the amount of charcoal than the richest income group (3.5 versus 9.6 tonnes per year per producer), but for fuelwood there is no significant difference between income groups, which might be explained by the lower entry threshold (less investment needed) and the lower value of the product. Woodfuel revenues contribute to both poverty mitigation (as they are used for basic needs, such as food, education and healthcare) and poverty elimination (providing capital for investment in other economic activities), and hence livelihood diversification for 72% of the fuelwood producers and 91% of the charcoal producers.

In conclusion, woodfuel revenues contribute to poverty reduction, but not in a way that reconciles development and conservation efforts. Woodfuel production in the Kinshasa context has a considerable impact on the resource base. In the likely event that demand continues to increase in the near future, those households depending most on woodfuel revenues may end up in a particularly difficult situation, missing out on financial resources for daily needs and investments in other activities. The potential that woodfuel offers to the livelihood of producers in the supply zone of Kinshasa should be central to any future plans for guaranteeing energy supply while maintaining the natural resource base.

Acknowledgements

This study is part of CIFOR's contribution to the Makala 'Sustainable management of woodfuel in the Democratic Republic of Congo' project (2009-2013), an initiative coordinated by the French Agricultural Research Centre for International Development (CIRAD), with financial support from the European Union (see http://projets.cirad.fr/makala). The content of this chapter is the sole responsibility of the author and can in no way be taken to reflect the views of the European Union. The inputs and comments by the participants at the Poverty Environment Network (PEN) workshop (13-14 June 2011, Norwich, UK) are gratefully acknowledged.

References

Arnold, J.E.M. and Perez, M.R. (2001). Can non-timber forest products match tropical forest conservation and development objectives? Ecological Economics 39(3): 437-447.

Arnold, J.E.M. and Persson, R. (2003). Reassessing the fuelwood situation in developing countries. International Forestry Review 5(4): 379-383.

Arnold, J.E.M., Kohlin, G. and Persson, R. (2006). Woodfuels, livelihoods, and policy interventions: changing perspectives. World Development 34(3): 596-611.

Belcher, B.M. (2003). What isn't an NTFP? International Forestry Review 5(2): 161-168.

Belcher, B.M. (2005). Forest product markets, forests and poverty reduction. International Forestry Review 7(2): 82-89.

Belcher, B. and Schreckenberg, K. (2007). Commercialisation of non-timber forest products: a reality check. Development Policy Review 25(3): 355-377.

Blaikie, P., Cannon, T., Davis, I. and Wisner, B. (1994). At risk: natural hazards, people's vulnerability, and disasters. Routledge, London, UK, 284 pp.

Daurella, D.C. and Foster, V. (2009). What can we learn from household surveys on inequalities in cooking fuels in sub-Saharan Africa? World Bank, Washington, DC, USA, 38 pp.

De Beer, J.H. and McDermott, M.J. (1989). The economic value of non-timber forest products in Southeast Asia with emphasis on Indonesia, Malaysia and Thailand. Netherlands Committee for IUCN, Amsterdam, the Netherlands, 175 pp.

Division Urbaine de l'Intérieur, Décentralisation et Sécurité (2010). Tableau statistiques 2009 des services de la population de 24 communes de la ville de Kinshasa. Division Urbaine de l'Intérieur, Décentralisation et Sécurité, Kinshasa, DRC, 6 pp.

Eckholm, E. (1975). The other energy crisis: firewood. Worldwatch Paper 1. Worldwatch Institute, Washington, DC, USA, 22 pp.

Ellis, F. (1998). Household strategies and rural livelihood diversification. Journal of Development Studies 35(1): 1-38.

Food and Agriculture Organization (FAO) (2004). UBET Unified bioenergy terminology. Food and Agriculture Organization, Rome, Italy. Available at: http://www.fao.org/DOCREP/007/j4504E/j4504E00.HTM.

Food and Agriculture Organization (FAO) (2010). Foresterie urbaine et périurbaine en Afrique. Quelles perspectives pour le bois-énergie? Document de travail sur la foresterie urbaine et périurbaine no. 4. Food and Agriculture Organization, Rome, Italy, 95 pp.

Food and Agriculture Organization (FAO) (2011). State of world's forest 2011. Food and Agriculture Organization, Rome, Italy, 264 pp.

Girard, P. (2002). Charcoal production and use in Africa: What future? Unasylva 53(211): 30-34.

Khundi, F., Jagger, P., Shively, G. and Sserunkuuma, D. (2011). Income, poverty and charcoal production in Uganda. Forest Policy and Economics 13(2011): 199-205.

Killmann, W. (2009). Non-wood forest products for livelihoods and sustainable development. In: Spathelf, P. (ed.) Sustainable forest management in a changing world: a European perspective. Springer, Dordrecht, the Netherlands, pp. 83-91.

Marien, J.N. (2009). Forêts périurbaines et bois énergie: quels enjeux pour l'Afrique centrale? Les Forêts du Bassin du Congo. In: De Wasseige, C., Devers, D., De Marcken, P., Eba'a Atyi, R., Nasi, R. and Mayaux, Ph. (eds.) Les forêts du Bassin du Congo – Etat des Forêts 2008. Office des publications de l'Union Européenne, Luxembourg, Luxembourg, pp. 217-230.

Nepstad, D.C. and Schwartzman, S. (1992). Introduction: non-timber product extraction from tropical forests evaluation of a conservation and development strategy. In: Nepstad, D.C. and Schwartzman, S. (eds.) Non-timber products from tropical forests: evaluation of a conservation and development strategy. Advances in Economic Botany 9, New York Botanical Garden, New York, NY, USA, pp. vii-xii.

Neumann, R.P. and Hirsch, E. (2000). Commercialisation of non-timber forest products: Review and analysis of research. Center for International Forestry Research, Bogor, Indonesia, 176 pp.

Openshaw, K. (2010). Biomass energy: employment generation and its contribution to poverty alleviation. Biomass and Bioenergy 34(3): 365-378.

Paumgarten, F. and Shackleton, C.M. (2009). Wealth differentiation in household use and trade in non-timber forest products in South Africa. Ecological Economics 68(12): 2950-2959.

Peters, C.M., Gentry, A.H. and Mendelsohn, R.O. (1989). Valuation of an Amazonian rainforest. Nature 339(6227): 655-656.

Plotkin, M.J. and Famolare, L.M. (1992). Preface. In: Plotkin, M.J. and Famolare, L.M. (eds.) Sustainable harvest and marketing of rain forest products. Island Press, Washington D.C., USA, pp. xiii-xv.

Ros-Tonen, M.A.F. and Wiersum, K.F. (2005). The scope for improving rural livelihoods through non-timber forest products: an evolving research agenda. Forests, Trees and Livelihoods 15(2): 129-148.

Schure, J., Marien, J.N., De Wasseige, C., Drigo, R., Salbitano, F., Dirou, S. and Nkoua, M. (2012). Contribution du bois énergie à la satisfaction des besoins énergétiques des populations d'Afrique Centrale: perspectives pour une gestion durable des ressources disponibles. In: Les Forêts du Bassin du Congo – Etat des Forêts 2010. Office des publications de l'Union Européenne, Luxembourg, Luxembourg, (in press).

Shackleton, S., Campbell, B. and Lotz-Sisitka, H. (2008). Links between the local trade in natural products, livelihoods and poverty alleviation in a semi-arid region of South Africa. World Development 36(3): 505-526.

Shackleton, C.M., Shackleton, S.E., Buiten, E. and Bird, N. (2007). The importance of dry woodlands and forests in rural livelihoods and poverty alleviation in South Africa. Forest Policy and Economics 9: 558-577.

Sills, E., Shanley, P., Paumgarten, F., Beer, J. and Pierce, A. (2011). Evolving perspectives on non-timber forest products. In: Shackleton, S., Shackleton, C. and Shanley, P. (eds.) Non-timber forest products in the global context. Tropical Forest Series. Springer-Verlag, Heidelberg, Germany, pp. 23-51.

Sunderlin, W.D., Angelsen, A., Belcher, B., Burgers, P., Nasi, R., Santoso, L. and Wunder, S. (2005). Livelihoods, forests, and conservation in developing countries: an overview. World Development 33(9): 1383-1402.

Trossero, M.A. (2002). Wood energy: the way ahead. Unasylva 53(211): 3-12.

United Nations Development Programme (UNDP) (2009). Province de Kinshasa: Pauvreté et conditions de vie des ménages. Programme des Nations Unies pour le Développement, Unité de Lutte Contre la Pauvreté, Kinshasa, DRC, 19 pp.

United Nations Development Programme (UNDP) (2010). Human development report 2010. The real wealth of nations: Pathways to human development. United Nations Development Programme, New York, USA, 227 pp.

Wunder, S. (2001). Poverty alleviation and tropical forests – What scope for synergies? World Development 29(11): 1817-1833.

Zulu, L.C. (2010). The forbidden fuel: charcoal, urban woodfuel demand and supply dynamics, community forest management and woodfuel policy in Malawi. Energy Policy 38(7): 3717-3730.

Part II. Community-based natural resource management

6. Discourses of community forestry

Wil de Jong
Center for Integrated Area Studies, Kyoto University, 46 Shimodachi-cho Yoshida, Sakyo-ku, Kyoto, 606-8501 Japan; wdejong@cias.kyoto-u.ac.jp

Abstract

This chapter uses the discourse concept and theory to explore discourses of community forestry. The concept of community forestry (CF) was introduced by the UN Food and Agriculture Organization (FAO) in the 1970s and promoted by them since then. Three major shifts in CF discourses are described: the *forest resource scarcity CF discourse*, the *tropical forest conservation CF discourse* and the *community enterprise CF discourse*, leaving the question unanswered whether they should be considered as distinctive discourses, or rather as one discourse with different key elements. These three are the most prominent manifestations of the CF discourse, but other more specific discourses could also be identified as CF discourses or modifications of the CF discourse, such as the forest common property discourse and the adaptive forest management discourse. I analyse each of the three CF discourses, reviewing their main arguments, their main protagonists and their public manifestations. The chapter also reviews wider societal trends and how they link to the shifts in CF discourses. The forest resource scarcity CF discourse emerged shortly after the beginning of the international concern about whether the world's limited natural resources will be sufficient to meet future economic growth and related consumption. The tropical forest conservation CF discourse appears to have been a result of international concern about tropical deforestation, which has dominated international environmental agendas since the 1980s. Finally, the community enterprise CF discourse appears closely linked to a worldwide shift towards neoliberal economic policies, a reduction in the role of government in non-public affairs and a focus on poverty reduction in international development assistance. The three examples suggest a close relation between CF discourses and practice and other wider societal environmental and economic concerns and discussions. Given the emerging concern about climate change issues, it can be predicted that a future CF discourse will likely incorporate climate change mitigation elements.

Keywords: forestry discourses, community forest management, non-timber forest products, tropical deforestation, forest livelihoods

6.1 Introduction

The term 'community forestry' (CF) came into use in the 1970s, when the UN Food and Agriculture Organization (FAO, 1978) initiated activities and programmes related to rural communities and their forest-related activities. In contemporary forest development circles, the term is used to refer to practices used by rural dwellers to manage tree vegetation. It also refers to a development goal pursued by governments, international development agencies and non-governmental organisations (NGOs) in which such local management of tree vegetation is promoted to achieve a number of environmental and social objectives. Finally, CF is an area of applied academic inquiry, rather like a specialised field within forestry sciences.

Over the years, related terms have been suggested, such as communal forestry, community or communal forest management, community-based natural resource management and others. Numerous publications contain an equally large number of proposed definitions of CF. For example, Arnold (1991: 2) defines CF as 'any situation which intimately involves local people in a forestry activity'. However, from this early definition, quite divergent additional definitions have been suggested. For instance, McDermott and Schreckenberg (2009: 158) emphasise in their definition the 'power and influence' that local people exert over 'decisions regarding management of forests'. Pokorny *et al.* (2008: 84) propose a definition of CF as 'commercial local forestry that is actively being promoted by external agents, emphasising the protagonist role of NGOs, development agencies, or national, regional and local governments'.

The aim of this chapter is to identify and, to the extent that this is possible, analyse different discourses on CF that have appeared over the last few decades. Discourse analysis is a new field in social science research that has become prominent since the 1990s. Discourse analysts hold that in our modern, information-focused society, multiple actors each construct discourses around issues that are of interest to them, and that it is possible to analyse these discourses. The starting point of this chapter is that since CF gained prominence as an area of interest, multiple actors have developed their own CF discourses.

The next section of this chapter briefly dwells on the concept of discourses and discourse theory, which is followed by a review of the most prominent CF discourses. These are then linked to other major societal issues, debates or discourses that may have influenced the CF discourses reviewed here. I then go on to reflect on these links and the implications for our view on CF discourses, and the CF programme in general.

6.2 Environmental discourses

Hajer and Versteeg (2005: 175) define a discourse as 'an ensemble of ideas, concepts and categories through which meaning is given to social and physical phenomena, and which is produced and reproduced through an identifiable set of practices'. According to Foucault, the pre-eminent theorist on discourses who developed much of his theory since the 1970s, discourses 'construct time and location-specific realities' and thus define 'how we perceive those realities' (Winkel, 2011: 2). Foucault's ideas on discourse and knowledge, and to what use such knowledge be put, are also closely linked (Winkel, 2011). The construction of realities through discourses, of which academic debates are an intrinsic part, is, according to Foucault, an exercise of power. This can be interpreted in the sense that discourses, and the actors who produce these discourses, pursue agendas or try to achieve objectives that are in agreement with the interests that these actors share.

If a discourse is viewed from such a perspective, this implies that a discourse analysis reveals the ensemble of ideas, concepts and categories that are being produced and reproduced, plus the actors who produce the discourse, their agenda, their objectives behind the discourse, as well as their shared world-views, ideologies and core values. A discourse analysis can demonstrate different, parallel or consecutive debates related to a certain issue. It may also expose multiple, often competing interests and shifting views (constructions) on one single issue, or disclose the powers and strategies actors use to achieve objectives that best match their interests, or reveal profound differences in world-views, ideologies and core values.

Hajer and Versteeg (2005) have pointed out the increased prominence of discourse analysis related to environmental issues since the second half of the 1990s. They believe that environmental issues are appropriate topics for discourse analysis because environmental policy making and implementations are typically 'messy and complex processes' (Richardson and Sharp, 2001: 194). As a result, different actors may engage in creating their own discourses in order to advance their own interests or point of views. Arts *et al.* (2010) confirm this for forest sciences and provide an overview of the multiple 'forestry discourses' that have appeared over the last fifty years or so.

A review of the international literature on CF shows that the term gained international prominence, but that this prominence has fluctuated over the past few decades as different actors have taken an interest in the topic and have each 'produced' their own CF discourse. Reviewing these different discourses will provide a useful view of how different actors have perceived CF, what they have seen as its potential and what major societal objectives could possibly be addressed through CF.

A last note on discourses and academic debates is warranted here. The concept of discourse is not always easy to apply, given its broad and somewhat open-ended definition. According to discourse theory, an academic discussion – which is mostly a stream of academic papers or other written published material – is also part of a particular discourse. In most cases, however, a discourse includes additional means of information exchange, like position statements, advocacy papers or leaflets, articles in the public media, radio and TV programmes, internet blogs and websites. Arts *et al.* (2010) make a distinction between *meta* discourses, *regulatory* discourses, and *forest* discourses. They suggest a hierarchy of discourses, meaning that global meta discourses (on economic, political or cultural issues; see Arts *et al.*, 2010) and influence, or even shape more specific sectoral discourses, as for instance forest related discourses.

I believe it is useful to make a distinction between a particular academic debate, i.e. a set of academic papers that discuss a certain topic, and the wider discourse of which these papers form a part. That discourse would in addition to the academic papers also be characterized by other multiple information streams. This distinction makes much sense if a particular issue is discussed not only in academia, but also through other means of information dissemination. Separating the two allows us to view the debate on the issue itself as a discourse, which usually implies not commenting on the argument of the discourse, but more on the way in which the debate was shaped, its dynamics, its influence and the like. Separating the academic debate that is linked to a particular discourse makes it possible to contribute to the academic debate itself. This is also relevant for the present chapter, which intends to be a contribution not only to academic debates on discourses, but also to an academic debate on CF itself.

6.3 Discourses of community forestry

The question of whether the CF debate can be viewed as a discourse deserves a fully affirmative answer. Debates on CF represent an ensemble of ideas, concepts and categories which are produced and reproduced to give meaning to a physical reality: communities and their forestry activities (Elands and Wiersum, 2001). Furthermore, these CF ideas are reproduced through multiple media, although maybe not always with exactly the same terminology. Community forestry has been discussed in numerous academic papers, but also in other printed material intended to mobilise actors who might be interested in carrying forward the CF programme initiated by FAO. As I discuss below, for a brief time during the late 1980s in particular the public media (mostly in the USA, but also in Europe and Asia) reported different ideas about tropical forest conservation and the perceived role of local communities, or CF, in contributing to this objective. Publications in public media, rather than specialised discussion outlets, are often considered a more appropriate indicator of a discourse, as understood by discourse theory (Kleinschmit, 2011).

Maybe a more intriguing question, and certainly one more difficult to answer question, is: are there multiple CF discourses? There is reason to view the CF discourse as one discourse with a single overarching theme, but whose more detailed content has changed over the years. Another view is that it is possible to distinguish between subsequent, and even partly concurring, CF discourses, because their core arguments and the actors who carry the discourse have changed substantially from one to the other.

This chapter distinguishes between three major shifts in the CF discourse, identified here as the *forest resource scarcity CF discourse*, the *tropical forest conservation CF discourse* and the *community enterprise CF discourse*. These three CF discourses are largely sequential: the first 'started' during the 1970s, with the introduction of the term by FAO. Since about the mid 1980s, tropical forest decline became an issue of international concern and the international CF project became more linked to this issue, while the resource scarcity issue moved to the background. Somewhere around the early 1990s, forestry reforms adjusted to a worldwide neoliberal economic policy shift and this was again reflected in the information stream related to CF, which emphasised exploitation of forest products with national and international market demand and a related organisational adjustment inside forest communities.

This proposed distinction is, of course, debatable. To begin with, the first and the third discourses referred to above clearly are CF discourses, because many of their manifestations actually use the term community forestry, or similar terms. This is less the case with discourse number two, where in my view a CF discourse was phrased in different terms, for instance as a 'non-timber forest products' discourse. Second, are these three discourse *different* discourses, or are they the same discourse that has adapted itself to the flavour of the time? Here I propose making a distinction between the three, because the protagonists of the discourses changed, the dominant arguments of the discourse changed and the media in which the discourse was displayed also changed. While a periodisation of discourses may be convenient, it is not easy to identify clear-cut transitions between the three: to some extent, the discourses coincided and even overlapped, and they continue to do so today.

Finally, there are several other themes related to CF that have been debated, especially during the 1990s. These include 'forest devolution', 'common property regimes' and 'adaptive co-management'. Each of these subjects can be viewed as a CF discourse or subdiscourse on their own. I limit my analysis to the three mentioned CF discourses, as in my view they are the most important ones that have emerged during the last 40 years or so.

The forest resource scarcity CF discourse

The initial CF discourse, the one that was stimulated especially by FAO, focused on how a perceived danger of forest resource scarcity resulting from unsustainable use would eventually cause a decline in rural welfare. Arnold (1991) pointed out that when FAO initiated its CF focus, development theory and international development cooperation was shifting its focus toward the rural population and agriculture, and to meeting the 'basic needs' of rural dwellers. This followed a general concern about limited local resources and the constraints this might put on satisfying basic needs. This shift in focus towards a rural context gave the importance of forests and trees in rural livelihoods more prominence. The perceived woodfuel scarcity

(Eckholm, 1975) was an important influence on how this link was viewed and the view reproduced. The concern about fuelwood shortage was also linked to the first 'oil crisis', when in 1973 crude prices jumped to previously unknown levels. Together with the woodfuel crisis, it was believed that rural population growth and higher per capita consumption would deplete forest stocks, reduce soil fertility and progressively lead to a scarcity of resources available to rural populations. The practical steps taken in response to this forest resource scarcity CF discourse mostly focused on reforestation and afforestation projects (Arnold, 1991; De Jong, 2010). From its inception, it was recognised that CF required close cooperation with the targeted beneficiaries. Participation became a core term of the initial CF vocabulary, although in practice it remained an ideal rather than a reality (Arnold, 1991; Hickey and Mohan, 2004).

The first CF discourse was, therefore, very much inspired by a more general natural resource depletion discourse (Arts *et al.*, 2010) that dominated the debates, and was diffused by the departments of the multiple agencies that considered these issues as their core mandate (FAO, World Bank, donor-country international development cooperation agencies). While some reproduction of the core facts and arguments of this resources scarcity CF discourse did take place in the public media, this was relatively limited. Most of the written manifestations of the discourse took place in various outputs produced by the specialised agencies, as mentioned above. Numerous international conferences where held which were attended largely by forestry and rural development specialists.

The tropical forest conservation CF discourse

Tropical deforestation became a worldwide concern in the early 1980s, although forestry experts had been commenting on the issue long before. The movement of the issue to the forefront of expert, and to some extent public, attention around that time had important consequences: established international organisations took up the quest to help stem tropical deforestation and large conservation NGOs joined the multiple actors that had tropical deforestation as a main agenda point. The ensuing 'tropical deforestation discourse' seriously affected the CF discourse, which was until then centred on forest resource scarcity, fuelwood shortage, reforestation, afforestation and agroforestry (Arnold, 1991) and pushed it into the tropical forest conservation CF discourse.

The multiple studies that appeared in the 1980s on how communities treated forests were probably the first serious announcement of a shift from the forest resource scarcity to the tropical forest conservation CF discourse. These studies were mostly carried out in Asia and Tropical America (see De Jong *et al.*, 2008 for an overview), and confirmed that not only forests and their resources are important in household economies, but also that forest dwellers engage in active forest management, be it of standing forest, swidden fallows, forest gardens or scattered trees in smallholder estates. A highly relevant offspring of this local forest management awareness became the 'non-timber forest product' (NTFP) discourse, a discourse that I place within the tropical forest conservation CF discourse.

The NTFP debate co-evolved with the attention to local forest management in the 1980s, which demonstrated the multiplicity of forest products used by forest dwellers. A monumental paper by Peters *et al.* (1989) in the journal Nature is the benchmark study that supports the proposition that sustained harvesting and marketing of NTFP yields a net present value of standing tropical forests which exceeded that of agricultural uses or timber extraction. For much of the second half of the 1980s and until well into the 1990s this became a core argument of the tropical forest conservation CF discourse: 'community forest management through sustained harvesting and sales of NTFP is a more profitable use of tropical forests and at the same time contributes to tropical forest conservation'. International development cooperation embarked on projects in many tropical forest locations to turn the core assumptions of this discourse into practice. NGOs emerged that specialised in working with forest communities to identify marketable NTFPs and establish the related value chains. In addition, the so-called conservation and development projects emerged – projects that aimed primarily to ensure forest conservation, or biodiversity conservation in general (mostly in conservation areas), but that tried to achieve this through various means, including the generation of income for resident communities.

A particularly relevant aspect of the tropical forest conservation CF discourse is its presence in the public media. For several years, especially during the late 1980s, the quest to conserve tropical forest was taken up not only by forestry specialists but also by various entertainment celebrities. A well known example is the 'Don't bungle the jungle' concert, which was held in New York in 1989 and tried to raise public awareness and funds for the tropical forest conservation cause. Prominent artists like Madonna and The B-52s participated in the concert. Sting, the singer and guitarist of the rock band Police, became a voice against tropical deforestation and his wife appeared on numerous public events with Brazilian Indian leaders to do the same. Tropical deforestation and the need to stop it, but also the option for profitable sustainable NTFP extraction, were widely published and commented on in newspapers, but equally in recreational periodicals for a non-specialised, popular audience. While the NTFP propaganda machine was in full swing, forest dwellers' capacities to manage forests and their actual customs of doing so were documented worldwide.

The community enterprise CF discourse

A subsequent round of CF discourse once again saw a change in the main concepts and arguments that were disseminated, as well as in the actors involved. I call this the community enterprise CF discourse. This latest CF discourse is mostly a post-2000 phenomenon, although it was already visible in the 1990s. It emphasises, much more than the two previous discourses, CF as a programme designed and implemented by specialised agencies: development cooperation, forestry agencies and rural development NGOs. In this latest discourse, CF is seen as an abstract objective of communities who manage forest estates, mostly to produce forest commodities. The principal commodity is timber, or in exceptional cases other forest products that are marketed internationally, such as Brazil nuts in Bolivia. Communities are to organise themselves, preferably as communal forest enterprises, along western enterprise models, with their related hierarchy, business management approach and the like. Forest management is to be laid down in forest management plans and annual operation plans, which have to be prepared according to legally prescribed norms.

The community enterprise CF discourse does not have the same prominence outside the specialised literature as its forerunner, the tropical forest conservation CF discourse, which received much more attention in the non-specialised public media. Interestingly, however, it coincided with the widely commented upon neoliberal revival, which started in the 1980s, but which was progressively implemented (also in the forestry sector) during much of the 1990s. Quite a few commentators have argued that many of the forestry reforms that took place during the 1990s, and that seemed to favour forest communities, were essentially part of a neoliberal forestry reform agenda (Pacheco *et al.*, 2010). Since the late 1990s, the idea of communal enterprise appeared in the debates among CF experts, mostly in development cooperation agencies or their associates. This coincided with, or possibly followed, a trend that began around the same time to shift the CF focus from resource security or sales of NTFP to an exclusive focus on the more valuable forest products such as timber, as mentioned above. This, in turn, can be linked to the shift in legal forest ownership towards communal ownership in many tropical forest regions in the world since the 1990s.

In my view, the CF enterprise discourse was aggressively promoted by development cooperation agencies and their partners, even though these ideas were viewed with more reservation among academics and national forestry agencies. The discourse, in turn, found resonance among decision makers who decided on the funding of the CF development cooperation projects within national governments or international bodies. The discourse was mostly reproduced in quite technical and specialised media, and very much directed at a specialised public.

6.4 Assessing the CF discourse

There are two relevant questions that can be posed in relation to the various CF discourses that can be identified over the last four decades or so. The most obvious one is: what are the origins of the discourses or changes of discourse? Can they be linked to other international political economic trends or debates, or 'meta discourses issues' (Arts and Buizer, 2009)? The second, more complicated, question is: What are the implications of a discourse view of CF for the 'CF project', or the accumulated efforts to identify, demarcate and mobilise a forestry practices among rural dwellers who share a number of similar characteristics to achieve wider societal goals?

While discussing three of the most prominent CF discourse of recent decades, it is possible to make links with wider societal issues. The forest resource scarcity CF discourse emerged during the years that the first international debates on the limits to growth, future resource scarcity and sustainable development ideas emerged. Arts *et al.* (2010) identify this as the 'limits to growth' discourse, which was a backlash against the 'modernisation discourse' that preceded it. Relevant here is that after the 1960s, which were dominated by a focus on economic expansion, technological innovations, increasing consumption, etc., gradually a concern arose about the limits of natural resources and the capacity of the earth to sustain a projected indefinite growth.

This basic line of thinking was transferred to the communal forestry debate. As previously stated, the first CF discourse focused on perceived or anticipated local resource scarcities. Reports on local communities and their use of forests and local forest knowledge had emerged from before that era. Terms like 'minor forest products' or 'secondary forest products' had also been proposed earlier. These insights suggested that CF could become a way to address rural resource scarcity issues or even become attractive national economic sectors. Hence, the emergence of the forest resource scarcity CF coincided closely with the worldwide concern about the limits of the world's environments to be able to accommodate the increased production and consumption patterns.

The tropical forest conservation CF discourse coincided with the worldwide concern for tropical deforestation, as indicated above. Arts *et al.* 2010 locate the beginning of this tropical deforestation concern in the mid 1970s, and the issue dominated the agendas until well into the 1990s. During this period an important number of publications appeared on the various direct and underlying causes of tropical deforestation (e.g. see Angelsen and Kaimowitz, 1999 for an overview). According to Arts *et al.* (2010), forest decline (in Europe as a result of acid rain, among other causes) and tropical deforestation were the two predominant forestry discourses of the 1980s. As in the case of the forest resource scarcity CF discourse, the tropical forest conservation CF discourse coincided with a larger issue that circulated in global debates and was on the agenda of a large number of environmentalists and the media. A difference with the previous resource scarcity CF discourse is that whereas this was linked to what Arts *et al.* (2010) identify as a meta discourse, the tropical forest conservation CF discourse was linked to a sectoral 'forest discourse'. While this is the case, the tropical forest conservation CF discourse was certainly more prominent in the public media.

The community enterprise CF discourse, as mentioned, coincided with a revitalisation of neoliberalism, including a reduction in the role of governments in public life and a shift towards poverty reduction in international development assistance (De Jong *et al.*, 2008). Arts *et al.* (2010) locate the beginning of neoliberalism in the 1970s, even though in many tropical countries international lending institutions began to impose structural adjustment programmes only in the late 1980s. This practically marks the beginning of neoliberal reforms in the forest policy and forestry development agendas. Until that time, the latter was largely a sector that national governments viewed as an economic growth sector, or a sector that served clientist political agendas (De Jong and Ruiz, 2010; Pacheco *et al.* 2010).

Although with appropriate modification the community enterprise CF discourse essentially advances the same model, in many locations either communities have either been given collective property rights over communal lands, including forest lands, or the land where they live has been given a special land use status that restricts its use by outsiders and gives certain exclusive rights to communities. Where CF is an option, communities can legally exploit forest products that are in demand on national or international markets by essentially following similar procedures to those imposed on corporate or other commercial actors, which means preparing forest management plans and annual operation plans. Communities need to acquire a legal status, which means they need to register themselves as some kind of organised structure, a communal enterprise. The external support received by communities is predominantly to support these kinds of CF (Sabogal *et al.*, 2008).

6.5 Community forestry discourses: *quo vadis?*

The discussion so far suggests some important insights about CF if the latter is viewed through a discourse lens. This chapter identifies three CF discourses, which have appeared since the concept of CF was first introduced in the technical forestry literature in the 1970s. In summary, the actors that carried each of these discourses were different and the three discourses were not equally prominent; the tropical forest conservation CF discourse in particular was widely publicised in the public media.

More relevant has been the shift in the core postulates and discursive argument of each of the three CF discourses. These postulates are not obvious and only make sense if they are linked to other issues of wider societal interest. We could identify links between the forest resources scarcity CF discourse and the dominant concerns of the limits to economic growth, which followed an era of belief in limitless technological innovation and economic expansion. Equally, we could link the tropical forest conservation CF discourse to a worldwide concern for tropical forest decline that became prominent from the mid 1980s. The community enterprise CF discourse can be related to the worldwide shift towards neoliberal economic policy and other related economic policy adjustments. This had an impact first in the forestry sector and subsequently on how CF was viewed by the actors who then dominated the CF discourse.

There are two opposite viewpoints on the divergence of CF discourses. On the one hand, it is not surprising that CF discourses correspond in some ways to meta discourses, or even the dominant forest discourses at a given time. Modern society constantly has to address multiple challenges, and each is usually resolved after intense debates. In issues related to economics, politics and culture that are of wide societal interest, certain concepts and postulates are typically dominant. Certain ideas emerge and for some time they become the accepted truth for a large number of participants in public debates or people who have executive power. These widely known and accepted ideas subsequently filter down into more specific, sectoral discussions, policies or programmes. From the evidence collected here, this seems to be a pattern that also applies to CF and explains the vagaries of CF discourses identified here.

On the other hand, it might be considered an anomaly that the postulates and discursive arguments of the CF discourses appear not to have been driven by those actors who have most at stake, the forest communities themselves. The CF programme, as well as the CF discourse formation should, from a political fairness point of view, be driven by the actors most affected by the practical outcomes. Although somewhat an over-generalisation, forest communities have so far not really seriously participated in the production and reproduction of concepts and categories. The discourse is about them, but they are not serious participants in the formation of the discourse.

From a discourse theory point of view, the limited participation of forest communities in shaping discourses that are about themselves is not that strange. Discourses are, after all, a reflection of preferences, needs, interests and underlying core values. Discourses are usually dominated by actors who have resources and political advantage, and who hold skills that allow them to influence the kind of media that carry the discourse. Forest communities have had tools to enable them to participate in shaping CF discourses, such as counter-mapping and indigenous rights movements. In recent decades, they have gradually become more vocal in manifesting their views and it can conceivable be argued that an 'indigenous CF discourse' now exists.

Given the above analysis, it is worth dwelling for a moment on the possible future trends of the CF discourse. If the general principle remains that the CF discourses adjust to wider, meta or sectoral discourses, then it can be expected that forest-based climate change mitigation is likely to become the next meta discourse to influence CF discourses. To some extent, this trend is already apparent. Reducing Emissions from Deforestation and Degradation in developing countries (REDD+) and Clean Development Mechanism (CDM) (reforestation and afforestation) measures, the two principal forestry methods to mitigate climate change, are part of a CF discourse as they are both methods that can be adopted by forestry communities. If they do so, it will probably increase their bargaining power within the climate change mitigation mechanism deals. However, since the stakes are higher under a climate change mitigation CF scenario, different discursive arguments may eventually emerge that will change how forest communities are politically positioned in the forest climate-change mitigation debate.

The participation of forest communities in shaping future CF discourses is likely to be influenced by the expected changing political position that forest communities will take under a climate-change mitigation forest sector. In recent years forest communities have, through their representative organisations, already and increasingly become active actors in shaping the CF discourse. Latin American indigenous communities, for instance, have proposed an 'Indigenous REDD' (AIDESEP, 2011), which includes modifications of the REDD mechanism and implementation methods to address the realities and needs of indigenous communities.

6.6 Conclusions

This chapter took a discourse perspective on CF debates and analysed in detail three CF discourse variants, while recognising that it might be possible to distinguish additional CF discourses and variants of the CF discourse. The three discourses analysed are the forest resource scarcity CF discourse, the tropical forest conservation CF discourse, and the community enterprise CF discourse. The discourses differ in terms of the actors who shaped and carried the discourse and the intensity of production and reproduction of the ideas and arguments. The postulates and core discursive arguments of the three discourses could clearly be linked to other prominent and broadly debated issues. This raises the question of whether the CF programme, which is still largely driven by a number of specialised actors, also produces and reproduces CF discourses. It is likely that the climate-change mitigation agenda will have an important influence on the shape of future CF discourses. When that happens, it is likely that forest communities – the protagonists in these future discourses – will become more active participants in producing and reproducing them.

References

AIDESEP (2011). Construyendo REDD Indigena. Available at: http://www.forestpeoples.org/.

Angelsen, A. and Kaimowitz, D. (1999). Rethinking the causes of deforestation: lessons from economic models. The World Bank Research Observer 14(1): 73-98.

Arnold, M. (1991). Community forestry: ten years in review. Food and Agriculture Organization, Rome, Italy.

Arts, B. and Buizer, M. (2009). Forests, discourses, institutions. A discursive-institutional analysis of global forest governance. Forest Policy and Economics 11: 340-347.

Arts, B., Appelstrand, M., Eba'a Atyi, R., Enters, T., Hamakers, I., Kleinschmit, D., Pülzl, H., Visseren-McGinley K. and Yasmi, Y. (2010). Discourses, actors and instruments in international forest governance. Embracing complexity – meeting the challenges of international forest governance. IUFRO World Series 28, pp. 57-73. Available at: http://www.iufro.org/science/gfep/forest-regime-panel/report/.

Eckholm, E. (1975). The other energy crisis: firewood. Worldwatch Paper 1, Worldwatch Institute, Washington, DC, USA.

De Jong, W. (2010). Forest rehabilitation and its implication for forestry transition theory. Biotropica 42(1): 3-9.

De Jong, W. and Ruiz, S. (2011). Strangers among trees: policies and politics for foreign residents in northern Bolivian forests. Forest Policies and Economics, in press. DOI: http:/dx.doi.org/10.1016/j.forpol.2011.02.004.

De Jong, W., Louman, L., Pokorny, B., Sabogal, C., and Stoian, D. (2008). Antecedentes, realidad y oportunidades del manejo forestal comunitario en América Latina. In: Sabogal, C., De Jong, W., Lauman, B. and Pokorny, B. (eds.) El manejo forestal comunitario en América Latina: experiencias, lecciones aprendidas y retos para el futuro. CIFOR, CATIE, Belem, Brazil, pp. 33-74.

Eckholm, E. (1975). The other wood energy crisis, firewood. Worldwatch Institute, Washington, DC, USA.

Elands, B. and Wiersum, F. (2001). Forestry and rural development in Europe: an exploration of socio-political discourses. Forest Policy and Economics 3(1/2): 5-16.

Food and Agriculture Organization (FAO). (1978). Forestry for local community development. Forestry Paper 7, FAO, Rome, Italy.

Hajer, M. and Versteeg, W. (2005). A decade of discourse analysis of environmental politics: achievements, challenges, perspectives. Journal of Environmental Policy and Planning 7(3): 175-184.

Hickey, S. and Mohan, G. (2004). Participation: from tyranny to transformation? Exploring new approaches to participation in development. Zed Books, London, UK.

Kleinschmit, D. (2011). Confronting the demands of a deliberative public sphere with media constraints. Forest Policy and Economics, in press. DOI: http://dx.doi.org/10.1016/j.forpol.2010.02.013.

McDermott, M. and Schreckenberg, K. (2009). Equity in community forestry: insights from North and South. International Forestry Review 11(2): 57-170.

Pacheco, P., De Jong, W. and Johnson, J. (2010). The evolution of the timber sector in lowland Bolivia: Examining the influence of three disparate policy approaches. Forest Policy and Economics 12: 271-276.

Peters, C.M., Gentry, A.H. and Mendelsohn, R.O. (1989). Valuation of an Amazonian rainforest. Nature 339: 655-656.

Pokorny, B., De Jong, W., Louman, B., Pacheco, P., Porro, N., Sabogal, C. and Stoian, D. (2008). Experiencias y retos del manejo forestal comunitario en América Latina. Recursos Naturales y Ambiente 54: 81-98.

Richardson, T. and Sharp, L. (2001). Reflections on Foucauldian discourse analysis in planning and environmental policy research. Journal of Environmental Policy and Planning 3: 193-209.

Sabogal, C., De Jong, W., Pokorny, B. and Lauman, B. (eds.) (2008). El manejo forestal comunitario en América Latina: experiencias, lecciones aprendidas y retos para el futuro. CIFOR, CATIE, Belem, Brazil.

Winkel, G. (2011). Foucault in the forests: a review of the use of 'Foucauldian' concepts in forest policy analysis. Forest Policy and Economics, in press. DOI: http://dx.doi.org/10.1016/j.forpol.2010.11.009.

7. Conservancies in Namibia: a discourse in action

Maaike de Vette[1,2], *Rose-Mary Kashululu*[3] *and Paul Hebinck*[4]
[1]*Aqua for All, P.O. Box 93218, 2509 AE The Hague, the Netherlands;* [2]*Women for Water partnership, Benoordenhoutseweg 23, 2596 BA The Hague, The Netherlands;* [3]*Ministry of Lands and Settlement, Omusati District, Namibia;* [4]*Wageningen University, Sociology and Anthropology of Development group, Hollandseweg 1, 6706 KN Wageningen, the Netherlands; paul.hebinck@wur.nl*

Abstract

Using case material collected in three conservancies in Northern Namibia, this chapter explores how the global discourse of community-based natural resource management (CBNRM) is translated into action through state interventions and how these initiatives are enacted in Namibia. Our research, which focused on the conservancy level, shows how the various subtle, often indirect, mechanisms make and break conservancy communities, creating fluid, overlapping and constantly changing boundaries between communal institutions. We therefore problematised what constitute communities as a way to unpack the working of the technologies of rule in conservancies and the making of subjects. Conservancies were riddled by attempts to include and exclude certain groups. One of the outstanding community dynamics in relation to the institutionalisation of conservancies is that these form the basis of new power relations in the communal areas that compete with the existing traditional hierarchies of power. This shift has, however, not been able to guarantee a fair distribution of benefits. The conservancy committees often served as a vehicle for the emergence of new local elites closely associated with the modern global commodity economy. Increased reliance on external methods in wildlife conflict mitigation and the subcontracting of key tasks, such as hunting and tourism development, challenge assumptions about local capacity development. Also, conservancies were streamlined by regulations that are designed, fine-tuned and monitored by the state and by experts. This research confirms statements on how CBNRM discourse is clouded by various romanticised statements, which leads to a range of paradoxes and ambiguities in its approach, generating processes of social change with problematic outcomes at the local, community level. While decentralisation is the discourse, recentralisation seems to be the practice. This discursive shift requires a detailed analysis of the state and the role of local and international elites in the commoditisation of environmental resources.

Keywords: CBNRM, governmentality, environmentality, community, de(re) centralisation, institutions, benefit distribution

Maaike de Vette, Rose-Mary Kashululu and Paul Hebinck

7.1 Introduction

This chapter narrates and explores how the global discourse of community-based natural resource management (CBNRM) is translated into action through state interventions and how these initiatives are enacted in Namibia. CBNRM stands for the devolution of rights to manage natural resources from the state to local communities. CBNRM programmes, spearheaded by the famous CAMPFIRE experience in neighbouring Zimbabwe (Alexander and McGregor, 2000; Balint and Mashinya, 2008), have been extensively promoted throughout Southern Africa. Community-based approaches have since become increasingly integrated into development programmes. Conservancies are the institutional expression of CBNRM in Namibia and have been established since the country gained its independence in 1990. By law they are managed by local conservancy committees to reap and distribute benefits from nature.

However, CBNRM is clouded by various romanticised statements concerning its ability and potential to answer the globally pervasive concerns about environmental degradation and resource depletion. Failures attributed to state management and market-oriented policies have made the concept of community – and thus community-driven development – attractive to many policymakers and practitioners as an alternative actor for managing game, forests, pastures, water and fisheries (Agrawal, 2003; Fabricius and Koch, 2004; Nelson and Agrawal, 2008). Case material collected in three conservancies in Namibia supports this, but shows how the approach is loaded with contradictions and complexities and generates processes of social change with problematic outcomes at the local community level with regard to sharing benefits and the use of power. These are indications that the broadcasted CBNRM discourse generates a range of paradoxes and ambiguities.

The findings of our studies presented in this chapter support the criticism elaborated in the academic literature, a field of study in which Freerk Wiersum has contributed, worked and published substantially over the years. Although focusing specifically on forestry and forest environments, he unveiled many critical issues in CBNRM. Over his long career he has witnessed various shifts in both theoretical approaches as well as management ideas. One of his many concerns is the shift to commercialisation, which demands the opposite of decentralisation: recentralisation of natural resource policies, and thus the need to rethink institutional arrangements for the governance of natural resources.

In this chapter we critically examine this aspect of Wiersum's work (1997, 2009) by exploring how institutions in CBNRM contexts actually work and are reworked in practice. Contemporary CBNRM debates zoom in on the tendency of local and national authorities to recentralise decision-making. For example, Bologna (2008: 123) pointed to one problematic aspect of this, arguing that 'there is a persistent reliance on paradigmatic models that are unable to accommodate the complexities of local-lived realities'. He added that CBNRM seems to have shifted from challenging and opposing more stable hegemonic discourses towards becoming such a discourse itself. Extremely high expectations of CBNRM, as Schiffer (2004) has already argued, overburden projects and lead to frustration when intended outcomes are not met, increasing the gap between discursive theory and practice.

Further, we focus on what happens at the conservancy level, showing how various subtle, often indirect mechanisms make and brake conservancy communities, creating fluid, overlapping and constantly changing boundaries between communal institutions. To do this we have to focus on the institutions that mediate people-environment relationships. Like Leach *et al.* (1999) and Ostrom (2005) we problematise notions of community and understand institutions as regularised patterns of behaviour between individuals and groups in society. We argue that this is analytically superior to the idea of community-level organisations prevalent in most CBNRM approaches, which assumes consensus and harmony among the members of such organisations.

We first provide a brief description of the empirical setting before launching into a more theoretical exploration of how to unpack conservancy, followed by an analysis of our data.

7.2 The empirical setting

The analysis originates from two field studies in three different conservancies in Northern Namibia: Uukwaluudhi conservancy in Omusati (Kashululu, 2009) and Sobbe and Balyerwa conservancies in Caprivi (De Vette, 2009). Registered in 2003, the Uukwaluudhi Conservancy (UC) is the second largest of the 52 in Namibia in terms of population, representing approximately 30,000 eligible members of the Uukwaluudhi community. The UC incorporates an entire existing kingdom and covers an area of 1,437 km^2. Uukwaluudhi is an administrative district in northern central Namibia inhabited predominantly by Ovambo, but there are also Herero, Himba, Dhemba and San people. Uukwaluudhi district is headed by the king, who is assisted by several headmen. The land remains state property and the Tribal Authority (TA) only issues usufruct rights. The region is one of the poorest and most densely populated regions in Namibia. Livelihoods revolve around the use of natural resources (notably harvesting the mopane trees, which is regulated by the TA) and a mixed farming/livestock system that is mainly oriented towards satisfying own consumption needs and local barter. Livestock, however, is widely sold and purchased. The commodity economy of Northern Namibia has recently started to flourish and expand. This is partly fuelled by increasing labour migration, but also by the peace settlement in neighbouring Angola, which has generated an increased demand for food and other products (Dobler, 2009; Kreike, 2009). The UC has fenced in some 6,000 hectares with an electrical fence as a core conservation area where small and large wildlife species are kept and trophy hunted. Some small animals and elephants are also found outside this core area as the fenced area is too small to house all the wildlife.

Sobbe and Balyerwa conservancies are situated in the Caprivi region. In contrast to the UC area, this region has a more complex tribal structure and there have been several tribal conflicts recently. Until 2004, some groups in Caprivi had also attempted to separate from Namibia. Caprivi is geographically, politically and economically a more excluded region in Namibia and has the highest poverty rate in the country, but there are signs of a booming border-trade economy (Zeller, 2009) with visible effects on local employment opportunities. The formation of the two conservancies in Caprivi in 2006 was supported by various local non-governmental organisations (NGOs), the most important being the Integrated Rural Development and Nature Conservation (IRDNC) agency. Sobbe Conservancy (SC) consist of woodland and grassland and covers 404 km^2. Its population of about 2,000 people are spread over five villages, of which 681 adults were registered members in 2009. The people are from the Mafwe tribe, but they adhere to two different TAs, the Mayuni TA and the Chinchimane TA. The two TAs, however, see themselves as two separate tribes and some land disputes remain unsettled. The southern part of the conservancy is almost entirely dedicated to wildlife conservation, although people practice agriculture and graze cattle as well. Balyerwa Conservancy (BC) covers 223 km^2 and lies along the Kwando River. It borders Mudumu National Park in the north and Mamilli National Park in the south, making the whole conservancy part of an existing wildlife corridor between the two national parks. The landscape consists mainly of woodland and grassland, with some floodplains in the east bordering the Kwando River. The population comprises approximately 1,500 people, including children, dispersed widely across several small villages. All the inhabitants belong to the Mayeyi tribe and only one TA governs the area of the conservancy.

7.3 The theoretical setting: analysing conservancies

Following Agrawal (2003, 2005), we interpret conservancies as ideal manifestations of the 'technologies of government', which describe the subtleties of various indirect mechanisms and discourses that encourage people to self-regulate their behaviours. Agrawal coined the term 'environmentality' to study 'the enormously interesting, complex, and crucial relationship between changes in government and related shifts in environmental practices and beliefs' (Agrawal, 2005: 162). Environmentality implies applying the Foucauldian 'governmentality' principle to studying environmental management and is rooted in the belief that politics, institutions and subjectivities together characterise government (Agrawal, 2003, 2005). Governmentality captures the workings of disciplining institutions that, without any apparent coercion, manage to make people act in certain ways (see also Li, 2007). Moore (2005: 4) characterises an important aspect of governmentality called 'control at a distance', which does not hinge on force, but on streamlining, encouraging and ordering 'actions among subjects whose agency

[is] deployed – rather than destroyed – by government'. Governmentality, in essence environmentality, entails looking at power relations unfolding between and among actors and their institutions, and examines how power works through the constitution and making of subjects (such as citizens, civil servants, communal farmers, conservancy members, and so on) through certain discursive rituals and administrative practices (see also Nuijten, 2005).

Conservancy committees, we argue, play a key but varying role in the making and disciplining of environmental subjects. Conservancies represent new institutions enacting change set in motion by specific social actors operating at local and global levels. Relevant political actors perceive gains from institutional change, and the transfer of authority (i.e. the devolution of rights to manage natural resources to communities) involves a danger of elite groups being formed. Agrawal (2005) and Platteau (2004) also point to elite capturing processes. We will show that elite capture also takes place in the conservancies we examined, which in turn indicates the problematic nature of the notion of community. The emergence of new institutions is therefore a highly political affair (Agrawal, 2003) that jeopardises the success of CBNRM, which in turn depend on the degree to which the interests of these actors intersect or overlap with those of the larger collective.

The conservancy model in Namibia has introduced new sets of actors that have begun to link the communal areas more firmly to the global economy in different ways than before. Ecotourism and trophy hunting have become important over the years and represent new avenues for the region's integration into the regional and global commodity economy. These which continue to coexist with the classic forms, such as labour migration and the sale of cattle. In turn, the networks thus created are the seedbed for new hierarchies of power in the communal areas, which revolve around the contestation of the commons and its traditional organisational model of Tribal Authorities. Nelson and Agrawal (2008) summarise this nicely as follows:

> CBNRM is fundamentally premised on institutional reforms that decentralize authority over – and benefits from – land and natural resources to local actors. In the absence of such reforms, the incentives for local groups of people to collectively invest in natural resource management are unlikely to exist or emerge. The lack of downwardly accountable decentralization or devolution is increasingly recognized as the principal barrier to CBNRM in sub-Saharan Africa (Nelson and Agrawal, 2008: 558).

However, we must not ignore the counter tendencies (see Arce and Long, 2000) that are at work in conservancies: conservancy committees and 'localised' networks within and around conservancies actively, but not necessarily openly, reshape and redesign conservancies' technologies of rule. In other words, the conservancy policy discourse becomes enacted or unpacked in ways that are shaped by local socio-political rituals and repertoires. This is evidenced by conservancy committees making decisions independently of experts, the use of local 'indigenous' methods to scare game and to prevent elephants from grazing in the arable fields, employment opportunities distributed along patron-client relationships, witchcraft accusations, and so on. It is the analysis of this kind of reworking and enacting that provides a more complete and fuller picture of the dynamics at work in conservancies. By showing how the conservancy model has landed in everyday life in Namibia's communal areas, we ascertain that the formation of conservancies continuously shapes and reshapes communities inside and around conservancies. This does not imply that we like or agree with the outcomes that are generated at the local level. We find it important for the policy process to contrast this analysis with the image of the harmonious and static conservancy community that forms the core of state policies and CBNRM discourses.

7.4 The community question: conservancy dynamics

The question of what constitutes community has various dimensions. We found a key dimension to be membership. The mechanisms of the distribution of benefits tell us something about the interrelations between conservancy members and are expressions of power and authority. The theoretical issue we address is that community needs to be understood as process in which the actors forming the community negotiate benefits and control over territory; status, power and their use play an important role in the process. The perspective we will outline does not necessarily reject the notion of community, but, as Leach *et al.* (1999: 230) state, this perspective 'rather contextualise[s] it by describing a more or less temporary unity of situation, interest or purpose'. This, we argue, will allow us to unravel the complexities of community and the making of subjects in conservancies. We will consider aspects of membership and some aspects of exclusion, decision-making and benefit sharing to underline the complexity of notions like community.

Membership versus ownership and inclusion

It is important to highlight the difference between conservancy membership and conservancy residency, as not all residents are members of the conservancy. The Namibian CBNRM legislation (Amendment Act of 1996) requires that a conservancy must have a membership register. The members of the conservancy constitute the community. Membership in UC, for instance, was formed by kinship, marriage relations and cultural affiliations, and is clearly based on geographical location and residence. The members are every man and woman from the age of 18 years born in Uukwaluudhi or living permanently in a homestead within the boundaries of Uukwaluudhi district, and who pays an annual administrative levy to the Uukwaluudhi TA as required by custom and tradition. This includes those who reside outside the boundaries because they have full-time jobs elsewhere in Namibia or South Africa. Labour migrants, whose remittances are economically important for their families, hardly spend time in the conservancy and occupy an ambiguous position. Immigrants from elsewhere in Namibia and from Angola who live in the conservancy maintain strong social links with their family members back home. As such, the community spans geographical and social boundaries in multiple ways. Migrants we spoke to were generally unaware of the conservancy and most of them were not in favour of the ban on hunting. Half of the immigrant in Sobbe and Balyerwa were not registered as members. Moreover, during the registration procedures, signatures were collected from residents, but more than half of them said they did not know what they signed for.

The role and position of the elderly in the conservancies show that membership does not automatically mean inclusion in the conservancy project. Elderly people are important as custodians of customs and hold a considerable body of knowledge about indigenous environmental practices. Furthermore, they receive old age pensions, which are an important source of household income. Sporton and Mosimane (2006) showed that the Omusati region has the highest proportion of pensioners (9%) and the highest level of household livelihood dependence on pensions (22%). Most elderly people were registered by their children and when interviewed they said they receive few benefits from the conservancy. In a village in BC, for example, the joint growing of chilli peppers as an elephant deterrent excluded an elderly couple as they could not help with growing them, which is a requirement for participation. Due to the migration of their children, elderly people often lack a social support network that would help them to benefit from conservancy programmes. Discussions revealed that they found it hard to adapt to the new regulations of the conservancy. They were used to making up their diet with bushmeat, which they were not allowed to do any longer because of conservancy policies. This is similar to the CAMPFIRE experiences in Zimbabwe (Alexander and McGregor, 2000). The growing influence of the new conservancy structures created tensions and induced resistance among the elderly residents because their ways of doing had been replaced by 'modern' ones. Elderly people often referred to the conservancy as a body that followed the rules of 'white men'. Where the elderly were trying to hold on strongly to traditional methods – and therefore often accused of witchcraft – the younger members were more likely to get pulled in the direction of the conservancy, increasing the generation gap.

Lastly, membership does not mean ownership. The conservancy programme measures its success by growth in membership, thereby automatically assuming commitment, ownership and participation. There were many signs of active participation among conservancy members, suggesting that people were attempting to become involved rather than rejecting the conservancy model. However, the continuing wildlife conflicts, the perceived passiveness of the conservancy committees and stricter enforcement of laws against local hunting (which is also a way for people to defend themselves) contributed to growing discontent, but did not lead to a decline in membership. The non-registered residents of the conservancy also had to contend with the stricter enforcement of government legislation by the conservancy rangers, which they were not happy with.

Benefit distribution

Nelson and Agrawal (2008: 567) concluded that the Namibian CBNRM system is designed 'to reduce the incentives that state wildlife authorities in Namibia possess to resist devolution of wildlife management to local communities'. This was derived from the belief that the 'transparency' of conservancy institutions was favouring the conservancies in terms of benefit distribution and downward accountability. The policy-makers thereby assumed equality and decentralisation at the local level. This is confirmed by Namibia's 'broader macro-political standing as one of sub-Saharan Africa's wealthier and less corrupt countries' (World Bank, 2006, cited in Nelson and Agrawal, 2008: 567). Although Nelson and Agrawal (2008) did emphasise the relative nature of their findings, it is important to acknowledge that their study took a national and regional perspective. For example, they mentioned 'hunting revenues to have risen from NAD 19.6 million in 1992 to NAD 80 million in 2000' (Agrawal and Nelson, 2008: 566) and concluded that through its transparency, this system provided valuable contributions to the conservancies. In doing so, they assumed values to be evenly distributed among local residents. We would like to emphasise that while the model might intend equal distribution of benefits, local dynamics in conservancy formation cause unintended structures in benefit distribution that prevent even distribution at the level of the conservancy. This is also an indication that membership does not automatically translate into democratic decision-making in the conservancies.

The prevailing situation at the UC is similar to what Balint and Mashinya (2008) found in CAMPFIRE, where proceeds distributed to the wards by the district council in Nyaminyami went only into the pockets of the local political elite. In the UC, members are left in the dark with regard to conservancy operations and proceeds. The chairperson has suspended all monthly and even quarterly planning meetings, and started working in isolation and taking all major decisions alone, without the annual general meeting (AGM) as provided for by the constitution. All financial proceeds from the conservancy were handled by the chairperson, who makes unilateral announcements on its use during the AGM. He negotiated with trophy hunters and lodge operators and approved hunting of wildlife, disregarding quotas without providing any reasons. The chairperson also collaborates closely with the King of Uukwaluudhi and the constituency councillor, and performs tasks at their request. For example, a gemsbok, which was not part of a quota in 2007, was culled on the demands of the councillor, and 5% of the financial proceeds, as well as meat from the (TA) quota, were delivered to the king's palace instead of to the offices of the Traditional Authority Council (TAC).

In Balyerwa and Sobbe meat distribution followed a strict route through the conservancies, which gave priority to chiefs and other influential families. The poorest families were often the last in line, and remote villages were usually left with only the carcasses. Other forms of benefit distribution were also not perceived as fair. The conservancy received a lot of money from trophy hunting activities.

A BC resident said: 'We have no idea where this money goes to, but we do see the amounts paid to the conservancy on the financial status. The conservancy keeps a lot of money for themselves'. Conservancy members did not get this money returned to them in compensation for their loss of livelihood owing to increased wildlife damage and stricter enforcement of the ban on hunting. Some members did not trust the conservancy's financial reports. They said that receipts were not kept and the scheduled investments were not actually carried out. As a result, a group of more educated people in SC had proposed starting an independent monitoring committee to monitor the conservancy's finances. Others formed informal (illegal) groups to resist the ban on hunting. Many residents confirmed that underground networks of poachers were operating and selling meat to illegal traders. The conservancy did not have enough rangers to cover the whole area and a lot of poaching still went undetected.

The increased enforcement of hunting restrictions also shows how the government reinstated some control in the area through the conservancy. Although this lowered poaching levels and led to an increase in wildlife populations, an unintended consequence was that this also reduced people's control over their livelihood resources. At the time of our studies, the Human Animal Conservancy Self-Insurance Scheme (HACSIS) ran a pilot project to compensate for the loss of cattle and funeral costs in some conservancies in Caprivi. This scheme aims to match the actual losses caused by wildlife by compensating registered member victims with conditional stock, crop or funeral compensation (IRDNC, 2008). In Caprivi, the conservancy assigns part of its income to the HACSIS, and IRDNC adds 50% to the budget. Anticipating the implementation of the HACSIS, some members had started reporting damage to the conservancy committee. Since 2006, wildlife management data from conservancy rangers and reported damage has been monitored by the conservancy event book system introduced by IRDNC. When we examined the event book in 2009, we found that only a few cases of damages had been reported. Our interviews with members revealed that many people had stopped reporting damage, blaming the conservancy committee for not giving priority to the HACSIS, despite the fact that the committee had prioritised the scheme in 2007.

During the interviews, conservancy members ventilated their growing discontent with the conservancy, particularly with the lack of support for tackling wildlife problems. They had little confidence in the scheme and some of them stopped participating in the conservancy. When asked, the conservancy committee, IRDNC and the Ministry of Environment and Tourism (MET) did not associate the decline in eagerness to participate in the conservancy with the wildlife problems and the failure of the compensation scheme. Referring to the minutes of the quarterly planning meeting of January 2009, they argued that the reduced number of cases of damage could be attributed to better wildlife measures and reduced conflicts between people and wildlife. The reduction in reported cases of damage was treated as an example of successful management by the conservancy committees, and thus a boost for the CBNRM discourse. However, an in-depth cross check of the reporting books with the interviews showed that the conservancy rangers had regularly failed to register the damage brought to their attention by conservancy members. Our findings are largely in agreement with the research done by Schiffer (2004) and Jones (1999), which clearly shows the increasing gap between conservancy committees and local residents. These processes are indicative of conservancy committees being or becoming trapped between a discursive and a realistic CBNRM.

Local institutions and decentralisation

Decision-making in the conservancies, both traditional and modern, was hierarchical in nature. The most notable example is the procedure for shooting an animal that poses a potential threat to people, cattle and crops. The procedure can take longer than a month, and the decision then still has to be approved by the government department before it can be carried out in the field. The need to align actions with international policy plays a key role in this. Experiences with conservancies in Namibia underline the dominant influence of state and international experts in shaping conservancy policies and activities, despite the devolution of rights and responsibilities to the local level. One of the most important rights that conservancies have is the right to acquire wildlife 'off-take' quota from the government, some of which can be tendered to safari companies as trophy animals, sold as live capture or used for local consumption purposes. The final decision on when and what species of game to harvest is taken by state-appointed experts. However, there are still problems regarding the consistency of policy interpretation and application. Conservancies in communal areas are required to apply for permits to utilise huntable game on their lands, but farmers on freehold lands need not apply (Skyer and Saruchera, 2004). In turn, the state appoints trophy hunters to harvest the listed species and numbers of game. Hunting has to be performed under the guidance and supervision of a number of registered Namibian hunting guides, who must be certified by MET. Since the introduction of the CBNRM discourse in Namibia, hunting has become an exclusive practice, regulated by government and carried out by appointed trophy hunters. Conservancies are obliged to ensure that the ban on anything other than registered trophy hunting (e.g. hunting by residents for consumption, trade or poaching) is upheld. In doing so, they act towards residents as disciplinary institutions, making the rural population more answerable to state and expert regulations, and thus controllable. Conservancies therefore assist with implementing a rationalist, technocratic, modernisation approach to development.

The alignment of policy with the local situation is problematic as information and feedback from the bottom up was given less weight in the final decision-making. Moreover, responsibility for policy, legislation and regulations was often split between different tiers of government. Cirelli (2002: 2) describes the difficulty as follows: 'While the state remains responsible for the obligations to which it subscribed at the international level, harmonised compliance, even within a single country, may be made difficult by the inactivity of some local authorities or by possible differences in the implementation by them'. This is confirmed by other authors. For example, Murphree and Mazambani (in Martin, 2002) emphasise that 'often situations occur in which the state is unwilling to surrender its technical and proscriptive policy approaches whilst lacking the resources themselves to make them effective'. As a result, the local community lacks the authority, incentives and resources needed for local action. Martin (2002) builds upon Murphree's (1994) observations that planners in CBNRM tend to incrementally transfer power to local levels and argues that the transfer of power should be a prerequisite instead of a reward. He adds that 'bureaucrats have the tendency to position themselves as the "gate-keepers" for activities like the issuance of permits, licenses and concessions' (Martin, 2002: 20). This shows that even in a strategy where local expertise and ownership are emphasised, the state often looks differently at local situations than local actors, and transferring power to local levels proves to be difficult and ambiguous. Therefore, certain state regulations do not actually achieve the intended outcomes in the localities.

The government and major NGOs like WWF assume that in CBNRM the community is a 'natural entity outside or opposed to the state' (Li, 2002: 266). Scott (1998: 310) comments that 'by themselves, the simplified rules of the state can never generate a functioning community, city, or economy. Formal order, to be more explicit, is always and to some degree parasitic on informal processes, which the formal scheme does not recognise, without which it could not exist, and which it alone cannot create or maintain.' Whereas authors like Scott (1998) and Li (2002, 2007) focus on the interfaces between the state and locality, we must not forget that mismatches also occur horizontally, within government institutions and within and between local institutions. A substantial body of CBNRM literature points to the theoretical and political question of what constitutes 'community'. 'Community appears an elusive concept and is often a figment of imagination of planners and donors seeking for quick fixes' (Fabricius and Koch, 2004: 79). Also, Leach *et al.* (1997) complain that 'it is striking the degree to which simplistic notions of community are being reinvented in the context of practical efforts towards community-based sustainable development'. The notion of community is rather misleading as it assumes homogeneity and harmony – while often the opposite is the case – and complexities are ignored. Communities are not static, either. They are composed and constituted by people who actively monitor, interpret and shape the world around them and struggle for resources. Rules and norms that come to characterise community (or conservancies) are not only the product of people's practices, interactions and actions, but also of decisions and actions of actors elsewhere in the world (e.g. in Windhoek, or Germany). Their discourses may shape, but certainly not determine, social practices in conservancies. Actors cannot simply be reduced to an expression of their position within a system of social relations (as poor or rich, communal farmers or entrepreneurs).

Agrawal and Gibson (2001) stress the need to unpack the concept of community. Gender, age, social and economic inequalities, unequal access to resources and power differences all paint a picture of communities that are social and spatially differentiated, rather than equal and consensus driven. The analysis of case material in this chapter underlines this argument and contrasts starkly with the widely held view of eternal solidarity among community members. Solidarity is certainly not completely absent, but negotiations between villagers on using and sharing key resources often fail pitifully and end up in conflicts (often expressed in witchcraft accusations) and the exclusion of certain members, or attempts to exclude them. Communities are also viewed as natural entities based on common kinship, language, customs and authority. However, the case material contradicts this, revealing that community boundaries may in fact be delineated and strengthened by relationships between members of different communities. Agrawal and Gibson (2001) favour a focus on the multiple interests of actors within communities and how these shape decision-making processes. The recent literature shows a relative consensus on the value of focusing on institutions as a way of escaping from the simplistic notion and interpretation of community (Leach *et al.*, 1999; Ostrom, 2005; Poteete and Ostrom, 2004) and providing a more fruitful approach to understanding what constitutes a community.

The Namibian CBNRM legislation does not define 'community' as such, leaving it to the actors to define it themselves. This has created opportunities for manipulation and abuse of power by a few elites, who promote their personal interests and gains. The conservancies appear to be highly stratified and differentiated, with multiple interests posing particular but serious challenges to conservancies, resulting in varying incentives and disincentives for participating in CBNRM. This understanding of community is very much shaped by our field work, which indicates that specific local configurations of various networks or subnetworks exist that are differentiated by power. Within the conservancy area there exist groups of people who have some leadership roles within the community, such as civil servants, politicians, traditional leaders and businessmen. These are often trusted by the 'ordinary' members. As elsewhere, locally based dominant actors tend to hijack community-based processes and forcefully occupy the political space created by decentralisation. In Uukwaluudhi Conservancy, some members of the committee have started to form very close alliances with local politicians, members of the TA, army officials and safari operators. Other committee members remain mute and are afraid to criticise the chairperson, perhaps because the chairperson has close ties to the king and the councillor. The king appointed the current chairman, and the councillor chairs AGMs. Instead of raising and stressing environmental issues, what we see in Uukwaluudhi is a decline in members' interest in conserving wildlife. As a result, the committee now remains rather distant and is losing touch with their members, who they should be representing. This was confirmed in an interview with a government official: 'AGMs are not held as required or can be postponed with no concrete reasons. The chairperson is a sole decision-maker. Others are just there to implement his decisions. He runs the conservancy as his own business, not as a community business. He keeps the financial documents, such as the conservancy check book, and no proper records are kept, for example receipts, invoices, payment request sheets and all that. During the conservancy annual quota setting it was observed that some animals were hunted without permits. Some were hunted by the chairperson for no clear purpose, and it is not known if money was received for such utilisation or not'.

The devolution of authority in wildlife conservation in Namibia and elsewhere should not be treated as a fact, but rather as a process during which particular social relations unfold (Kayembe and Monela, 2000: 381). Conservancy formation is a process that needs to accommodate changes initiated by new actors. New actors mean new sets of social relationships that, whether we like it or not, also hinge on tribalism and related conflicts. People we spoke to in Caprivi, for instance, often referred to tribalism, which they felt was a source of conflict. In its active form, tribalism is more noticeable at the collective level than at the individual level and creates rather strong divisions in society based on tribal adherence. The underlying forces of division based on tribalism also influence institutional and individual relations. Caprivi experienced massive violent conflicts between 2001 and 2004, which involved acts of tribalism and dissent between tribes. This revealed that the conservancies were founded on a very fragile structure of traditional authority which has not helped to heal the many rifts in society. For example, Mashi, another nearby conservancy, became involved in a related dispute in which members of the conservancy wanted to split the conservancy in two parts based on chiefdoms. This was resisted by IRDNC and the MET because they feared such disputes would lead to donors withdrawing their support. The conservancy committee was convinced that it could settle the dispute, which indicates that dependence on donors and the national government discourages officials from facing up to local disputes: splitting the conservancy would have prejudiced the image of harmonious conservancy communities that the CBNRM discourse broadcasts.

7.5 New regimes of power?

Our case material goes beyond the difficulties and complexities at the level of the conservancy community per se. We argue that the CBNRM/conservancy discourse has not only provided a new avenue for elites to emerge, but has also provided them with a modern means for competing for status and power with the TA. Elites and conservancy management committees (CMCs) coexist and compete for power and status with existing structures in the communal area, such as the TA. Thus, in pursuing an environmentality agenda, the state devolved the legal ownership of rights over wildlife to conservancies, which by law are managed by the CMC. The power to decide does not reside directly in conservancy membership, but with the CMCs. Most members of the Uukwaluudhi CMC, for instance, belong to the local elite. In the communal areas in Namibia, the elite consists of individuals in professional positions (e.g. bank managers, supermarket managers, and to a certain extent teachers, healthcare workers and government employees) and private entrepreneurs. These elites are not only well connected professionally to the new modern economy, but also remain closely integrated into their communities through kinship ties and obligations. Elite capture may explain why community members decide to disengage from conservancy activities and decision-making, because of clientelist distribution of benefits. This dynamic has certainly stimulated the development of an unconstitutional style of management of the conservancies.

There are clear signs that conservancy dynamics cannot be disconnected from the expansion of the commodity and cash economy in the communal areas of Namibia. Joint venture ecotourism and trophy hunting has added a new dimension to the regional cash economy that involves external actors, in turn creating new patronage opportunities for the elites. One of the most important rights that the conservancy provides is the right to acquire a wildlife off-take quota from the government, parts of which can be tendered to safari companies as trophy animals. Both Sobbe and Balyerwa conservancies were in the process of contracting a large trophy hunting concession. Tourism and safari hunting in Namibia provide capital, expertise and, more importantly, international market access. Members of conservancy committees stated numerous times that local initiatives were impossible due to a lack of education, finance and skills, and that outside technical support is needed to establish such ventures. The conservancies rely heavily on outside investors as they do not have hunting guides, professional hunters and big game professional hunters to supervise hunting safaris, as required by law. Obtaining a full professional licence of the highest category (i.e. Big game professional hunter) requires passing a written examination and more than 30 years of experience. Without this professionalism, communal conservancy members cannot match the incomes commercial, large scale farmers in the freehold areas get from hunting. Grobler (2008) has given an indication of the kind of deals and money involved in trophy hunting royalties. Conservancies get about USD 10,620 per animal from commercial hunting safari companies, which in turn sell these to wealthy hunters for up to USD 30,171 per safari. In the joint venture contract, the conservancy is the landlord and the company manages the business and pays dividends and levies to the conservancy.

These encounters between the 'traditional' and 'modern' modes of regulation will eventually lead to a gradual shift in power. There are clear but conflicting views on the exercise of authority and the division of tasks and responsibilities. The decentralising ideal of including local structures within governance arrangements has reduced the power of the TAs to that of advisory bodies. In the process, the TAs have been transformed from owners and regulators to participants in a programme over which they had no direct control. This will eventually lead to new hierarchies of power and authority in the communal areas. The state plays an ambiguous role in this process of transformation of power. Conservancy members pointed to the close relation between the conservancy committees, IRDNC and the MET. Through liaising with them, the conservancy committees have acquired a stronger network, tying them more firmly into the global political and economic network. Nott and Jacobsohn (2004) have already shown how 'informal alliances between some politicians in the ruling party and tourism developers, opposed to new "fair" concessions required by conservancies for lodge operations, conspired to undermine and weaken some conservancies even though official government policy was supportive of these concessions'. Through the conservancies, the MET regained some local control over the enforcement of government regulations. In the case of hunting and poaching, the conservancy committees have become an extension of the MET policy. It is clear that the state and experts continue to play a key role in the regulation of resource harvesting and use within the conservancy, and have even strengthened their positions through the conservancy committees.

7.6 Conclusions

A few concise concluding comments are warranted. We have demonstrated the complexity of CBNRM in Namibia and have highlighted a range of ambiguities associated with, and triggered by, the formation of conservancies as the institutionalisation of CBNRM in Namibia. The analysis of the community question focused on contradictory processes, such as the formation of elites and resource capture versus the romanticised view of sharing benefits at the local level of 'community' organisation. We found these conservancies to be riddled by attempts to include and exclude certain groups. The conservancy committees, which are supposed to manage the affairs of the conservancy on behalf of, and for the benefit of, the community, often served as a vehicle for the emergence of new local elites closely associated with the modern, global commodity economy, which is gradually expanding into Namibia's communal areas. In addition, conservancies are subcontracting key tasks and activities to tourism and hunter associations, which challenges assumptions about local capacity to manage and control resource use. The conservancies are streamlined by regulations that are designed, fine-tuned and monitored by the state and experts. These regulations have shifted the right to control resource regulation from old to new institutions, and these have not been able to guarantee a fair distribution of benefits.

Agrawal (2003, 2005) associates environmentality with the disciplining of actors which we in turn have linked to the workings of conservancy. In the cases presented here one may wonder what environmentality has achieved and who is being disciplined. The analysis made clear that there is a strong need to differentiate between community members. The need to differentiate between subjects made by conservancy formation is heightened by the unequal participation of migrants and the elderly, who are members of the community but largely ignored. Instead, the conservancy committee facilitates the formation of elites through new forms of resource capturing (i.e. the conservancies). Conservancy formation also acts as a vehicle for the gradual reconfiguration of power in Namibia's communal areas. Conservancies will evolve further as a battleground for new power constellations in which 'old' forms of power will continuously be contested by new, modern forms of power which go beyond natural resource harvesting alone.

The role of the state is in the current setting rather ambiguous. Although decentralisation is the discourse, the practice is a recentralisation. This discursive shift requires a detailed analysis of the state and the role of local and international elites in the commoditisation of environmental resources. This brings us back to the work of Freerk Wiersum and the concerns he frequently expressed.

References

Agrawal, A. (2003). Sustainable governance of common-pool resources: context, methods and politics.' Annual Review of Anthropology 32: 243-262.

Agrawal, A. (2005). Environmentality. Technologies of government and the making of subjects. Duke University Press, Durham, NC.

Agrawal, A. and Gibson, C. (2001). The role of community in natural resource conservation. In: Agrawal, A. and Gibson, C. (eds.) Communities and environment. Ethnicity, gender, and the state in community-based conservation. Rutgers University Press, New Brunswick, pp. 1-32.

Alexander, J. and McGregor, J. (2000). Wildlife and politics: CAMPFIRE in Zimbabwe. Development and Change 31: 605-627.

Arce, A. and Long, N. (2000). Anthropology, development and modernities: exploring discourses, counter-tendencies and violence. Routledge, London, UK.

Balint, P.J. and Mashinya, J. (2008). CAMPFIRE through the lens of the 'commons' literature: Nyaminyami Rural District in post-2000 Zimbabwe. Journal of Southern African Studies 34: 127-143.

Bologna, S. (2008). Stakeholders and stickholders: power and paradigms in a South-African development context. Anthropology Southern Africa 31(3/4): 123-130.

Cirelli, M.T. (2002). Legal trends in wildlife management. FAO Legislative Study 74, pp 1-69.

De Vette, M. (2009). Emergent properties in community-based natural resource management: The case of conservancy formation in Caprivi, Namibia. Masters thesis, Wageningen University, Wageningen, the Netherlands.

Dobler, G. (2009). Oshikango: The dynamics of growth and regulation in a Namibian boom town. Journal of Southern African Studies 35: 115-131.

Fabricius, C. and Koch, E. (eds.) (2004). Rights, resources & rural development: community-based natural resource management in Southern Africa. Earthscan, London, UK.

Grobler, J. (2008). Namibia's quick-buck mentality. Mail & Guardian online, 10 August 2008. Available at: http://www.mg.co.za/article/2008-08-10-namibias-quickbuck-mentality.

Integrated Rural Development and Nature Conservation (IRDNC) (2008). A quick guide to the 'Human-Animals Conservancy Self Insurance Scheme' (HACSIS) for Kwandu, Mayuni, Mashi and Wuparo. IRDNC guide, Windhowk, Namibia.

Kayembe, G. and Monela, G. (2000). Empowering communities to manage natural resources: Where does the new power lie? A case study of Duru-Haitemba, Babati. Faculty of Forestry and Nature Conservation, Sokoine Unversity of Agriculture, Morogoro, Tanzania. Available at: http://www.tzonline.org/pdf/empoweringcommunities.pdf.

Kashululu, R.M. (2009). Conservancy as an arena of power struggle: a case study of Uukwaluudhi Conservancy in Namibia. Masters thesis, Wageningen University, the Netherlands.

Kreike, E. (2009). De-globalisation and deforestation in colonial Africa: closed markets, the cattle complex, and environmental change in North-Central Namibia, 1890-1990. Journal of Southern African Studies 35: 81-98.

Leach, M., Mearns, R. and Scoones, I. (1999). Environmental entitlements: dynamics and institutions in community-based natural resource management. World Development 27: 225.

Li, T.M. (2002). Engaging simplifications: community-based resource management, market processes and state agendas in upland Southeast Asia. World Development 30(2): 265-283.

Li, T.M. (2007). The will to improve: Governmentality, development, and the practice of politics. Duke University Press, Durham, NC, USA.

Jones, B.T.B. (1999). Policy lessons from the evolution of a community-based approach to wildlife management, Kunene Region, Namibia. Journal of International Development 11: 295-304.

Martin, R.B. (2002). Conditions for effective, stable and equitable conservation at the national level in southern Africa. Paper prepared for theme 4 workshop 'Local Communities, Equity and Protected Areas', 24-26 2002 Durban, South Africa. Available at: http://cmsdata.iucn.org/downloads/cca_rmartin.pdf.

Moore, D. (2005). Suffering for territory: race, place, and power in Zimbabwe. Duke University Press, Durham, NC, USA.

Murphree, M. (1994). The role of institution in community-based conservation. In: Western, D., Wright, R.M. and Strum, S.C. (eds.) Natural connections, perspectives in community-based conservations. Iland Press, Washington, DC, USA, pp. 403-427.

Nelson, F. and Agrawal, A. (2008). Patronage or participation: CBNRM reform in sub-Saharan Africa. Development and Change 39(4): 557-585.

Nott, C. and Jacobsohn, M. (2004) Key issues in Namibia's communal conservancy movement. In: Fabricius, C. and Koch, E. (eds.) Rights, resources & rural development: community-based natural resource management in Southern Africa. Earthscan, London, UK, pp. 194-200.

Nuijten, M. (2005). Power in practice: a force field approach to natural resource management. Journal of Transdisciplinary Environmental Studies 4: 1-15.

Ostrom, E. (2005). Understanding institutional diversity. Princeton University Press, Princeton, NJ, USA.

Platteau, J.P. (2004). Monitoring elite capturing in community-driven development. Development and Change 35: 223-246.

Poteete, A. and Ostrom, E. (2004). Heterogeneity, group size and collective action: the role of institutions in forest management. Development and Change 35: 435-461.

Schiffer, E. (2004). How does community-based natural resources management in Namibia change the distribution of power and influence? Preliminary findings. DEA Research Discussion Paper nr. 67, 35 pp.

Scott, J.C. (1998). Seeing like a state: how certain schemes to improve the human condition have failed. Yale University Press, New Haven, CT, USA.

Skyer, P. and Saruchera, M. (2004). Community conservancies in Namibia. An effective institutional model for commons management? PLAAS Policy Brief 14, 4 pp.

Sporton, D. and Mosimane, A. (2006). Population ageing and sustainable livelihoods in regions affected by HIV/AIDS. University of Sheffield, Sheffield, UK. Available at: htpp://www.shef.ac.uk/content/1/c6/06/20/28/BN1.pdf.

Wiersum, F. (1997). Indigenous exploitation and management of tropical forest resources: An evolutionary continuum in forest-people interactions. Agriculture Ecosystems and Environment 63: 1-16.

Wiersum, F. (2009). Community forestry between local autonomy and global encapsulation: Quo vadis with environmental and climate change payments? First Community Forestry International Workshop: 'Thinking Globally – Acting Locally: Community Forestry in the International Arena', 15-18 September 2009, Pokhara, Nepal.

Zeller, W. (2009). Danger and opportunity in Katima Mulilo: a Namibian border boomtown at transnational crossroads. Journal of Southern African Studies 35: 133-154.

8. REDD+: what's in it for community forest management?

Margaret Skutsch
Centro de Investigaciones en Geografía Ambiental, Universidad Nacional Autónoma de México (CIGA-UNAM), Antigua Carratera a Pátzcuaro no. 8701, Col. Ex-hacienda de San José de la Huerta, C.P. 58190, Morelia Michoacán, México; University of Twente, School of Management and Governance, Department of Political Science and Research Methods, P.O. Box 217, 7500 AE Enschede, the Netherlands

> Conflicts concerning forest use and management can be resolved through regulation, markets or negotiation. Conventional forestry has emphasized the first two categories of instruments, but at present the third category is gaining in importance (Wiersum, 1999, Proposition no. 4)

Abstract

As international REDD+ policy is being developed under the auspices of the United Nations Framework Convention on Climate Change (UNFCCC), many countries are proposing community forest management (CFM) as a major element of their national REDD+ strategies, for example in the readiness plans they are submitting to the World Bank's Forest Carbon Partnership Facility (FCPF) programme and the UN-REDD. This chapter first explores which elements of REDD+ can probably be tackled successfully by CFM. I conclude that its major niche will be in promoting forest enhancement, rather than in reducing emissions from deforestation, and that although it is likely also to reduce emissions from forest degradation, carbon credits for this will be difficult to claim as it is almost impossible to construct a reliable baseline for past forest degradation. The incentive systems being used to promote CFM under REDD+ national plans are then considered. Although at the international level, emission reductions achieved under REDD+ will be rewarded on an output basis (per tonne CO_2), it is questionable whether this same system is appropriate within national programmes. The case is made for input-based incentives, which will be easier to administer as they have much lower transaction costs. Moreover, an input-based incentive system will result in less conflict and higher participation. In other words, the argument is made that even if a market system is necessary at international level to encourage investment in REDD+, at the local level more traditional stimuli will be more efficient.

Keywords: community forest management, forest enhancement, degradation, REDD+, carbon credits, decentralisation

Margaret Skutsch

8.1 Introduction

Under the United Nations Framework Convention on Climate Change (UNFCCC), the international community is working towards a new policy for the mitigation of climate change: Reducing Emissions from Deforestation and Degradation in Developing countries (REDD+). As a global payment for environmental services system (for carbon), this could have considerable potential for encouraging communities to become involved in forest management, and indeed many countries have adopted community forest management (CFM) as part of their national REDD+ strategy. It is not yet clear whether REDD+ will work as a real market for carbon, but it is clear that it will have market-like characteristics; in particular, rewards will be based on outputs. Several interesting questions arise as a result, such as what type of carbon savings could be expected from CFM, how much carbon would be saved, and what the value of this might be. Another question is how communities would be incentivised and rewarded, given that international policy on REDD+ is essentially to be played out at national, not project, level. This leads to a question about whether REDD+ will lead to a recentralisation of control and a loss of the hard-won decentralisation achieved by CFM over the last two decades, as well as whether sufficient safeguards will be in place to ensure that community rights are upheld.

8.2 The nature of international REDD+ policy

Two reports have been influential in promoting the United Nations programme for REDD+: the Fourth Assessment Report by the Intergovernmental Panel on Climate Change (IPCC), which contains estimates that emissions from deforestation, primarily in the tropics, account for around 17% of all global greenhouse gas emissions (IPCC, 2007), and a report commissioned by the British government (Stern, 2007) which suggests that the cost of reducing forest emissions is low compared to most alternatives relating to reduction of fossil fuel use. Although not yet finalised, REDD+ policy is likely to be based on formal agreements by individual developing countries to try to reduce the rate at which their forests are being lost, compared with a reference level to be based at least in part on historical trends. At the end of a defined accounting period (e.g. 2012-2017), the difference between the (measured) 'with REDD+' case and the (hypothesised, counterfactual) 'without REDD+' case would be computed in terms of tonnes of carbon dioxide saved. The country would then be able to valorise these savings financially, either through an international carbon market (should one similar to that associated with the Kyoto Protocol come into being) or through an international fund set up for this purpose. As this policy has been one of the most popular of those proposed in the Bali Roadmap for the post-Kyoto period, both among the potential carbon credit purchasing countries and the developing countries who would supply the credits, it stands a good chance of being formally promulgated.

It is important to understand that the policy is directed in the first instance at the country level and, unlike earlier international funds for improvement of tropical forest management, the basis for the international policy REDD+ finance is strictly payment by results via carbon credits, whether the purchase of these goes through a true market or through a fund. Individual projects may be propagated, but at least in the long run the accounting for crediting will be in terms of net savings of carbon over the whole national forest territory. The first requirement for a country to qualify for this policy is therefore to construct a national reference emission level and a system of monitoring and reporting (MRV), both of which must be approved by the UNFCCC.

8.3 The prominence of CFM in national REDD+ plans

In preparing for this policy, a large number of countries have received capacity building funds from multilateral and bilateral sources (e.g. the World Bank's FCPF programme, UN-REDD and Norad) for the development of REDD readiness plans. These involve not just an explanation of how the reference emission level and MRV system will be developed, but also what kinds of policies and programmes will be promoted within the country to achieve the envisaged reductions and what safeguards (social and environmental) will be put in place. An interesting observation is that CFM is mentioned as a component in the proposed national REDD+ plans of almost every country. In a number of countries, such as Mexico, Nepal and Tanzania, CFM forms the central plank of this plan. Given that tropical deforestation is widely attributed either to poor regulation of the timber industry or to expansion of agriculture, particularly in the form of commercial plantations (Cropper *et al.*, 1999; Geist, 2001; Kaimowitz, 1997), this choice of strategy may seem surprising. The prominence of CFM in national REDD+ plans is, however, the result of a complex of political and economic factors.

Firstly, CFM is a relatively 'doable' option, compared for example to changing forest legislation and enforcing the laws that already exist but are flouted by commercial interests, often with government collusion. Even without corrupt practices, which are certainly present in many countries, the political pressure behind forest clearance and degradation by commercial selective logging is considerable, as the economic returns are often great. On the other hand, much CFM is carried out in dry tropical woodlands (savannah and cerrado) and temperate forests, which are of relatively low timber or agricultural value and are thus much less threatened by these forces. These forests are degrading because of increasing population pressure and the dependency of communities on the forests for their livelihoods. Where off-take as a result of cattle grazing in the forest and extraction of firewood, small timber, poles and fodder is greater than mean annual increment, and where the cycles of traditional swidden agriculture have been shortened to unsustainable levels, there are net losses of forest biomass every year, with corresponding carbon emissions. In many cases CFM has been able to bring forest off-take into balance with natural regeneration rates, and can be implemented at low cost to the national economy and without much opposition from the commercial sector. The biggest sticking point is usually tenure, but countries are increasingly dealing with this and making a variety of legal arrangements for benefit sharing with communities (Wiersum, 1999).

In recent years, new types of CFM have been developed in humid tropical forests, for example in the Amazon, mostly in connection with indigenous peoples' land rights (Wiersum, 2009). In these areas, population density is characteristically much lower and the forest areas assigned to each community much larger. Degradation of these forests is not usually the result of community activities, but of illicit timber sourcing by outside actors. The designation of forest as indigenous community reserves (ICRs) legitimates community ownership and makes it easier to exclude these external forces.

CFM therefore tends to have relatively low opportunity costs and can be presented as a people-friendly, 'win-win' option, not merely internally but also towards the international community, which is more and more concerned with social safeguards. International environmental NGOs and associations of indigenous peoples' organisations, such as the Forest Peoples Programme, have taken a strong lead in this, not least for fear that in trying to maximise carbon storage in natural forests, governments may be tempted to evict populations living in or near them who have customary, but not official rights (FPP, 2011). CFM offers an opportunity to involve local people in a positive way and to anchor their rights within REDD+. Moreover, as is the case with many CFM programmes, considerable optimism has been voiced about the potential poverty alleviation benefits. Since CFM under REDD+ would seem to involve payments for the (carbon) environmental services delivered, the potential income-generating aspects of this approach are appreciated (Peskett *et al.*, 2008; Scherr *et al.*, 2003).

8.4 The niche of CFM within REDD+

Evidently, CFM could inhabit a particular niche within a national approach to REDD+. It is unlikely to be strong enough to resist the market forces behind deforestation, since the financial returns to CFM are small. As noted above, CFM is generally effective at reducing degradation processes engendered by community uses (through managing the forest commons in an agreed and planned way), and it may also deter illicit degradation (selective logging) through the establishment of ICRs. In both cases, it is primarily the assignment of tenure and the establishment of clear rights to forest products which underpin the effectiveness of these schemes, and hence the REDD+ impacts.

The distinction here between deforestation and degradation is important. In UNFCCC terms, deforestation refers to a complete change in land use, from forest to non-forest, where forest is defined as having a minimum canopy cover of 10-30%, with trees which at maturity have a minimum height of 2-5 m, and covering a minimum area of 0.1-0.5 hectares (the selection of the threshold value on each of these variables is made individually by each country within these bounds). Degradation refers to a loss of biomass, and hence of carbon stock, in forests which remain forests. Most attention in the literature has focused on deforestation, since clearing a hectare of forest in one go emits anything from 100 to 750 tonnes of CO_2, depending on the forest density. Losses from degradation may be in the order of 5 to 20 tonnes CO_2 per hectare per year, depending on the density of the forest and the extent of the off-take. However, such degradation is usually on a year-in, year-out basis. Moreover, the areas of forest subjected to degradation in any one year may be far greater than the areas subject to deforestation. In addition, when sustainable practices are brought in to reduce rates of degradation, for example under CFM regimes, there is often also a forest restoration effect ('forest enhancement') as stock actually starts to increase. Under REDD+ policy, this sequestration may also be credited. The annual increase of biomass in forests recently brought under community management may in fact be three to five times more than the annual degradation losses avoided (Skutsch and Solis, 2011).

8.5 The commercial value of the carbon services delivered by CFM

Optimists have put the international market price of carbon at USD 10 per tonne CO_2 (WBCSD, 2009), which at face value would make the earning potential for communities involved in CFM enormous. However, as with most beautiful ideas for reducing poverty in rural areas, the validity of this one needs to be examined very critically. The price which REDD+ carbon will fetch on the international market is in fact completely unknown. More importantly, only a small fraction of the final selling price would be likely to reach the producers, as is the case for any commodity produced by rural labour in developing countries; even in Fair Trade deals, the producers rarely receive more than a few per cent of the final sales value. In the case of carbon credits, even though they are not perishable and have zero storage, packing and transportation costs, the transaction costs are likely to be high, because of the need for monitoring, verification and brokerage, in which the intermediaries will take a percentage cut at each step. Moreover, to deal with potential leakage and permanence issues, a proportion of the credits are likely to be deducted as insurance.

Other factors lowering the earning potential may be less obvious. Although reducing degradation can in principle be credited, in practice this will be very difficult, because establishing the reference scenario requires data on historical rates of degradation, whereas there is a chronic lack of systematic forest inventory data in almost all developing countries. While past deforestation rates can be estimated reasonably well by comparing a time series of Landsat or SPOT images to obtain statistics on forest area change, it is very difficult to quantify the changing stock volumes within the forests that remain forests from this kind of remote sensing. In most cases it will be impossible, as the error of estimate in the image interpretation is larger than the typical changes in the stock levels (Van Laake, 2011). Moreover, developing a reference level for this would entail projecting such historical losses forwards, which involves many unknowns. Forest enhancement, however, can be measured in situ, at ground level, over the accounting period and forms a much more robust indicator of carbon savings. It is probable, therefore, that in most cases only the forest enhancement factor could be credited.

8.6 Organising payments for communities

Many proponents of REDD+ support the idea of independent CFM projects generating their own credits and directly placing these on the international market, and are wary of government intervention, citing poor governance and inefficiencies (e.g. Cortez *et al.*, 2010). However, there are two major impediments to this approach. First, the transaction costs of marketing credits are strongly affected by economies of scale, and the number of credits likely to be generated by any one CFM project, or even by a small group of related CFM projects, would be too small to make this worthwhile. Moreover, the communities themselves may not have the skills to navigate the international system effectively, and the purchasers may not trust the validity of credits from such small entities (Munden Project, 2011). Second, it is hard to see how such a system could be sanctioned, given that international REDD+ is based on net savings of carbon over forests throughout the whole country. Although a country could agree that each project is entitled to sell its own credits, these credits first have to be squared up against the national reference level. This is not just a question of individual projects reporting on their carbon achievements to a national registry, but of ensuring that any losses in other parts of the country are taken into account when the project credits are calculated. If a project achieves carbon savings but the country has an overall net loss, the credits from the project will be invalid.

From a practical point of view, therefore, it is it likely that credits will need to be bundled for marketing at the national – or in cases of federal countries, perhaps at the state – level, with authorities at this level either acting themselves or appointing private sector agencies as local brokers. At this scale, the balance could be made between losses and gains over complete territories, and projects could in principle be assigned a proportional share of the net gains. However, deciding what a fair proportional share of the net gains would be is anything but simple. We are faced here with a number of dilemmas, including the classic 'angels and sinners' problem. Communities which have in the past allowed their forests to be heavily depleted are likely to benefit from fast regeneration rates and rapid stock increases once they introduce more sustainable management practices. Communities which have guarded their forests carefully in the past have much less room for carbon to grow. Communities in more favourable climatic areas may be able to increase their stock much faster than communities struggling in low rainfall regions. Clearly, making payments strictly proportional to the amount of carbon sequestered per hectare per year is unjust, and a payment system based on this principle may be highly demotivating (Skutsch *et al.*, 2011). Moreover, payment by results may encourage communities to try to falsify their carbon estimates.

The question is whether payment by results is really necessary to encourage participation; quite possibly people would respond equally well to financial stimuli which simply encourage improved management in general. This is borne out by the fact that most existing systems of payment for environmental services (for water services, conservation, biodiversity, etc.) are in fact based not on payments for *outputs*, but essentially on payments for *inputs*, or surrogates for these inputs (Skutsch *et al.* 2011). Most commonly, a set of management practices is prescribed and payments are made on a flat rate per hectare basis (Wunder, 2005). Outputs are rarely measured (the difficulties of quantifying outputs like water quality and quantity and biodiversity enhancement more or less preclude this in any case); evaluation is usually a question of checking whether the requisite management practices have been implemented. Flat-rate payments are easy to administer, less politically divisive than output based incentives, and transparent.

A flat-rate payment structure for general management activities carried out is likely also to accord better with community notions of the purpose of forest management. As Karky (2008) has shown, if community forests in Nepal were to be managed solely to maximise carbon stock growth, all off-take would have to be stopped. However, communities are heavily dependent on their forests, particularly for fodder and firewood, and Karky shows that the opportunity costs of these products exceed the likely financial rewards for the equivalent carbon by several times. In other words, carbon only becomes profitable if communities are allowed to extract their livelihood needs first. Communities are well aware of the multiple goods and services provided by forests and are often said to have a holistic view of their natural resources (Nelson, 2003; WRM, 2004). An agreed management plan which satisfies a variety of objectives may therefore be more appropriate in their view than a monofunctional one.

8.7 The dangers of recentralisation and the establishment of safeguards and rights

The underlying change which has made it possible for CFM to function is the recognition by national authorities (particularly the forest department) that centralised rules and regulations on their own are not usually effective in controlling the use of forest by local inhabitants. CFM rests on the premise of shared rights and responsibilities; the community is legally permitted to benefit from forest products, but in return is obliged to carry out some management activities and maintain off-take at sustainable levels. To a large extent this is bound up in a concept of ownership, even if the community does not legally own the forest, but merely has an extended contract to use it. Decentralisation is thus a key element in this approach to forest management.

REDD+, as negotiated by the Parties to the UNFCCC, is centralised in the sense that rewards (carbon credits) will be calculated on the basis of net carbon savings at national (or state) level. Concerns about recentralisation of forest management have as a result been raised in the literature (Phelps *et al.*, 2010) and are reflected in attempts by environmental NGOs, such as The Nature Conservancy, to promote what they call a 'nested approach' to REDD+ (Cortez *et al.*, 2010). In this approach, individual projects would be divorced from national programmes and credited entirely independently, although this is a strategy which is unlikely to be accepted by most governments or to the UNFCCC as a whole. Such projects would in most cases require foreign assistance to get set up, and particularly to market their credits. The REDD readiness plans, however, tend to envisage national programmes of support to community management, similar to those that have been operating in Nepal, Tanzania and a number of other countries for some years. Hence, most REDD readiness plans show that although countries are working towards centralisation of the accounting system for carbon, the implementation of REDD activities is understood to require a decentralised approach in which rewards are distributed among the many stakeholders, such as communities managing their forests in more sustainable ways. Moreover, there is a thriving exchange of ideas going on at present on how to ensure that community rights are safeguarded (e.g. Greenpeace, 2011; UN-REDD, 2010).

8.8 Regulation, markets or negotiation?

It is evident that the success of REDD+ will depend on a carefully designed balance between regulation, markets and negotiation. Regulation at international level is required to set the terms of the agreement and create the rules for the market. At national level, regulation may of course be required to strengthen forest law, but more particularly, and in the short run, to establish the rules of the game for CFM under REDD+ and to standardise the reporting and monitoring requirements for local participants. Such regulation is essential to support the credibility of claims of carbon savings and to make them acceptable at international level. Although frequently portrayed as a market mechanism, REDD+ is more likely to involve a heavily regulated quasi-market. The motivation for countries to participate in REDD+ at the international level may be the promise of financial benefits from the sale of carbon credits, but for very practical reasons of transaction costs and political legitimacy, it is unlikely that such a market could be propagated internally within the countries themselves, as argued above. The participation of stakeholders such as communities will depend on whether the rewards offered by governments for improved management and for monitoring the carbon stocks are perceived as covering the costs of such activities, including the opportunity costs of activities and forest products foregone. The rewards need not necessarily all be in cash, of course; as with current CFM programmes, rights to forest products and secure tenure may themselves be a considerable incentive.

Organisations propagating REDD+, including the World Bank's FCPF and the UN-REDD, have, in line with UNFCCC policy on this, been influential in promoting consultative processes at national level. Many community umbrella organisations and local NGOs are now well aware of what REDD+ has to offer and are actively pushing for community rights under this new initiative. It is likely, therefore, that most governments will find themselves under some pressure to negotiate the terms of their national REDD+ plans with these stakeholders, including the way in which rewards are made and their monetary value. Without the active participation of local stakeholders, there is unlikely to be any decrease in the rate of forest degradation or increase in the rate of forest enhancement. Moreover, without their involvement in monitoring these changes at the local level, there will be little data on which to base national carbon credit claims.

REDD+ is thus a very contemporary policy initiative, reflecting underlying power shifts in society as well as ideological movements. Although it certainly contains elements of a market-based system, it cannot survive without both regulation and negotiation. And while it would be wildly idealistic to suggest that forest communities all over the world will be liberated from poverty by REDD+, there is hope that their role in managing the forests will gradually receive more recognition and that they will receive at least some reward for their efforts in this regard.

References

Cortez, R., Saines, R., Griscom, B., Martin, M., De Deo, D., Fishbein, G., Kerkering, J. and Marsh D. (2010). A nested approach to REDD+. The Nature Conservancy, Arlington, VA, USA.

Cropper, M., Griffith, C. and Mani, M. (1999). Roads, population pressures, and deforestation in Thailand, 1976-1989. Land Economics 75(1): 58-73.

Forest Peoples Programme (FPP) (2011). Submission to the SBSTA regarding information for safeguards in REDD+. Available at: http://www.forestpeoples.org/topics/un-framework-convention-climate-change-unfccc/publication/2011/submission-sbsta-regarding-sys.

Geist, H.J. (2001). What drives tropical deforestation? A meta-analysis of proximate and underlying causes of deforestation based on subnational case study evidence. LUCC Report Series 4, LUCC International Project Office, University of Louvain, Louvain, Belgium.

Greenpeace (2011). REDD: Safeguards. Greenpeace International, Amsterdam, the Netherlands.

Intergovernmental Panel on Climate Change (IPCC) (2007). Fourth assessment report: Climate change 2007.

Kaimowitz, D. (1997). Factors determining low deforestation: the Bolivian Amazon. Ambio 26(8): 537-540.

Karky, B.S. (2008) The economics of reducing emissions in community managed forests in Nepal Himalaya. PhD Thesis, University of Twente, Enschede, the Netherlands.

Munden Project (2011). REDD and forest carbon: market-based critique and recommendations. Available at: http://www.redd-monitor.org/wordpress/wp-content/uploads/2011/03/Munden-Project-2011-REDD-AND-FOREST-CARBON-A-Critique-by-the-Market.pdf.

Nelson, A. (2003). Community forest management and action research: emerging models, emerging paradigms. Paper for the XII World Forestry Congress, Quebec, Canada.

Peskett, L., Huberman, D., Bowen-Jones, E., Edwards G. and Brown, J. (2008). Making REDD work for the poor. Overseas Development Institute, London, UK.

Phelps, J., Webb, E. and Agrawal, A. (2010). Does REDD+ threaten to recentralize forest governance? Science 328(5976): 312-313.

Scherr, S., White, A. and Kaimowitz, D. (2003). A new agenda for forest conservation and poverty reduction: making markets work for low income producers. Forest Trends, Washington, DC, USA.

Skutsch M. and Solis, S. (2011). How much carbon does community forestry save? In: Skutsch, M. (ed.) Community forest monitoring for the carbon market: Opportunities under REDD. Earthscan, London, UK.

Skutsch, M., Vickers, B., Georgiadou, Y. and McCall, M. (2011). Alternative models for carbon payments to communities under REDD+: a comparison using Polis model of stakeholder inducements. Environmental Science and Policy 14(2): 140-151.

Stern, N. (2007). Review on the economics of climate change. HM Treasury and Cabinet Office: Cambridge University Press, Cambridge, UK.

UN-REDD (2010). Monitoring governance safeguards for REDD+. UN-REDD Newsletter Issue 9.

Van Laake, P. (2011). Information requirements for national REDD+ programmes. In: Skutsch, M. (ed.) Community forest monitoring for the carbon market: Opportunities under REDD. Earthscan, London, UK.

Wiersum, K.F. (1999). Social forestry: changing perspectives in forestry science or practice? Wageningen Agricultural University, Wageningen, the Nethrlands, pp. 211.

Wiersum, K.F. (2009). Community forestry between local autonomy and global partnership: Quo vadis with environment and climate change payments? In: Proceedings of the community forestry international workshop 'Thinking globally – acting locally', Community Forestry in the International Arena. 15-18 September 2009, Pokhara, Nepal.

World Business Council for Sustainable Development (WBSCSD) (2009). A price of $10 per tonne could help save forests. Available at: http://www.wbcsd.org/Plugins/DocSearch/details.asp?ObjectId=MzY5MjM.

World Rainforest Movement (WRM) (2004). Community forests: equity, use and conservation. World Rainforest Movement, Montevideo, Uruguay.

Wunder, S. (2005) Payments for environmental services: some nuts and bolts. CIFOR Occasional paper no. 42, Center for International Forestry Research, Bogor, Indonesia.

9. Learning from the actors: the rise and demise of a CBNRM initiative in Mexico

Nasim Musalem and Gerard Verschoor
Wageningen University, Rural Development Sociology Group, Hollandseweg 1, 6706 KN Wageningen, the Netherlands; nasim.musalem@wur.nl; gerard.verschoor@wur.nl

Abstract

Advocacy and support for community-based natural resource management initiatives offers a possible solution to the impasse between conservation and development goals. This has resulted in a mushrooming of communal initiatives – yet most of them seem to fail. To explain these failures, experts resort to a variety of theoretical frameworks that bias the understanding of the dynamics underlying such initiatives from the outset. In this chapter we refrain from doing this and stick to the events leading to the rise and demise of a community-based producer organisation in Mexico as narrated and explained by the actors themselves. We do not pass judgement on what may retrospectively be considered to be a 'success' or 'failure', and invite the reader to associate the story with his or her own experiences. In so doing we hope to attract attention to the fundamentally uncertain nature of community-based natural resource management initiatives.

Keywords: community-based conservation, organising practices, *Chamaedorea*, small-scale producer organisations, Mexico

9.1 Introduction

Community-based natural resource management (CBNRM) aims to realise both community development and sustainable environmental management. The general theory behind CBNRM claims that the best way to do this is for local people to use their knowledge, skills and technologies. However, this theory does not correspond to practice: most CBNRM initiatives fail.

How can these apparent failures be explained? Attempts to do so have given rise to a tug-of-war between those who claim that local people, for one reason or another, have not made best use of their knowledge, skills or technologies (the 'blame the victim' argument underlying institutional and social capital approaches) and those who argue that local people have been hampered by the activation of causal mechanisms of power outside their control (the 'blame the system' argument underlying political ecology and postcolonial approaches).

There are three problems underlying this state of affairs. The first, and most obvious one, relates to debates about how to evaluate success and failure. People involved in CBNRM may often be disappointed with a specific process, and initiators of such schemes (especially when these concern extra-communal institutions) may call the outcome a fiasco. Yet, from the point of view of local actors, specific CBNRM initiatives are points in a series of projects that affect them. What looks like a failure from the point of view of the architects of a CBNRM scheme might be seen as a success from the perspective of those who are targeted, who may see the initiative as contributing to their overall livelihoods. It is all in the eye of the beholder.

The second problem is of a methodological nature and is more counterintuitive. Often, if not always, successes and failures are explained by retrospective explanations – that is, through an appeal to factors that are only known at the end of a given initiative. If, for example, *at the end* of a process it becomes clear that a lack of organisation can be blamed for the failure to achieve specific goals, the explanation that is given often invokes a lack of organisational capacity *from the start*. Likewise, if context can be shown to affect (often negatively) the content of a CBNRM programme, those who evaluate it will tend to point to the context from the outset. As Akrich *et al.* (2002: 190) warn, we should not believe for a moment those edifying stories which retrospectively invoke a cause; that is, one should not mistake the *explanans* (that which explains) for the *explanandum* (that which should be explained). Doing so overlooks the fundamentally uncertain character of CBNRM initiatives: for the actors involved, success or failure is only known at the end.

The third problem relates to the question of who provide the theoretical interpretation of the success or failure of a given CBNRM initiative. They are usually scientific experts. But this calls into question the relationship between experts' and laypersons' explanations and, by extension, the role accorded to the critical capacity of the latter by the first. By and large, experts (especially those in the social sciences) adopt a condescending attitude when faced with the need to explain success or failure, believing that their account is better than that of the actors they study. The implicit justification for this is that social scientists have been trained to see and explain processes (or structures, or forces of power and domination, or the invisible hand of the market, and so on) which are not normally evident to laypersons. Here, the argument seems to be that laypersons are so busy with their daily affairs that they are at a disadvantage when trying to explain why they do what they do. Social scientists, by contrast, have the time (and the resources) to reflect on all that laypersons fail to mull over. The critical capacity of non-scientists is, as it were, ignored.

To avoid falling into these traps, in this chapter we try to present a CBNRM initiative 'in the making' by sticking to three principles. First, in our account we refrain from making statements concerning the degree to which the initiative failed or was successful (though we do speak of the rise and demise of it). Second, since controversy tends to reveal processes that otherwise remain hidden (Law and Callon, 1992), we do not resort to elements which only became known at the end of the initiative. In doing so, we refrain from passing judgement on the different actors' standpoints and arguments at the time important decisions were made. Finally, we do not use the case to illustrate a specific (social) theory to account for what happened. Rather, we take in earnest the explanations of our *dramatis personae*: we learn from the actors themselves how we should think and write about them, rather than telling them how they should think about themselves. In so doing, we have been inspired by Freerk Wiersum's lectures – an aspect of his career that has been overshadowed by his extensive writings on people-forest interfaces. One of us (Verschoor) witnessed this for many years; it was mainly in his lectures that Wiersum emphasised the need to stay close to actors' experiences and explanations in order to learn from them how to make success or failure intelligible.

The setting of our case is Veracruz, Mexico. There, tropical forests are the source of many non-timber forest products (NTFPs) which, when collected and sold, complement the incomes of community members. In the past decade, demand for one of these products (*Chamaedorea,* a palm whose fronds are used to embellish flower arrangements in both Mexican and export markets) has risen – and so has its extraction. Since *Chamaedorea* is an endangered species, Mexican state institutions have devised strategies that combine biodiversity conservation and poverty alleviation in forested rural areas to create 'win-win' situations by supporting CBNRM initiatives that focus on this product. The implementation of these strategies, however, assumes that community members have common interests and goals, and will therefore naturally form producer associations or cooperatives and develop the management skills necessary to operate dynamic or 'green' markets – in spite of evidence to the contrary (Belcher, 2003; CEC, 2002; Marshall *et al.*, 2003; Trauernicht, 2005). It is against this background that we follow the trajectory of the life and demise of *Productores Integrados de Tezonapa* (PROINTE), a community-based organisation established to market exotic *Chamaedorea* palm leaves.

9.2 Constructing the organisation

The organisational process we describe spans a six-year period (2001-2007) and is set in the small rural community of El Paraíso (1,945 inhabitants), which forms part of the municipality of Tezonapa in the southern Mexican state of Veracruz. The community is located at the highly biodiverse point of convergence between the Sierra de Zongolica mountains and the coastal plains which border the Gulf of Mexico. The region's main economic activities are related to agricultural production and the processing of coffee, latex, timber and sugar. El Paraíso is one of the many rural settings on the agricultural frontier where forests have been cleared for the extraction of natural latex since the late 1960s and, later, for the establishment of sugarcane plantations (INEGI, 2000). This situation has led to the awareness about the importance of biodiversity and environmental conservation derived from the loss of tropical ecosystems in forested areas, which in turn has triggered locally-based, alternative, or 'green' economic initiatives. One of these, revolved around the production and marketing of *Chamaedorea* fronds, which are potentially profitable as they command high prices in the US and Europe. In what follows, we provide a chronological description of the crucial events related to four stages in the life of PROINTE: the origin and expansion of an idea, the planning of activities, their implementation, and the demise of the organisation, as well as the organising practices and politics that went along with these. Figure 9.1 presents a timeline with these four stages and the main events and decisions taken during the life of PROINTE.

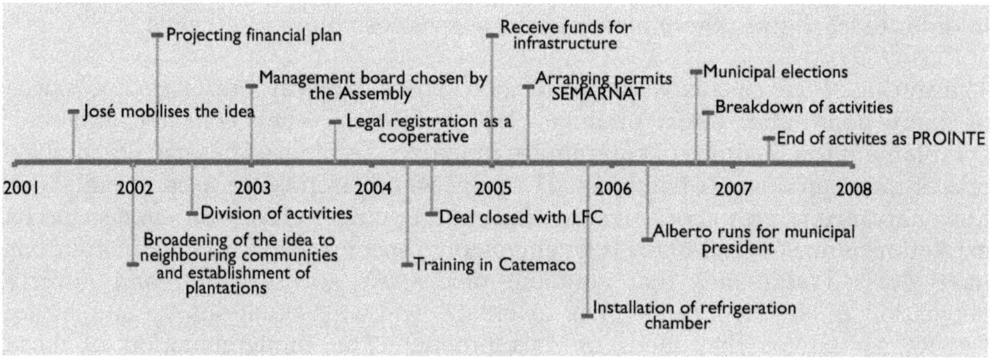

Figure 9.1. Timeline of events and decisions at PROINTE.

Origin and expansion of an idea

Although a convoluted series of discussions preceded the formal proposal to form an organisation, the initial steps can be traced back to the spring of 2001, when José, aged 56, a peasant and local fruit and wood trader from El Paraíso, set forth a plan to establish a cooperative for the commercialisation of tropical palm fronds in his community. As the key actor in the development of this construction, José attributes his ability to identify potential in starting businesses to what he calls an 'entrepreneurial spirit'.

José recalls starting 'doing business' early in his childhood, trading everything from buttons, threads and needles as a young boy, and later mangoes for the Mexican fruit mega-processor, Jumex. Although José dropped out of elementary school, his 'vision' for market opportunities went uncontested locally and, as a result, he is greatly respected throughout the community. Besides working his land – coffee intercropped with latex and timber – José has been active in many representational positions in his community, serving as community treasurer, secretary and president of the local coffee producers' cooperative between 2001 and 2006. During this time, he started mobilising ideas to establish a formal market for *Chamaedorea* in the community. This opportunity arose due to the sudden appearance of *coyotes* (intermediaries) from the nearby cities of Orizaba and Córdoba who wanted to commercialise *Chamaedorea* fronds found growing naturally on farmers' lands. Instinctively, José concluded from these visits that there was an opportunity for him to market tropical fronds at nearby wholesale markets, and this triggered his curiosity about the costs and profit margins. The calculations represented a positive margin of 40-50% that could be obtained by selling directly to wholesalers.

Another reason for José's actions lay in the general anxiety caused by the state's gradual abandonment of rural areas since the late 1980s. Direct government intervention in agriculture had been the hallmark of Mexico's development policy since the mid 1930s, yet reform, restructuring and the closure of parastatal enterprises (e.g. CONASUPO) in the 1990s, as part of market 'liberalisation' processes, were met with deep concern by rural communities (Yunez-Naude, 2003). In Tezonapa (the municipality of this study), as in many of Mexico's rural areas, the liberalisation of agricultural markets affected the way in which peasants were organised to access markets and 'pushed' them to organise themselves into new community or regional cooperatives to obtain greater bargaining power. In Tezonapa, most of the peasants linked themselves to sugar mills and coffee cooperatives (González-Jacome, 2004; Singelmann, 2003); so too the coffee cooperative of El Paraíso, which survived after the disappearance of the Mexican Coffee Institute (IMECAFE).

The *Chamaedorea* association slowly turned into a reality when José reached out to others in his community through regular visits to his closest friends and neighbours, with whom he often discussed his ideas and whom he tried to enrol in his projects. Because of José's position at the coffee cooperative, this effort was relatively straightforward as he was in touch with most of the inhabitants of El Paraíso at the monthly coffee-bean delivery to the coffee cooperative. José made the most of these encounters to interest others in joining forces locally to collect and market *Chamaedorea* fronds. These casual meetings revolved around the possible economic benefits and the perceived need to seek new markets. These opportunities were seen to contribute to improving local living conditions, particularly as coffee prices were falling.

Enrolling those not related to the coffee cooperative was also easy. *Chamaedorea* grow profusely in the canopy forests that cover the mountains surrounding El Paraíso. The plants require relatively little maintenance and harvesting them would not compete with existing local agricultural production systems. Furthermore, José presented *Chamaedorea* as an economically viable, environmentally friendly development option, raising the interest of several conservation organisations, especially when associated with coffee in agroforestry systems (Baumer, 1990; CEC, 2002).

By the end of 2001, discussions about forming a *Chamaedorea* association crystallised with the drafting of an affidavit supported by 15 landowners close to José. The association's goals were gradually defined, and, by the end of 2002, José and the 15 landowners were determined to set up an organisation that would guarantee a large base supply of *Chamaedorea* leaves, allowing them to seek better prices for their fronds than the ones paid by the *coyotes*. José's actions at this moment included studying prospective markets at the wholesale markets of nearby Córdoba and Orizaba. The real problem at this point was *how* to achieve this.

José's persistence had its roots in his awareness that the main comparative advantage of El Paraíso was its geographical location – at the foot of the Sierra de Zongolica mountains – with its nearly perfect ecological setting for *Chamaedorea* production (high elevation, good weather conditions). The fact that most landowners in El Paraíso knew that the amount of labour required to harvest the fronds would be very low supported José's assumption that the potential supply of the fronds would be very high. However, José and the other interested landowners were confronted by the fact that there were not enough *Chamaedorea* plants to guarantee sustainable production. The association therefore proposed establishing plantations to cultivate the fronds. Although environmental conditions were favourable and plant material was readily available, little was known about the cultural requirements of the species. Fortunately, such expertise was available at the National Institute for Forestry, Agricultural and Livestock Research (INIFAP) just 20 km away from El Paraíso. Staff at the research centre were fascinated by the El Paraíso initiative and provided the necessary information and tools to propagate *Chamaedorea* palms to quickly establish plantations.

During the development of the idea to establish a *Chamaedorea* cooperative, José needed to link the group of interested landholders with someone with a visionary mind who could help solve problems and constraints as they arose. This is why he joined forces with Alberto (48), a charismatic entrepreneur and aspiring politician from the neighbouring city of Tezonapa. Having known each other for years, Alberto and José formed a good team and were able to capitalise on their recognition and respect earned as leaders of the regional coffee cooperative. Together they visited representatives of the neighbouring communities. The rationale behind this was that since the project started as a local idea, the solidification of a larger supply base for the fronds would give them regional recognition, granting them a stronger voice and negotiating power with banks and municipal and state governments. Their motivation, willpower and commitment to constant work would open doors, or so they reckoned, which too often remained closed to small, unknown groups. As the number of members started to increase, rumours and expectations surrounding the project also grew, making the project attractive to many as it was generally felt that it would provide a better income to the inhabitants of the region.

The rest of the members-to-be of the initiative immediately welcomed the notion of joining forces with a large number of small-scale producers. Due to the distinctive way in which the commercial chain for *Chamaedorea* is organised – handled by few regional buyers, who then mainly sell to a single company, with prices offered to producers set at very low levels – membership of the organisation would offer prospects of obtaining better prices through collective marketing, rather than selling individually through local *coyotes*. This advantage for obtaining better prices gave José and Alberto a strong argument to convince small-scale producers that a communally sustained buying structure would be beneficial to all. In addition, the possibility of accessing credit and the wider benefits of municipal and state development programmes, such as FONAES and INDEVER, to finance the establishment of small-scale enterprises were attractive reasons that encouraged many to join the project. These programmes are aimed particularly at individuals and group initiatives with scarce resources in an attempt to support productive projects through social financing schemes (FONAES, 2009).

As the number of people interested in the project grew, so did the new responsibilities and tasks faced by the incipient organisation. By 2002, most of the El Paraíso landowners – approximately 80 – were part of the project. In an attempt to define the boundaries of the organisation, a division of tasks was agreed upon. During this phase, José and Alberto, who searched for 'good management characteristics', scrutinised the abilities of each of the members participating in the project. The definition of the desirable roles within the organisation started to take shape and become clear: it was decided that José would lead and coordinate most of the activities; Alberto would be in charge of member endorsement and expansion of the project to neighbouring communities; and the remaining 13 founding members would take on a number of different management tasks and set up nurseries to expand the potential for *Chamaedorea* cultivation. This was a crucial event, because it would later affect access to permits for sustainable use.

With this new division of tasks in place, Alberto and José's first administrative targets were the municipal authorities, who would inform them about the financial opportunities available in programmes for regional commercial development. At different stages, a series of tasks was required to comply with the rules of operation of the programmes (e.g. formulation of a financial plan, providing a census of producers involved, identifying the infrastructural needs). Alberto was in charge of this 'privilege' because his brother was the municipal president of Tezonapa. The state authorities were also approached to initiate the request for obtaining a sustainable use notice, which required regular visits to Jalapa (the state capital) to conduct meetings at the state government's rural development offices.

Planning of activities

To achieve solidity in the deployment of activities and start the actual work, money was required to cover the transport expenses and get things moving. However, the financial resources available to José and Alberto were limited. Furthermore, they divided their efforts and time between taking care of their agricultural land

and visiting neighbouring communities to reach agreements to get more farmers interested. Additionally, the paperwork was costly and reaching communities was very time-consuming. In view of this, during this time José sought someone to assist with the paperwork and help with errands related to administrative matters. Margarita (30) was asked to join them in a non-remunerated position, but with the agreement that once the cooperative was on its feet she would occupy one of the main management positions. Margarita is a young and outspoken woman trained as a business manager. In close coordination with José and Alberto, she took on the task of writing and identifying the needs for the envisioned organisation and developing a monitoring/control system for it. Margarita was also in charge of drafting a feasibility study, a financial forecast and an executive summary for potential funding institutions. Once finished, these documents (PROINTE 2003) became the organisation's 'business card' and would be used in lobbying to obtain financial support from the municipal authorities of Tezonapa.

Apart from presenting boundaries and objectives (and, to a certain extent, the organisational room for manoeuvre), the documents Margarita drafted also yielded an important projection: the required capital and costs of marketing. According to the documents, it would be necessary to obtain an initial capital of MXN 1,687,699 (approximately USD 135,000) to start activities. This ballpark figure was based on a rough estimate of supply by the members of 4,000 to 5,000 *gruesa* (bunches of 12 dozen stems each) per week, for an average price of MXN 20 (USD 1.50). For members, this meant an increase of MXN 5 per *gruesa* (USD 0.5) on the prices paid by the *coyotes* as well as a long-term buying contract with the organisation. It was believed that this pricing strategy would cut off the supply of fronds to the *coyotes*.

From this point on, weekly management meetings were scheduled to track progress with activities related to the establishment of the organisation. Late in August 2003, the first General Assembly as an 'organised group' took place. The most relevant issues discussed focused on three issues: difficulties with obtaining the starting capital; the form of supervision needed to reach and sustain the projected supply of fronds; and the importance of keeping an updated record of all producers occupied in the establishment of plantations.

More than 20 communities were represented in the Assembly, and 20 more would follow within the next couple of months. This reflected the work of Alberto, who had been very successful in convincing and offering neighbouring communities a place in the project. The first communities to be enrolled were the ones neighbouring El Paraíso. The communities of Limonestitla and Caxapa – known to be highly organised – approached the project directly, requesting participation. Tolentino Méndez (leader of the community of Limonestitla) explains why:

> Before all this organisation fuss started, INIFAP from Jalapa had already told us that the fronds were a good business! ... but before the constitution of the cooperative, we wanted to wait until we could

> establish plantations in our communities and see how this little palm behaves. Only then would we join fully. They [José and his team] also told us that a large amount of money was going to be given to the cooperative to start operations, and that we would not need to put down any money from our own pockets yet.

For the management team, the success of the project depended on its capacity to obtain the initial capital needed to initiate operations and build the necessary infrastructure (mainly a selection and refrigeration area). Most of these initial efforts were centred on lobbying municipal and state government authorities, carried out during a six-month period in 2003, during which the anticipated marketing projections were presented. As expected, the request for the full amount for the infrastructure (selection and refrigeration area) was welcomed and supported by Tezonapa's municipal government. As Alberto mentioned after a meeting with the municipal authorities:

> We presented the idea of the organisation with a very clear and understandable overview from plantation to commercialisation, objectives, goals; they [the authorities] evaluated it and thought it viable. In the region, there are already some projects for coffee, sugarcane, citrus fruits - but none for fronds. Therefore, because it was an innovative idea, the authorities decided to support it. Our next step is the legal registration of the cooperative.

Accordingly, the Registration Assembly took place during the first few days of September 2003. In a general vote it was agreed to form a cooperative, to expand its area of influence to all neighbouring communities in the municipality of Tezonapa, and to establish a joint business with El Paraíso's coffee cooperative (with which it would share basic infrastructure, such as office space and land to establish the refrigeration chamber). A Board of Directors was elected, with José as president, Alberto as vice-president, Samuel (aged 38) as treasurer, Margarita as manager, and five members who had been involved early on as board members or *vocales*.

On 7 September 2003, the cooperative was legally registered in Córdoba, Veracruz. The name chosen was *Productores Integrados de Tezonapa*, S.C. de R.L. (PROINTE), sharing the legal address with the coffee cooperative. It is important to mention that PROINTE was established as an independent cooperative, but was closely associated with the coffee cooperative of El Paraíso (an older structure derived from the government withdrawal of the coffee regulation programme in the late 1980s). While José continued to represent the coffee cooperative, he simultaneously worked and took decisions for both cooperatives.

9.3 Putting the CBNRM initiative into practice

Once PROINTE was legally registered, with José serving as president, and having obtained the necessary financial support, it was time to put their business plan into practice and start building the infrastructure. Although this looks like a late start, the 'organising work' took more time than they expected. For José, the installation of the refrigeration chamber was a step into an unknown world. There was considerable activity at this time. José invested several months in the ins and outs of the installation, which required the largest expense of financial resources and specialised skilled labour. This new challenge needed a different plan of action and was beyond José's direct reach. The specification was unknown to José. It required technical knowledge of refrigerated warehouses, knowledge of the needs of the international markets and about how the fronds were classified in commercial sizes. Since practical knowledge about standards and grades was an unknown field for him, José contacted La Flor de Catemaco (LFC), the largest and most successful wholesaler of *Chamaedorea* palm fronds in Mexico, with an estimated export of two million fronds per year and over 30 years of experience in the field (La Flor de Catemaco, 2008).

The main concern for José at this point was how to get the attention of LFC. By the sheer number of producers it represented, PROINTE was potentially a substantial supplier of fronds and thus an interesting partner for LFC. For this reason, while recognising that their room for negotiation was limited, as PROINTE was not yet producing fronds, a team headed by José scheduled a meeting in the summer of 2004 in Catemaco (the Mecca of tropical foliages and headquarters of LFC) to introduce themselves to the directors of the company. Although, according to José, they were very nervous, the meeting was productive. The willingness of the team to prove that they could do more than just get PROINTE off the ground impressed the managers of LFC. Both parties reached an agreement to set up a training programme for PROINTE's key personnel, which included the handling and selection techniques and procedures for classifying the *Chamaedorea* fronds, and to hold a series of talks to increase awareness of the importance of quality for products to be marketed internationally.

In order to design the infrastructure needed at El Paraíso, José sought assistance from an independent technician specialised in refrigerated warehouses. This resulted in a blueprint for the selection and refrigeration chambers (340 m^2) that fulfilled technical requirements, could store the expected supply of fronds for four days at -3 °C, and had sufficient storage capacity to load one refrigerated trailer-truck a week with a capacity of 1000 *gruesas*. In the blueprint, the premises would also be equipped with a 17 °C climate controlled selection area that could accommodate around 15 employees for the selection, bundling, packaging and sorting of 5,000 fronds a week. José and the rest of the management board were certain that PROINTE could not start working as planned without the refrigeration chamber. They would not be able to cut out the intermediaries and would not be in a position to solidify their agreements with LFC in terms of quality and volumes. Therefore, the installation of the huge metallic structure stands out as a major achievement, because it required the heaviest mobilisation of actors, resources and economic investment, as well as planning that played, at that time, a crucial role in the decisions and strategies of PROINTE. In the summer of 2004, the refrigeration chamber was installed inside the premises of the coffee cooperative as stated in the registration of PROINTE. This location seemed obvious to José as he was president of this cooperative.

Being able to comply with the specifications for commercialisation in the international markets is what had led LFC to be a successful enterprise in its field. It required a high level of quality awareness from all the staff involved in the production and handling of the fronds. Therefore, six months after the visit to LFC, a group of 20 of PROINTE's employees travelled to Catemaco to receive training for a full week. The training programme incorporated the specifics about quality standards and personnel management. The employees were also taught how to handle the fronds. The broader experience of LFC in the business implied a strict quality control system as well as clear-cut definitions of specific species of *Chamaedorea* that complied with international market demand.

The next bottleneck for PROINTE was the establishment of standards to select fronds according to the commercial names, as well as a detailed route planning for the collection of fronds at farmers' fields. The latter aspect at first seemed to be simple, but turned out to be problematic because it entailed exclusion of those PROINTE members who lived in hard-to-reach, mountainous areas where no paved roads existed, or where roads were impassable during adverse weather conditions. This was very much against the spirit of PROINTE (which emphasised that all members had equal rights), caused irritation among members, and jeopardised the initial idea of expanding the organisation. Indeed, during a meeting in Limonestitla in late 2004, members threatened to withdraw from the project should their community not be included in the collection route.

Following this dilemma, the newly established cooperative faced new challenges of an altogether different nature: implementing a totally new management model in the production process. Indeed, PROINTE needed to start working as a collective, taking into consideration the opinion of all members, and not just of specific individuals - as during the early stages of the project.

As mentioned earlier, since *Chamaedorea* is an endangered species, its use is a hotly debated issue in conservationist circles, and also plays an important role in the formulation of policies and regulatory frameworks pertaining to these products. Thus, as expected, its market in Mexico is heavily regulated, requiring producers to obtain notices and permits to market them in domestic and international markets. These are granted by the Ministry of Environmental and Natural Resources (SEMARNAT) and supervised by the Federal Environmental Protection Office

(PROFEPA). Obtaining these permits is a complex matter, as all the aspects stated in them have to be taken into consideration for the legal management and marketing of the fronds. PROINTE managed to obtain legal permits for the extraction of *Chamaedorea* for all participating communities. These permits had to be requested at SEMARNAT's federal offices, which required regular trips to the state capital, Jalapa, and to Mexico City.

9.4 Demise

Everything seemed to go well for PROINTE until, unexpectedly, a group seeking control over the cooperative caused polarisation within PROINTE's Assembly. This came up during the election campaign for the municipal government in Tezonapa – held every four years – in which Alberto was as candidate for the ruling party. The schism was a result of Alberto's decision to run for municipal president.

This decision seemed to be a tactical one in that, should Alberto be elected to office, it would open up new opportunities for PROINTE. Both José and Margarita gave up their personal time, and some of the time that would otherwise have been dedicated to PROINTE's establishment, to help Alberto in his political campaign. As his main campaign discourse, Alberto used the 'achievements' in bringing together more than 400 producers in PROINTE and his knowledge and awareness of rural areas to secure the vote of the population. In the following weeks, after the start of the campaign, he dedicated his time purely to promoting his candidacy. During this time, José and Margarita multitasked and combined work for the management board of PROINTE with the political campaigning of Alberto. This was done by using the same resources (e.g. the office was used to hold political meetings), people (chauffeurs were delivering pamphlets; Margarita was paid to do the accountancy for PROINTE while also managing funds for the campaign). Some members of the cooperative warned Alberto about combining these activities and the dangers involved in this – above all, the risk of confusing/mixing financial resources. However, Alberto continued on the same course and never separated his political campaign from his management tasks.

The support given to Alberto's campaign bore fruit as the local Partido Acción Nacional (PAN) Assembly chose Alberto as their official candidate. Subsequently, Alberto moved his operational centre to the PAN office and continued campaigning in most of the communities in the municipality of Tezonapa and organising political gatherings to promote his image. A crucial turning point in Alberto's campaign occurred when Martín, a member of PROINTE who was also in the race for the municipal presidency and Alberto's main political opponent, accused Alberto of 'inheriting' the municipal presidency from his brother. Martin and his supporters broadcast this accusation by painting many of Tezonapa's walls around the main plaza with claims that 'power should not be inherited'. The wall paintings referred not only to political inheritance, but also to the continuity of Alberto's brother's economic interests through Alberto's future administration, as they shared various projects (e.g. ownership of sugarcane plantations, *Chamaedorea*, coffee, and representational positions in public-private projects with Nestlé and INIFAP, and in the regional farmers' committee). The episode was followed by an article published in the local newspaper that criticised Alberto's political campaign and insinuated that the resources – more than MXN 1.5 million – obtained for the establishment of PROINTE's infrastructure were also being used to finance Alberto's political campaign. This was a big setback for Alberto's aspirations, but despite this adversity his team continued with the political campaign. José explains:

> There was this article claiming that we had gained more than a million pesos from the state government, and that hurt us a lot. Everyone in the region read that article and immediately attention focused on our activities; we were under a microscope. Everyone thought there were millions involved and thought we were using it for our personal benefit! But this was not the case! With the money we had to make several trips to Jalapa to visit communities to organise the people. You had to drive your car everywhere ... So things are not what they seemed, right? Those funds were used entirely for building the infrastructure, but those are the things that people in general do not know.

The newspaper article triggered many reactions and a general feeling of unease among the members of the new cooperative because it fuelled gossip about the source of Alberto's campaign funding. Alberto (and also José) countered that the most important issue was that Alberto should win the municipal presidency. Through his presidency, so they argued, it would be relatively easy to access funding from a variety of the state's regional development programmes and use this to solidify PROINTE's efforts. Winning the elections could be achieved only with the vote of the large number of PROINTE's members. From José's and Alberto's point of view, PROINTE's members were thus exchanging votes for financial support. However, members' worries were not appeased by this logic. From their perspective, Alberto (and, by extension, José) would indeed use the municipal presidency to access funding, but it was not at all guaranteed that the funds would find their way exclusively to PROINTE as Alberto, as explained earlier, was involved in other economic activities parallel to PROINTE. From the members' point of view, they were exchanging votes for the further expansion of Alberto's parallel businesses.

The members' point of view was, as it turned out, partially the outcome of active politicisation by a group within PROINTE led by Martín. The group backed up the rumours about the misuse of funds obtained for the installation of the refrigeration chamber and the delays in starting operations. José challenged these criticisms and rumours by backing Alberto's candidacy, but his efforts proved unsuccessful as the Assembly was already divided on the issue, and his leadership had been strongly eroded. José did his utmost to reverse the situation by visiting all of the participating communities and present his take on the issue: the necessity to have cash flow (by means of 'securing' the municipal presidency) to start commercial activities (buying the first batch of members' fronds, paying salaries to selection and transport personnel).

José's efforts to soothe members' anxieties (and especially those of his closest collaborators) took him the best part of six months. Obviously, José did not want to give up the dream that had taken him so much time and effort to build, especially as the commercialisation of fronds seemed around the corner. His efforts notwithstanding, José sensed that the collective spirit had gone. Therefore, as a last resort, in late 2006, José and Alberto called for a general meeting of PROINTE during which they tried to regain faith in the project.

Much to their surprise, however, a common feeling of distrust and anger took over the room when members of Martín's faction accused Alberto and José of misusing the political campaign to further their personal interests. José and Alberto, for their part, tried to divert the accusations by presenting clear financial statements, a full description of expenses incurred so far (including those related to the installation of the selection and refrigeration areas), but the members did not seem to care. What seemed to be on their minds was to remove José and Alberto from their functions at PROINTE.

A couple of weeks later, by mid 2007, the Assembly chose a new management board. José, Alberto and the rest of their team were replaced by farmers belonging to Martín's faction. After their replacement, José and Alberto reacted by restricting entry to the refrigeration chamber, which was built on the coffee association's premises. In turn, this was seen by members as an illustration of the problems that could arise from Alberto's and José's mixing their parallel economic activities, as well as of the need to increase transparency by having a clear-cut division of enterprises.

The new board did not improve the situation and was not able to initiate commercial activities. This was due mainly (as in the previous board's administration) to a lack of cash flow, but also because the agreement with LFC had been a personal one between José and the company. Martín and his followers faced the same difficulties and uncertainties that José's group had to tackle before, but smoothed over their failure to start activities by stating that José and Alberto had left the organisation's bank account empty. To act on the situation, Martín called for a meeting of the General Assembly to discuss the future direction PROINTE should take. Because a quorum could not be achieved, the meeting never materialised and so the project came to a halt.

José, Alberto, Samuel, Margarita and others continued with their other businesses. In all the communities involved, former members referred to the whole process as a 'big loss' caused by a mixing of political and personal economic interests.

9.5 Conclusion

In this account, we have followed the actors and described 'how things get done': how, in practice, so-called community initiatives are not 'collective' in the traditional sense of the word, but turn out to be the result of entanglements between different entities with different goals. What we have seen is that the trajectory of a 'communal initiative' obeys the passionate, personal interests and political manoeuvrings of specific individuals who – for a while – are able to enrol others into their projects and act as collective spokespersons. But 'speaking for others' is a political act, and ultimately political opposition to the project turned out to be stronger. This is not to say, however, that the PROINTE initiative was 'successful' or 'unsuccessful'. Rather, we have kept to our principle of impartiality regarding the veracity of actors' arguments, and not ignoring the likelihood of there being a multiplicity of interpretations. Furthermore, we have been deliberately atheoretical; that is, we have refrained from using a set of a priori concepts in order to frame or understand the struggles and negotiations that led to the rise and demise of PROINTE. We take the view that the value of cases is not limited to serving as a specific instance of a more general theoretical argument. In fact, we would argue that there are many more ways of marshalling specificities.

As Mol and Law (2002) claim, case studies may do other kinds of work. For example, they might make the reader aware of events and situations elsewhere that have not been recognised so far. Our case can serve to put readers' knowledge of similar cases into relief; that is, it may help them to search for differences, repetitions, commonalities. The case also invites the reader to look further and ask 'So what happened afterwards?' In our case, activities were eventually picked up once again and the initiative obtained a second chance (even though this was much later – sometime in 2009). The case may also serve to undermine cherished assumptions, such as the idea that policies to support conservation and development can automatically trigger the right responses. Or it may work allegorically (like a parable) and trigger all sorts of other associations. It may convey messages that are hard to express concretely, but from which one nevertheless learns. Coincidentally, or by design, this was the virtuosity of Freerk Wiersum's lectures: one could literally see how students' souls started wandering once he treated them to one of his famous, vivid, captivating stories of how people relate to forests.

References

Akrich, M., Callon, M. and Latour, B. (2002). The key to success in innovation Part I: the art of interessement. International Journal of Innovation Management 6: 187-206.

Baumer, M.(1990). The potential role of agroforestry in combating desertification and environmental degradation technical centre for agricultural and rural cooperation, CTA, Wageningen, the Netherlands, 250 pp.

Belcher, B.M. (2003). What isn't a NTFP? International Forestry Review 5: 161-168.

Commission for Environmental Cooperation (CEC) (2002). In search of a sustainable palm market in North America, Commission for Environmental Cooperation, Montreal, Canada.

Fondo Nacional de Apoyos para Empresas en Solidaridad (FONAES) (2009). ¿Qué es FONAES? Available at: http://www.fonaes.gob.mx/index.php/fonaes/que-es-fonaes.html.

González-Jacome, A. (2004). Dealing with risk: small-scale coffee production systems in Mexico. Perspectivas Latinoamericanas 1: 1-39.

Instituto Nacional de Estadística Geografía e Informática (INEGI) (2000). XII Censo General de Población y Vivienda, Instituto Nacional de Estadística, Geografía e Informática, Aguascalientes, México.

La Flor de Catemaco (2008). La Flor de Catemaco Website. Available at: http://www.laflordecatemaco.com/productos.html.

Law, J. and Callon, M. (1992). The life and death of an aircraft: a network analysis of technical change. In: Wiebe, E. and Law, J. (eds.), Shaping technology/building society. Massachusetts Institute of Technology, Cambridge, MA, USA, pp. 21-52.

Marshall, E., Newton, A.C. and Schreckenberg, K. (2003). Commercialisation of non-timber forest products: first steps in analysing the factors influencing success. The International Forestry Review 5: 128-137.

Mol, A. and Law, J. (2002). Complexities: an introduction. In: Law, J. and Mol, A. (eds.), Complexities. Social studies of knowledge practices. Duke University Press, Durham, CT, USA, pp. 1-22.

Productores Integrados de Tezonapa (PROINTE) (2003). Proyecto de Comercialización de Palma Camedor, Tezonapa, Veracruz, México.

Singelmann, P. (2003). La transformación política de México y los gremios caneros del PRI, Revista Mexicana de Sociología Instituto de Investigaciones Sociales, México, D.F., pp. 117-152.

Trauernicht, C. (2005). The effects of non-timber forest product cultivation on the plant community structure and composition of a humid tropical forest in southern Mexico. Forest Ecology and Management 219: 269-278.

Yunez-Naude, A. (2003). The dismantling of CONASUPO, a Mexican State trader in agriculture. The World Economy 26: 97-122.

Part III. Biocultural diversity

10. Perceptions and values of local landscapes: implications for the conservation of biocultural diversity and intangible heritage

Michelle Cocks[1] and Tony Dold[2]
[1]Institute for Social and Economic Research, Rhodes University, P.O. Box 94, 6140 Grahamstown, South Africa; [2]Selmar Schonland Herbarium, Rhodes University Department of Botany, P.O. Box 94. 6140 Grahamstown, South Africa; m.cocks@ru.ac.za

Abstract

South Africa is widely recognised as a highly diverse country with regard to its people, culture, landscapes, biological resources and ecology. The importance of policies on protecting cultural diversity and how they relate to the environment is only just emerging at an international level, but in South African we find there is a complete lack of institutional support for rural communities' attachment to local landscapes. The main objective of this case study is to improve our understanding of the meanings and values that Xhosa people (amaXhosa) of the Eastern Cape Province of South Africa attach to their natural landscapes. It is clear that the amaXhosa are still intrinsically connected to their natural environment, in particular the thicket biome that is called *ihlathi yesiXhosa* (Xhosa forest). Access to places and spaces within the natural environment provides a sense of wellbeing, a link to ancestral spirits, a location for religious rituals, plus a wealth of culturally-inspired uses of specific species and sacred places in the landscape. We suggest that local cultural values could be incorporated into the formal conservation process.

Keywords: biodiversity, cultural values, perceptions, rural communities, South Africa, thicket biome

10.1 Introduction

South Africa is widely recognised as a highly diverse country with regard to its people, culture, landscapes, biological resources and ecology (Cadman *et al.*, 2010). Indeed, South Africa is often referred to as the 'rainbow nation', a metaphor introduced by Archbishop Desmond Tutu (Evans, 2010) which refers to the cultural and ethnic diversity of its population, as well as a desire to embrace rather than suppress or homogenise this diversity. Eleven languages are officially recognised and several more are spoken.

South Africa has been identified as the third most biologically diverse country in the world, with 80% of its plant species being endemic (Cadman *et al.*, 2010), but also with the highest known concentrations of threatened plant species in the world (Wynberg, 2002). Economic development, agriculture, rapid urbanisation, invasive alien species and wildlife trade are some of the major drivers of biodiversity loss in the country (DEAT, 2006). Calls to curb this ecological demise are reflected in the Millennium Development Goals, which clearly state the importance of conserving biodiversity to ensure the safety and health of our ecological systems as well as human life (Millennium Ecosystem Assessment, 2005). Furthermore, there is a rapidly emerging acknowledgement of the need for an integrated approach to the conservation of biological and cultural diversity to ensure sustainable development (Knight *et al.*, 2010; Pretty *et al.*, 2010). Communal lands in South Africa comprise a significant proportion of the country's land surface and represent areas where people are still directly linked to biodiversity through subsistence livelihoods and cultural practices. Two of South Africa's largest former communal homelands, Ciskei and Transkei, both in the Eastern Cape, are within the Maputaland-Pondoland-Albany Biodiversity Hotspot (Mittermeier *et al.*, 2004). However, policy responses to this integrated paradigm have been slow to emerge in the communal areas of South Africa.

As sustainable use and conservation of biodiversity is seen as being embedded in societal and cultural systems (Zweifel, 1997), international policy-makers are increasingly considering how to incorporate social and cultural values into conservation practices, as noted in the Convention on Biodiversity Conservation (CBC) of 1992, for example (McNeely, 2000). Attention is being given to the traditional use and management of species for either utilitarian or cultural purposes (e.g. Berkes *et al.*, 2000; Cunningham, 2001), but the concept of biodiversity concerns not only species, but also landscapes.

For most of the 20th century the concept of 'landscape' was imbued with nationalistic, religious and scientific associations, particularly within Europe and North America. For example, the concept of landscape became associated with wilderness or wild nature, something apart from people. In the late 1990s, in response to the growing 'cultural landscape' movement (Taylor, 2008), landscapes were recognised as providing a sense of time and continuity. Today, landscapes have become a symbol of the fundamental links between local communities and their heritage, between humankind and its natural environment (Rössler, 2006). This interpretation had evolved to such an extent that a diversity of living cultural places, sacred sites and cultural landscapes had been included on the UNESCO World Heritage List (Rössler, 2006).

With respect to studying landscape values, the term 'value' may be interpreted in different ways. For example, the ethnobotanical focus on landscapes tends to focus specifically on the cultural values of plants (Berkes *et al.*, 2000; Posey, 1999). This approach is mostly descriptive, with attention being given to various types of values underlying specific cultural practices. Environmental psychology studies tend to focus on specific emotional processes, such as the potential of natural areas to stimulate a person's level of arousal or to act as a repository of location specific narratives. This approach has mostly focused on the landscape values for recreationist and tourist venues in western countries. Reunala (1999) identifies cultural values as being specific to a certain cultural group, whereas psychological values are individual based.

An understanding of the complexities of different cultural perceptions of landscapes has been recognised by the Millennium Ecosystem Assessment (2005) as contributing to more effective strategies for ecosystem management and socioeconomic development. To date, however, little systematic attention has been given to the study of local landscape values as representations of people-environment interactions. The need for a better understanding of these relations to improve the relevance and effectiveness of conservation research, planning and implementation has nevertheless been identified as an important issue. The importance of cultural protection policies and how they relate to the environment is only just emerging at an international level (Pretty *et al.*, 2010). Within the South African context we find that there is a complete lack of institutional support for rural communities' attachment to local landscapes. The protection and preservation of cultural heritage in South Africa is the mandate of the South Africa Heritage Resource Agency (SAHRA). Their focus, however, is predominantly on tourism and they give very little attention to intangible components of cultural heritage. Tourism has been identified as one of the fastest growing economic sectors in South Africa. It currently contributes 7.1% (ZAR 53.9 billion) of the total GDP, resulting in many natural and rural spaces being developed for tourism and recreation related activities (Puren *et al.*, 2007). South Africa's spatial planning policies and legislation give no formal recognition to intangible values of landscapes (Puren *et al.*, 2007).

The main objective of this case study is to improve our understanding of the meanings and values that Xhosa people attach to their natural landscapes. These meanings and values were informed by narratives recorded in the field. Narrative is recognised as providing a sense of order and meaning within and across time. It also conveys a group's norms and values and plays an important part in people's identity and sense of belonging (Rapport and Overing, 2000). This increased understanding contributes to the numerous calls that have been made within the conservation literature (Amo-Rodrigues *et al.*, 2010; Fabricius *et al.*, 2004; Goebel *et al.*, 2000) to improve our understanding of the relationship between people and biodiversity.

10.2 Methods

Field research was conducted in several peri-urban villages and townships in the Grahamstown, Alice and Peddie districts of the Eastern Cape Province, South Africa. Villages include Benton and Tharfield in the Peddie district and Llangolan and Victoria Post near Alice, all in the former Ciskei Homeland. Further investigations concerning initiation schools and resettlement of farm workers were undertaken in Joza, Phumlani and Extension 7 in Grahamstown (Figure 10.1).

Homelands were established by South Africa's apartheid government to divide the population into ethnic groups, each with its own territory and inherent potential that would be developed into separate sovereign nations (Boonzaire and Sharp, 1988). More than 3.5 million people were shifted to mostly unproductive lands, with most of them living in 'homelands' to work as labourers in nearby or distant 'white' industries and urban areas. Since the end of apartheid in 1994 the homelands have been incorporated into the South African provinces. The Eastern Cape Province incorporates the former homelands of Ciskei and Transkei and is the second biggest province in South Africa (Buffalo City Municipality, 2009). The area still suffers the effects of apartheid policies as it remains characterised by high levels of poverty (Ainslie *et al.*, 1997; Bank, 2002), a lack of infrastructure, high population densities, inferior education opportunities and families that have been divided though migrant labour and the continuing rural-urban migration (De Wet and Whisson, 1997; Palmer, 1997).

All the study sites are within close proximity to areas of thicket vegetation (units within the thicket biome) called *ihlathi lesiXhosa* (literally, Xhosa forest). For years,

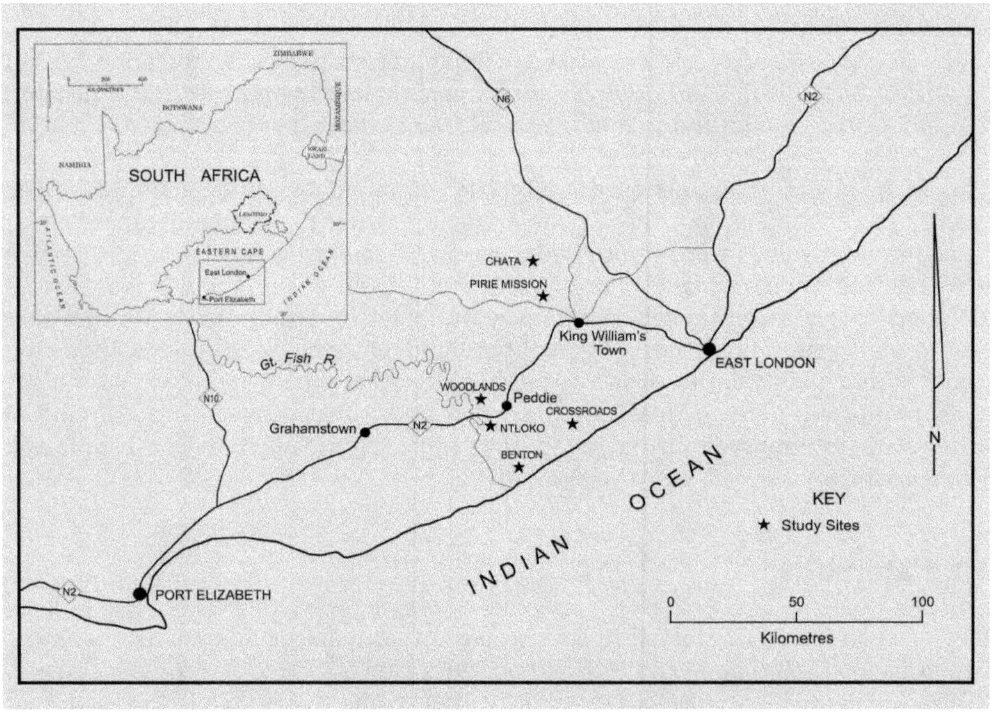

Figure 10.1. The Eastern Cape Province of South Africa showing the study sites.

thicket vegetation was denigrated as a 'false' or transitional vegetation type, and as a result had been neglected and poorly understood by ecologists and scientists. In 2002 its status (in part) was elevated to a new and unique biome called Albany Thicket (Mucina and Rutherford, 2006), giving long-overdue scientific acknowledgment to its importance. Experts estimate that there are approximately 6,500 plant species in the Albany Thicket and an impressive number of mammals, including 5 species of tortoise, 48 large mammals, 25 ungulates and 421 birds (Cowling and Pierce, 2009).

Following the identification of a number of central themes from previous research in the same study sites (Cocks, 2006; Cocks *et al.*, in press), a multidisciplinary research approach was used to collect further data on local perceptions and meanings of natural landscapes. A combination of ethnographic, sociological and botanical methodologies were employed, including 44 informal in-depth interviews with participants in their homes, 2 participatory mapping exercises, a botanical specimen collecting excursion and 6 'walk in the woods' interviews. In addition, one of the authors (MC) accompanied groups of women on fuelwood gathering excursions on three occasions and attended a number of informal games and play activities with children to record riddles in a game known as *rayi-rayi* (guess-guess). One of the authors (TD) visited three initiation schools to provide insight into preferred sites of seclusion lodges, a traditional hunt with dogs in the 'forest', and a religious ritual to present gifts to the ancestors residing in the forest. In addition, a singing competition with primarily women and children was sponsored to record songs that express meanings of nature.

10.3 Results

The overall analysis of the data collected suggests that the cultural meanings of landscapes expressed by the amaXhosa are complex and embody several themes. These have been categorised into three interrelated components as defined in Taylor (2008): the physical features or appearance of the landscape, the observable activities that take place within it, and the functions, meaning and symbols they provide.

Physical landscape

A clear distinction is made between domestic and natural landscapes, and a number of identifiable zones are distinguished. For example, domestic landscape includes homegardens, cultivated fields, cemeteries, dipping tanks and stock pens. Five natural landscapes are recognised by a combination of values and attributes, such as vegetation, dominant plant species, accessibility, distance from village, safety, grazing and browsing value, availability of fuelwood, fruit, medicinal plants and honey, and cultural and religious significance. Culturally important sites occur primarily within non-village landscapes and include the ocean, estuaries, river pools, caves, dense *Euphorbia* groves and mountain peaks. The importance of these sites is attributed to their association with the ancestors and the sites are therefore held sacred.

Perceptions of forested landscapes: *ihlathi lesiXhosa*

Thicket vegetation, known locally as *ihlathi lesiXhosa,* is considered to be a sacred place where the ancestors communicate with their living descendants by means of messengers (*izithunywa*) in the form of birds, animals, insects or even the wind. Local people distinguish between 'cool air' in the forest and 'hot air' in the town or village. In this instance, hot and cool are euphemisms for negative and positive, and air can be translated as 'spirit'. *Ihlathi lesiXhosa* is considered to be positively charged, while the village environment is negatively charged. The forest is therefore understood to be a place that bestows spiritual health and wellbeing (*impilo*). The following poignant quotes recorded in this study convey a significant appreciation of *ihlathi lesiXhosa*:

> When I breathe the forest air and I take in the smell of the flowers and medicines I feel good. My troubles seem insignificant. When I return from the forest I feel happy, I know that I have received a blessing.
>
> When I breathe the forest air and I take in the smell of the flowers and medicines (*amayeza*) I feel good. My troubles seem insignificant.

Notwithstanding the negative environmental impact of continuous harvesting from the forest, particularly of medicinal plants to supply growing urban markets and fuelwood for sale, it is believed by many that the ancestors take care of the forest and that any destruction will always recover if the ancestors intervene. Unlike the western comparison of 'unspoilt nature' versus 'transformed nature', even heavily transformed *ihlathi lesiXhosa* is still appreciated for its recreational and spiritual value. Nevertheless, most people do understand the concept of environmental degradation and consider it to be disrespectful to the ancestors.

Activities

Five activities shape the values of and relationships with *ihlathi lesiXhosa*: religious activities, fuelwood collection, forest sojourn, hunting and initiation.

1. Religious activities. Several religious rituals, facilitated by diviners (*amagqirha*), take place in the *ihlathi lesiXhosa*. The diviner is shown the site in a dream beforehand and gifts of traditional beer, cooked maize and tobacco are left there for the ancestors. It is said that 'We visit the forest because this is where the ancestors are and we must talk with them from time to time.'
2. Fuelwood collection. Despite the backbreaking work involved in daily fuelwood collecting (*ukutheza*), village women enjoy the chore and look forward to the opportunity to get out of the house, chat with friends and spend time in

the forest under the benevolent eyes of the ever-present ancestors. Despite increased electrification in the study area over the last decade, wood is still preferred for cooking certain foods, such as *imifino* (pot-herbs), because of the particular flavour it adds. Food prepared for religious rituals must always be cooked with fuelwood. Collecting fuelwood is considered to be the task of a hardworking housewife, who takes great pride in the size and neatness of her *igoqo* (woodpile).

Urban women who no longer have access to these resources often say that wood gathering is the one activity that they miss the most. One middle-aged woman stated the following: 'When I was working in *eQonce* [King William's Town] I missed collecting firewood in the forest; I did not feel like I was *umama* [a mother; hard working adult woman].' Some women in towns still collect fuelwood on municipal commonages. It is with great pride and respect that a mother is referred to as *ihlathi lokuzimela* (forest of hiding), referring to her role as protector and provider for her family.

A sad but widely acknowledged aspect of fuelwood gathering is that some women find solace from abusive partners in the forest, and *ukutheza* is a socially acceptable way of escaping from this. The traditional Xhosa idiom *uThixo ulihlathi lam* (literally: God is my forest; figuratively: God is my refuge/protection), referring to the haven provided by forest/thicket, is particularly appropriate under these circumstances.

3. Forest sojourn. Men also enjoy their time in the Xhosa forest where they may spend hours with their cattle and goats or walking with their dogs. It is common for village men who are migrant workers to spend time alone in the forest a day or two before returning to their workplace. Besides the opportunity for personal reflection, this ritual forest walk, often to a remote place, is described as a spiritual sojourn and reports of encounters with clan totem animals (*isilo sasekhaya*) and messenger animals (*izithunywa*) are common. It is not unusual for a migrant worker to collect an object, such as a shiny stone or a beautiful feather, in the forest to take back to the city as a symbolic keepsake. Despite the attraction of urban life, village people still have strong cultural associations with nature and remain hopeful that their children will maintain these ties.
4. Hunting.Hunting small game with dogs is a common activity in the study area and is considered to be a traditional sport for young men. Traditional hunting is considered to be very different from poaching or commercial hunting and is a highly competitive sport that is said to sharpen the mind and body of both hunter and dog. Hunting dogs are skilled and disciplined and are known on occasions to be given charm medicines from diviners to increase their prowess, particularly for hunting competitions. Traditional hunting with dogs has long been considered by conservation authorities as reprehensible and has been an illegal activity for many years in South Africa. Nevertheless the young men interviewed in this study showed remarkable insight and understanding of the habitats and habits of the animals they hunt.
5. Initiation. The passage of a Xhosa boy from boyhood to manhood is characterised by the rite of circumcision (*ulwaluko*), followed by a period of ritual seclusion in

a temporary seclusion lodge. The lodge is strategically located out of view from the public and preferably in close proximity to Xhosa forest and a river or stream. The lodge, called *ibhoma*, is constructed from branches and covered with layers of grass to house the initiates (*abakhwetha*) during the period of seclusion. The central upright pole is preferably cut from an *umnquma* tree (*Olea europaea* L. subsp. *africana* (Mill.) P.S.Green) or an *umthathi* tree (*Ptaeroxylon obliquum* (Thunb.) Radlk.) as these are symbolically linked to the ancestors who, by this gesture, are called upon to take care of the initiates. The lodge site is considered to be sacrosanct and is often used by several generations of initiates. Men often revisit their lodge sites to reminisce and spend time alone in nature.

Symbols

There are five types of symbolic associations with the natural environment: animals, linguistic expressions of nature, expressions, riddles and times of day.

1. Animals in Xhosa culture. The Xhosa forest is believed to be home to a wide variety of wild creatures, including dangerous animals such as snakes and 'tigers' (leopards), which are greatly feared. However, certain animals are believed to be representatives of the ancestors (*izinyanya*). For example, swallows (*iinkonjane*) and wagtails (*iicelo*), when frequently seen around the home, nesting or entering the house, are believed to bring good luck and blessings from the ancestors. Messenger animals are those repeatedly seen around the home or entering the home. A diviner may be consulted to interpret the message from the ancestors and it may convey good or bad tidings. Some animals are considered taboo as they are clan totems (*izilo zekhaya*, literally 'animals of the home'), for example the *amaMfene* (baboon) clan will never kill or eat the flesh of a baboon. Similarly the *amaGqwashu* and *amaSukwini* clans will never kill their totem, the monkey, and the *amaNdlovu* (elephants) clan would not harm an elephant, nor would the *amaJwara* hunt a leopard.

   Plant and animal materials necessary for traditional medicine, customs and rituals to appease the ancestors are believed to be available only in *ihlathi lesiXhosa*. These plants and animals are considered sacred and they are imbued with the spiritual power of the ancestors.
2. Linguistic expressions of nature. Songs, proverbs and riddles form an important part of Xhosa narrative. A number of songs were recorded, transcribed and translated into English. Nine of the songs were specifically about animals and refer to elements of the natural world that connect people to the spirit world. For example:

   > *Uxam ulele ngesisu ngasemlanjeni, Uxam ulele ngesisu, Ndithi ulele ngesisu uxam, Uxam ulele emfuleni.* (The monitor lizard is sleeping on its stomach next to the river, The monitor lizard is sleeping on its stomach, I say it is sleeping on its stomach, The monitor lizard is sleeping next to the stream.)

Because of its riparian habitat, the Nile monitor lizard is strongly associated with *abantu bomlambo* – the river people (ancestors of the river). It is believed that the sighting of *uxam*, the monitor lizard, in or near the home is an ill omen. Despite this perception, the skin and fat of a monitor are considered to be powerful medicines. *Uxam* is a totem animal for some clans, such the *amaNtshilibe* clan.

3. Expressions. Many isiXhosa proverbs (*amaqhalo*) and idioms (*izaci*) refer to nature and specifically to plants. The wording and interpretation illustrate the complex relationship between culture/society and nature/environment. It is said that cultures and languages are defined by the natural environment in which they evolve, expressed here by the use of plant names in nuances of the Xhosa language. For example:

   Umthathi uyawuzala umlotha (sneezewood generates only ashes)

   This means that a worthy father may have a worthless son. *Umthathi* refers to the sneezewood tree (*Ptaeroxylon obliquum* (Thunb.) Radlk.), which has a strong, long-lasting valuable wood, but is resinous and burns quickly and easily, leaving only worthless ash.
4. Riddles. A number of riddles were recorded with groups of children playing the riddle game called *rayi-rayi*. Many of the riddles ascribe human features to plants and animals. The riddles function on two levels: first as a puzzle with an answer, and second as a metaphor for a human figure. The riddles often present subtle culturally significant symbolic meanings or associations. For example:

   Rayirayi, rayirayi. Ndinamfana wam, usoloko enxibe ibhulukhwe yekaki nebhatyi eluhlaza, ukuze ngamaxesha athile athwale isitshaba esibomvu – likhala xa lidubulayo. (Guess, guess. I have a young man of mine, he is always dressed in a pair of khaki trousers, a green jacket and sometimes he wears a red crown – it is the aloe when it blossoms.)

   This riddle is solved by identifying the young man as an anthropomorphised description of the aloe (*Aloe ferox* Mill.). On the metaphorical level, the aloe specifically suggests an *ikrwala* ('new man') who, during this period of initiation, wears khaki trousers, a jacket and red cosmetic paste on his face. The symbolic association between the *ikrwala* and the aloe, as highlighted by the riddle, resides in the fact that both are undergoing renewal, the one through natural processes and the other through a cultural practice.
5. Times of the day. The names and descriptions of as many as thirty times of day in *isiXhosa* are beautifully descriptive, often drawing on vivid images of rural life and nature. For example, *urhatya lwemivundla* (twilight of the hares) means the time after sunset when the *imvundla* (scrub hare) comes out of its shelter in the long grass.

10.4 Discussion

The results of this study clearly indicate that the amaXhosa are still intrinsically connected to their natural environment, in particular *ihlathi lesiXhosa*. Access to places and spaces within the natural environment provides a sense of wellbeing, a link to ancestral spirits, a location for religious rituals, plus a wealth of culturally-inspired uses of specific species and sacred places in the landscape. Many of the narratives collected from local people who interact with Xhosa forest portray an enjoyment of being in nature that highlights qualities, such as silence, beauty and tranquillity, the opportunity to observe wild animals and a chance to escape the worries of home. These qualities and opportunities are likely to contribute significantly to the wellbeing of local people. It is said that 'the spirit of the forest bestows one with wellness (*impilo*)' and it is therefore believed that without access to the *ihlathi lesiXhosa* one becomes spiritually and culturally impoverished. The numerous songs, riddles and expressions recorded also reflect aspects of nature and convey a deep connection with nature.

Unfortunately, this type of knowledge and experience is being eroded by factors which prevent people from spending time in nature (Alexander, 2011). Amongst these are increasing urbanisation, competing activities like schooling, crime and regulations restricting people's access to natural resources. There are concerns, especially among older people, that the youth and the ever-growing urban population are becoming increasingly alienated from nature, and many people expressed a wish that the younger generation should have access to nature so that they will stay in touch with their cultural roots. Again, this will only be possible if people have continued access to natural resources, including for extractive use.

10.5 Conclusion and implications

It has been proposed that resource management models in developing countries need to include both biological and cultural aspects which reflect the variety of knowledge, values and activities of local communities (Amo-Rodrigues *et al.*, 2010; Pretty *et al.*, 2010). One of the ways in which the values of local communities could be incorporated into the conservation processes in South Africa is to promote the conservation of culturally important plant and animal species in specific landscape niches, such as initiation sites and sacred sites, particularly within communal areas (Cocks, 2006; Cocks *et al.*, in press). The maintenance of such sites would ensure enclaves of biodiversity within communal lands and provide the opportunity for conservation to occur outside formally protected areas. The management of these sites should be incorporated into spatial biodiversity plans and mainstreamed into local government environmental decision-making processes as stipulated under the National Environmental Management Biodiversity Act of 2004. The adoption of such a process would actively assist in empowering local stakeholders to take responsibility for managing these areas (Knight *et al.*, 2006).

Existing policies still do not contain provisions for local people to take ownership of their cultural distinctiveness and connections to the landscape, despite the fact that these connections provide meaning and nurture the Xhosa people's existence (Cocks and Wiersum, unpublished data). Formal acknowledgement of cultural sites and activities would actively contribute to the protection and preservation of South Africa's intangible heritage. Rural communities would then be in a position to take ownership of their cultural distinctiveness and connections to the landscape (Puren *et al.*, 2007). It is therefore a matter of urgency that conservation efforts in South Africa incorporate local cultural values to protect both its rich cultural diversity and biodiversity – 'if we lose the one, we lose the other' (Persica and Martin, 2008).

Acknowledgments

This study was sponsored by SANPAD (South African Netherlands Programme on Alternatives in Development). The authors would also like to thank community members of Tharfield, Masakane, Woodlands and Ntkolo.

References

Ainslie, A., Cinderby, S., Petse, T., Ntshona, Z. and Bradley, P. (1997). Rural livelihoods and local level natural resource management in Peddie District. Land and Agriculture Policy Centre, Johannesburg, South Africa and SEI, Stockholm, Sweden.

Alexander, J.K. (2011). Stories from forest, river and village: exploring children's cultural environmental narratives and their implications for community conservation. MA thesis, Rhodes University, Grahamstown, South Africa.

Amo-Rodrigues, S.D., Vergara-Tenorio, M.D.C., Ramos-Prado, J.M. and Porter-Bolland, L. (2010). Community landscape planning for rural areas: a model for bio-cultural resources. Society and Natural Resources 23: 436-450.

Bank, L. (2002). Beyond red and school: gender, tradition and identity in the rural Eastern Cape. Journal of South African Studies 28(3): 631-649.

Berkes, W., Colding, J. and Folke, C. (2000). Rediscovery of traditional ecological knowledge as adaptive management. Ecological Application 10(5): 1251-1262.

Boonzaire, E. and Sharp, J. (1988). South African keywords: the uses and abuses of political concepts. David Philip, Cape Town, South Africa.

Buffalo City Municipality (BCM) (2009). Integrated Development Plan 2009/2010. Buffalo City, South Africa.

Cadman, M., Petersen, C., Driver, A., Sekhran, N., Maze, K. and Munzhedzi, S. (2010). Biodiversity for development: South Africa's landscape approach to conserving biodiversity and promoting ecosystem resilience. South African National Biodiversity Institute, Pretoria, South Africa.

Cocks, M.L. (2006). Wild resources and practices in rural and urban households in South Africa: implications for bio-cultural diversity conservation. PhD thesis, Wageningen University, Wageningen, the Netherlands.

Cocks, M.L. and Wiersum, K.F. (in press). Bio-cultural interactions amongst the amaXhosa in the Amathole region, South Africa. Conservation and Society, in press.

Cocks, M.L., Dold, A.P. and Vetter, S. (in press). 'God is my forest' – Xhosa cultural values provide untapped opportunities for conservation. South African Journal of Science, in press.

Cowling, R. and Pierce, S. (2009). East of the Cape conserving eden. Fernwood Press, Cape Town, South Africa.

Cunningham, A.B. (2001). Applied ethnobotany: people, wild plant uses and conservation. Earthscan, London, UK.

Department of Environmental Affairs and Tourism (DEAT). (2006). South Africa environment outlook. A report on the state of the environment. Department of Environmental Affairs and Tourism, Pretoria, South Africa.

De Wet, C. and Whisson, M. (1997). From reserve to region. Apartheid and social change in the Keiskammahoek district of (former) Ciskei: 1950-1990. Occasional Paper No. 35, Institute for Social and Economic Research (ISER), Rhodes University, Grahamstown, South Africa.

Evans, M. (2010). Mandela and the televised birth of the rainbow nation. National Identities 12: 309-326.

Fabricius, C., Koch, E., Turner, S. and Magome, H. (eds.) (2004). Rights, resources and rural development: community-based natural resource management in southern Africa. Earthscan, London, UK.

Goebel, A., Campbell, B., Mukamuri, B. and Veeman, M. (2000). People, values and woodlands: a field report of emergent themes in interdisciplinary research in Zimbabwe. Agriculture and Human Values 17: 385-396.

Knight, A.T., Cowling, R.M. and Campbell, B.M. (2006). An operational model for implementing conservation action. Conservation Biology 20(2): 408-419.

Knight, A.T., Cowling, R.M., Difford, M. and Campbell, B.M. (2010). Mapping human and social dimensions of conservation opportunity for the scheduling of conservation action on private land. Conservation Biology 24: 1348-1358.

Millennium Ecosystem Assessment (2005). Ecosystems and human well-being: current states and trends, Volume 1. Island Press, Washington, DC, USA.

Mittermeier, R.A., Hoffmann, M., Pilgrim, J.D., Brooks, T.B., Mittermeier, C.G., Lamoreux, J.F. and Da Fonseca, G. (eds.) (2004). Hotspots revisited: earth's biologically richest and most endangered ecoregions. Cemex, Mexico City, Mexico.

McNeely, J.A. (2000). Cultural factors in conserving biodiversity. In: Wilkes, A., Tillman, H., Salas, M., Grinter, T. and Shaoting, Y. (eds.) Links between cultures and biodiversity. Proceedings of the Cultures and Biodiversity Congress. Yunnan Science and Technology Press, Yunnan, China PR, pp. 128-142.

Mucina, L. and Rutherford, M.C. (eds.) (2006). The vegetation of South Africa, Lesotho and Swaziland. Strelitzia 19. South African National Biodiversity Institute, Pretoria, South Africa.

Palmer, R. (1997). Gender, households, and environmental changes in two rural communities in South Africa's former Ciskei. Unpublished manuscript for the Institute of Southern African Studies at the National University of Lesotho, Roma, South Africa.

Persica, A. and Martin, G. (eds.) (2008). Links between biological and cultural diversity – concepts, methods and experience. Report of an International Workshop, UNESCO, 48, Paris, France.

Posey, D.A. (1999). Introduction: culture and nature – the inextricable link. In: Posey, D.A. (ed.) Cultural and spiritual values of biodiversity. United Nations Environment Programme, Nairobi, Kenya and Intermediate Technology Publications, London, UK, pp. 3-18.

Pretty, J., Adams, W., Berkes, F., Ferreira de Athayde, S., Dudley, N., Hunn, E., Maffi, L., Milton, K, Rapport, D., Robbins, P., Sterling, E., Stolton, S., Tsing, A., Vintinner, E. and Pilgrim, S., (2010). The intersections of biological diversity and cultural diversity: towards integration. Conservation and Society 7(2): 100-112.

Puren, K., Drewes, E. and Roos, V. (2007). An exploration of sense of place as informative for spatial planning guidelines: a case study of the Vredefort Dome World Heritage site, South Africa. World Academy of Science, Engineering and Technology 28: 217-224.

Rapport, N. and Overing, J. (2000). Social and cultural anthropology: the key concepts. Routledge, London, UK.

Reunala, A. (1999). Forests and Finnish culture: 'There behind yonder woodland...'. In: Reunala, A., Tikkanen, I. and Äsvik, E. (eds.) The green kingdom. Otava Publishing Ltd., Keuruu, Finland.

Rössler, M. (2006). World heritage cultural landscapes: a UNESCO flagship programme 1992-2006. Landscape Research 31(4): 333-353.

Taylor, K. (2008). Landscape and memory: cultural landscapes, intangible values and some thoughts on Asia. In: 16th ICOMOS General Assembly and International Symposium: 'Finding the spirit of place' between the tangible and the intangible, 29 September – 4 October 2008, Quebec, Canada.

Wynberg, R. (2002). A decade of biodiversity conservation and use in South Africa: tracking progress from the Rio Earth Summit to the Johannesburg World Summit on Sustainable Development. South African Journal of Science 98: 233-243.

Zweifel, H. (1997). The gendered nature of biodiversity conservation. The National Women's Studies Association Journal 9(3): 107-123.

11. Biocultural diversity in the Netherlands: from ecologically noble savages towards biocultural creatives

Birgit H.M. Elands[1] *and C.S.A. (Kris) van Koppen*[2]
[1]*Wageningen University, Forest and Nature Conservation Policy Group, P.O. Box 342, 6700 AH Wageningen, the Netherlands;* [2]*Wageningen University, Environmental Policy Group, P.O. Box 8130, 6700 EW Wageningen, the Netherlands; birgit.elands@wur.nl; kris.vankoppen@wur.nl*

Abstract

Biocultural diversity is usually associated with traditional ways in which local or indigenous people interact with the natural environment. This chapter deals with biocultural diversity practices in a highly modernised society, the Netherlands. It introduces the concept of 'biocultural creatives': groups of people who, driven by an engagement with society and nature, create new cultural models and practices for interaction with biodiversity. Biocultural creativity aims to reconnect people with nature and is sought for in this chapter within the domains of recreation (use of nature) and financial support and volunteer activities (nature protection). Within recreational practices, creativity is found in the application or rejection of technology and the world of fantasy or contemplation; within financial support, creativity is found in traditional economic instruments applied to the world of nature conservation; and within volunteer work, creativity is found through the modernisation of traditional volunteer activities and new styles of citizen governance in the planning, design and management of conservation areas and landscapes. We conclude that although none of the biocultural creative initiatives have been adopted by large groups of Dutch citizens, they are important to the conservation of biodiversity because they re-create a constant awareness of the cultural roots of nature conservation and remind us that these roots link up to the dynamics in society and because they enrich our cultural connectedness with the natural environment. Therefore, we encourage policy-makers and nature managers to see biocultural diversity as an opportunity for the emergence of new perspectives and ideas, which can be meaningful for nature conservation. We also recommend researchers to do further studies into the dynamic and creative aspects of biodiversity that we touch upon in this chapter.

Keywords: biocultural diversity, cultural creative, nature conservation, people, activities

Birgit H.M. Elands and C.S.A. (Kris) van Koppen

11.1 Introduction

In this chapter we make a provisional effort to apply the concept of biocultural diversity to the Netherlands. As we discuss below, biocultural diversity is usually associated with traditional ways in which local indigenous groups interact with nature. But it does not have to be restricted to these situations. Biocultural diversity may also be an interesting concept in understanding how dynamic, industrialised and globalised societies deal with nature. In collaboration with Freerk Wiersum, Michelle Cocks has made a fascinating step towards a more dynamic interpretation of biocultural diversity in her studies of the way Eastern Cape Xhosa deal with nature (Cocks, 2006; Cocks and Wiersum, 2003; Cocks *et al.*, 2008). In this chapter we take this further to the extreme by investigating biocultural diversity in the Netherlands and by focusing on the creation, rather than the conservation, of biocultural diversity.

Some may feel such an effort is somewhat iconoclastic in a world where the protection of nature and cultural tradition is heavily under pressure. However, we would stress that, given the current processes of industrialisation and globalisation, our exploration is not meant to downplay the importance of protecting traditional, indigenous natural and cultural values. By presenting an out-of-the-box approach to biocultural diversity, we hope to contribute to a better understanding of the concept by illustrating some of its potential dynamics in a modernising and globalising world.

11.2 The cultural roots of biodiversity conservation

Among scholars studying society-nature relationships it has become widely accepted that nature and culture are closely related for at least two main reasons. Firstly, many parts of nature – including areas that have long been thought of as pristine – have been subject to human cultivation in one way or another. Nature protection itself very frequently involves specific new interventions in ecosystem functioning. The Netherlands, of course, is a good example. Dutch nature is a product of intensive interaction between society and its natural environment. Many of the traditional nature areas are marginal rural soils reforested for hunting and logging purposes or traditional cultural landscapes for dairy farming or sheep herding. The more recent 'nature development' (habitat creation/restoration) areas along rivers and in polders are often literally dug out or dredged into the landscape. The saying 'God created the world, but the Dutch created the Netherlands' applies to Dutch nature areas as well.

Secondly, and perhaps more fundamentally, culture and nature are closely linked because the way we perceive nature and what we value in it is rooted in culture. The studies by William Cronon (1995) and Simon Schama (1995), among many others, have investigated and illustrated many examples of this close relationship. In the words of Schama (1995: 61): 'Landscapes are culture before they are nature.' This culture is not just a separate domain of nature lovers, but it is deeply embedded in the traditions, the daily practices and the hopes and fears of the societies involved, whether they are traditional and rural or highly urbanised and industrialised.

The intertwining of biological diversity and the diversity of cultural practices and values is strongly expressed in the concept of biocultural diversity. UNESCO (2008: 8) defines biocultural diversity as 'the sum of the world's differences regarding biological diversity at all levels and cultural diversity in all its manifestations (ranging from individuals to entire cultures), and their interactions'. Taken literally, this definition is somewhat nonsensical, because things like Bach's cantatas, Chinese traditional operas or African drumming, all beautiful manifestations of different traditions in music, definitely belong to this sum. In fact, it would be hard to find something on earth that does not belong to it. However, if we isolate and focus on the interactions aspect, the definition becomes more illuminating. Biocultural diversity, then, is biodiversity that is understood to be interacting with culture: moulded and influenced by culture, perceived and valued through culture. The concept of biocultural diversity, in other words, focuses our attention on the 'inextricable link' between nature and culture (Posey, 1999: 3).

11.3 From 'ecologically nobles savages' to 'biocultural creatives'

As we stated in the introduction, the protection of biocultural diversity is often associated with more 'exotic', 'indigenous', 'native' and 'tribal' people living in remote and isolated communities, far away from modernity and urbanised areas, in close harmony with their natural surroundings. It is true that biocultural diversity is critical to indigenous communities that live around forest and natural areas, such as the Raute in western Nepal, both for their material and cultural needs (Wiersum and Shreshta, 2010). However, the many publications on biocultural diversity focus almost exclusively on the ways that indigenous and tribal groups attribute spiritual and cultural values to nature and consequently develop protective regimes towards their natural environment (see, for example the landmark volume on cultural and spiritual values of biodiversity edited by Darrell Posey, 1999). The notion of 'ecologically noble savages', as such peoples are sometimes called, has several weaknesses. While it is clear that many indigenous communities have developed sustainable ways of dealing with nature, the idea that all indigenous group have always acted as nature conservationists has come under dispute (Hames, 2007). Besides, the notion implies a static interpretation of culture, which does not allow for any modernisation. Research has also clearly shown that biocultural diversity is not only relevant to indigenous communities, but also to rural and urban communities in South Africa. Moreover, its relevance goes beyond poor households, as wealthy households in South Africa use wild plants extensively for cultural and material reasons (Cocks, 2006; Cocks and Wiersum, 2003; Cocks *et al.*, 2008). Finally, the notion seems to limit itself to non-Western societies, as if biocultural diversity values and practices are absent in the rest of the world.

When we move away from the static picture of the ecologically noble savage, however, we also need to rethink the role of culture. If biocultural diversity can be associated with dynamic social processes, then the issue is not only how to preserve or restore cultural practices and values, but also how to modify, adapt and create culture in ways that enhance the diversity of nature. This gives rise to an intriguing question. Can we think of processes which create biocultural diversity? And which groups in society take the lead in these processes? In identifying these groups, we took inspiration from the concept of 'cultural creatives', which gained popularity in the marketing community during the last ten years. According to its advocates, there is a group of people in society who 'are at the leading edge of several kinds of cultural change, deeply affecting not only their own lives but our larger society as well' (Ray and Anderson, 2000). These 'cultural creatives' take a critical view of society as it is, feel engaged with social and environmental sustainability, and believe in personal and spiritual growth. They are the ones that design new cultural models of working, consuming and socialising and apply them in their own lives. Drawing on this concept, we introduce for the purposes of this essay the concept of *biocultural creatives*: groups of people who, driven by an engagement with society and nature, create new cultural models and practices for interaction with biodiversity.

11.4 Our conceptual approach to biocultural creativity

Before we can arrive at an analysis of the roles of biocultural creatives in the Netherlands, we have to outline our conceptual approach in more detail. First we describe how we use the terms biodiversity and biocultural diversity. Then we explain the perspective of activity domains, which is central to our approach. The activity domains will guide us in our exploratory search for biocultural creatives.

Biodiversity refers to the totality of species, populations, communities and ecosystems, both wild and domesticated. In this broad sense, it conveys a similar meaning to 'nature'. We approach biodiversity mainly from a nature perspective. Moreover, we do not limit nature to nature conservation areas, but include urban parks, agricultural landscapes and outdoor recreational areas.

As explained before, we focus on the interaction aspect of biocultural diversity. Pretty *et al.* (2009) distinguish the following intersections of biological and cultural diversity: beliefs and worldviews, livelihoods and practices, knowledge bases and languages, and norms and institutions. As we are interested in the way Dutch citizens relate to nature in their daily lives, our approach can be positioned within the livelihoods and practices perspective.

Daily interaction with nature is highly varied: it ranges from visiting a nature area to watching a nature documentary to engaging in political action. Three activity domains can be distinguished, each with specific roots in the history of nature conservation (Van Koppen, 2002). The first domain, *nature conservation*, comprises individual and group activities for protecting nature in a broad sense. It includes practices such as making nature surveys, acting as a nature guide, voluntary nature management activities, taking care of nesting and feeding places for birds, membership of nature protection organisations, etc. These activities emerged with the rise of the nature conservation movement around 1900 and since the 1970s have spread across substantial sections of the Dutch population. The second domain is that of *nature consumption* activities related to nature. It includes walking and hiking in nature areas, nature-oriented outdoor sports and other forms of recreation in nature. Nature recreation, in a broad sense, has a long history in the Netherlands and other Western countries. Since the first half of the 20th century, nature recreation has become an explicit aim in nature policy and spatial planning. Other forms of nature consumption, such as watching nature documentaries, reading books and articles on nature or buying ecolabelled food produced in a nature-friendly way are also included in this domain. Finally, the domain of *nature politics* includes activities related to political decision-making on nature, such as voting, participation in policy processes or engagement in political protest. Since the institutionalisation of nature policy from the 1970s, public support for nature policy measures and participation in nature policy processes have become significant aspects of nature conservation.

Our search for biocultural creatives proceeds from the perspective of activity domains.

11.5 How Dutch citizens interact with nature

How are the types of activities in these three domains expressed across the Dutch population? To first obtain a broad view of the potential groups in the Netherlands, we take a representative public survey of support for nature and nature policy among Dutch citizens as our starting point (De Bakker *et al.*, 2007). This research explicitly focused on the varied ways in which Dutch citizens interact with nature. The three activity domains 'use', 'protection' and 'decision-making' were operationalised into a number of practical activities. On the basis of a cluster analysis, the survey identified four different groups of citizens based on these practical activities (Figure 11.1).

The largest group is the 'average citizens'. These people visit nature areas on a regular basis, but their participation in other nature-related activities is limited. They are concerned about nature, but do not express this in their behaviour and activities. The second group of people are the 'passive citizens'. These people are hardly involved in nature-related activities. They might visit a nature area once a

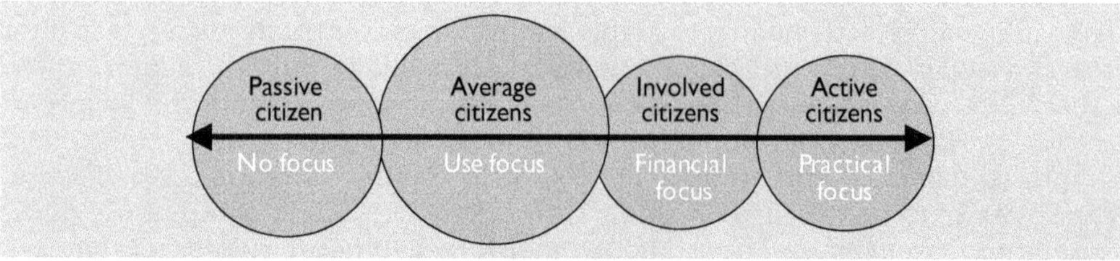

Figure 11.1. Four groups of Dutch citizens clustered according to their approach to nature protection and nature policy (adapted from De Bakker et al.*, 2007).*

year, but that is about it. The 'involved citizens' are more willing to donate money than to take part in active work. They tend to be members of nature conservation societies and are more willing than the other groups to pay for nature, both as consumers and as taxpayers. They also visit nature areas relatively frequently. Finally, the 'active citizens' are engaged in many volunteer activities and often visit nature areas for recreational purposes. They show their engagement through their participation in nature conservation activities and their consumption behaviour. If they are involved in decision-making, they do it at the local level.

If we consider these groups of Dutch citizens more carefully, we can conclude that there are three dominant modes of interaction with nature: use of nature, protection of nature through financial mechanisms and protection of nature through volunteer activities. We will describe these three interactions one by one in search for biocultural creativity.

11.6 Biocultural creatives

Use of nature

The degree of participation in recreational use of natural areas is high in the Netherlands. Almost three-quarters of the Dutch population visit a protected area of other nature area at least once a year (Bijl *et al.*, 2010). In 2006/2007, 384 million visits of at least an hour were made to Dutch forest and nature areas (including water and coastal recreation) (Goossen, 2008). People visit forest and nature areas for tranquillity, leisure, adventure or social reasons.

In urbanised societies, where people need to escape the regularity, bustle and harshness of the built environment, natural environments are considered to be places where people can relax and express themselves. Outdoor recreation is increasingly becoming an expression of lifestyle and identity. The way people express themselves, however, is as varied as people themselves. Consequently, how people experience nature varies from 'fast nature experiences' (adventure, fun, novelty) to 'slow nature experiences' (environmental responsible behaviour, contemplation and familiarity) (Elands and Kaae, 2010).

Biocultural creatives are involved in the construction of unique nature experiences. Creativity evolves predominantly around two themes. The first theme relates to technology as a key characteristic of the modernised world, either through its integration or its removal. Technology is a crucial component in some recently created forms of nature recreation. One of the best examples is the development of adventure forests or climbing forests. These forests usually include suspended trails in the forest canopy. A system of cables, wooden platforms and ropes allows users to move safely along canopy trails where balance and concentration (even more than physical dexterity and fitness) are essential (Elands and Kaae, 2010). The Netherlands has at least ten adventure forests (www.klimbos.info/), which are growing in popularity. In contrast, another increasingly popular type of nature experience involves the rejection of technology: wilderness survival trips in which subsisting in nature without the help of modern technology is the key challenge. Woodcraft and bushcraft are typical forest-related examples of wilderness survival experiences in which people are trained in ancient skills and traditional knowledge, such as hunting, fishing and shelter building (http://en.wikipedia.org/wiki/bushcraft).

The second theme relates to imagination, through which we can escape our daily duties and responsibilities and reconnect to ourselves. Reconnecting people to nature is currently a topical issue in the Netherlands. One of the most prominent Dutch biocultural creatives is Thomas van Slobbe, director of wAarde Foundation, a think-tank within the natural and environmental movement that develops innovative and inspiring visions and projects to reconnect people to their natural surroundings (Henstra, 2008). Van Slobbe is of the opinion that in the modernised world contact between human beings and the natural environment is seriously threatened and that 'nature deficit' is a worldwide problem. The wAarde Foundation develops many initiatives to reconnect people, especially children and young people, with nature. In 2010 a decision was made to create an Avatar forest in Amsterdam. Named after the successful movie of 2009, the forest will contain footpaths for rambling and places for dreaming away as, in the movie. The forest should attract teenagers, for whom virtual reality is a natural part of their daily lives. This reconnection is not only achieved through touching upon the world of fantasy, but also through the stimulation of sensory experiences, such as smell, touch, balance and acceleration, which are claimed to be essential in the wellbeing of human beings in general and the psychomotoric development of children in particular.

Imagination is also about spirituality, contemplation, meditation and reflection. Nature can facilitate this. People search for these kinds of experience individually, but it can also be organised collectively. The United Kingdom is the first Western country to develop 'faith woodlands', which are dedicated areas of woodlands that serve as spiritual and meditative space for a diversity of faith communities, such as Muslims, Christians, Hindus and Sikhs (Forestry Commission England, 2007). Another growing trend in Western Europe is the establishment of burial grounds in woodland sites, where the body is given back to earth, as an alternative to existing funeral practices. The first woodland burial ground was created in the United Kingdom in 1993 and at least 200 natural burial sites have been created since then. The designation of natural burial grounds in the Netherlands is still in its infancy, but it is attracting growing interest.

It is unclear how successful these biocultural creative initiatives are in terms of the numbers of Dutch people participating in them in their leisure time. This would be expected to lead to an increase in the number of visits to forests and nature areas. At present, it seems that the number of visits to forest and nature areas in the Netherlands is stable, perhaps even declining slightly (Goossen, 2008). More important than numbers, however, is the question of what sort of biocultural diversity is being created by these new developments. We cannot offer firm answers here, but we can tentatively point out some patterns. Canopy trails open up a new layer of landscape in forests, with its own scenery and species. They make participants more aware of the ecological and scenic richness of forests. Survival trails add experiences of harvesting, feeding, finding shelter and sleeping in nature to the array of sensual recreation experiences of walking through nature. Avatar forests, faith forests and nature burial grounds are all ways of linking – or re-linking – cultural and spiritual values to nature. What these have in common is that they all enrich and broaden the nature experience of people with new sensory or spiritual meanings. They connect biocultural diversity to a wider set of practices and values in the life of modern urban citizens.

Protection of nature through financial mechanisms

In the Netherlands, the most common financial mechanism is through membership of nature conservation organisations, which is generally high. In 2006, about 30% of all Dutch households were members of at least one nature conservation organisation (De Bakker *et al.*, 2007). The total membership of all nature conservation organisations in the Netherlands is more than two million.

People have a range of options for contributing financially to nature conservation. In the Netherlands there are many initiatives that allow people to adopt or sponsor elements of the natural environment, such as a cow, a tree or a meadow bird's nest. An instrument for raising capital for the protection of biodiversity that is attracting interest globally is the concept of the Landscape Auction. At a landscape auction people bid for diversity and landscape elements (Triple E, 2011). Some examples are the establishment of a wildlife corridor, the conservation of a watermill and the maintenance of heath in a particular area. Anyone can participate, from individuals and service clubs to companies. A bidder does not become the owner of the landscape element, but becomes a supporter of the preservation costs of this particular element. The principle of individuals paying for landscape conservation and maintenance is also the main focus of the Landscape Foundation. This instrument differs from the auction in the sense that it is mainly directed towards inhabitants of a certain area or people living nearby a particular landscape.

Dutch biocultural creative Tom Bade tries to link nature with economy through a variety of innovative ideas and products. He is director of Triple E (which stands for Economy, Ecology & Experience), which developed the Landscape Auction concept in cooperation with the NGO ARK and the ViaNatura Trust Fund (Wensing, 2008). Tom Bade has also raised the ideas of a 'landscape petrol tax' and a 'landscape panorama guarantee' to be included in the purchase price of a house.

So far, the financial mechanisms described above make only a modest contribution to nature conservation compared with public funding. In highlighting the initiatives by Triple E, we do not claim that such private financing mechanisms could replace the role of government in safeguarding nature as a collective good. According to Vreke (2011), who has studied the potential of private funding in nature and landscape management, private partners will only invest financially if they benefit from it and if the investments are feasible within their business or household context. For the purposes of this chapter, however, the experiments by Triple E are of interest because they probe new ways in which citizens can take responsibility for nature and the landscape. Financial engagement may well affect cultural engagement, and *vice versa*. What the mutual influence is, and whether it is positive or negative, needs further exploration.

Protection of nature through volunteer activities

The focus of volunteering is on 'doing'. In the Netherlands, many people are involved in some sort of volunteering work regarding nature: some people are active in keeping up small forest patches, some maintain hiking trails, while others are involved in data collection. Roughly, we can distinguish between three types of volunteer activities: nature and landscape management, mainly aimed at maintenance of the ecosystem, but also at the creation of visitor facilities; nature monitoring, focusing on collecting data on the distribution and abundance of flora and fauna; and nature education, which involves activities such as organising excursions, courses, exhibitions and lectures. In total, the number of volunteers for these activities has increased from 53,000 in 1993 to almost 77,000 in 2005. It should be noted that the total number of individuals doing volunteer work in these organisations is smaller, as some people are active in more than one organisation.

It is relevant to distinguish between nature-related activities initiated by professional or governmental organisations and those initiated by people themselves. One of the most successful nature and landscape management initiatives, organised by the Dutch NGO Landschapsbeheer Nederland (Landscape Management Netherlands), is the 'Nature Labour Day', which is held on the first Saturday in November and in 2011 attracted more than 13,000 volunteers at 406 different locations throughout the Netherlands (www.natuurwerkdag.nl). Children can also participate in the Nature Labour Day and many special activities are organised for them. This year, a provincial department of Scouting Netherlands was able to mobilise 650 scouts, representing about 15-20% of the scouts in that particular department, to join the Nature Labour Day (Van Dam *et al.*, 2011). As Landscape Management Netherlands is very active in organising volunteer activities and involving the public in decision-making about nature and landscape (participation), one could argue that the organisation itself is a biocultural creative.

Members of the public may also draw up nature and landscape plans. Anyone can be a cultural creative, from farmers, office workers, housewives and children to hobby clubs, service clubs and companies. These grass-roots initiatives take place mainly at local level. The creation and management of small inner-city natural gardens or parks, the design and maintenance of public grounds close to residential areas, and plans to improve the spatial quality of the countryside for its inhabitants are increasingly taken up by local people who care about their natural surroundings. Although biodiversity benefits, this is not the primary aim. On the contrary, most projects have multiple aims, including recreational use, education of children, quality of life, social cohesion and health.

A remarkable example is the Heimanshof, a botanical garden in the Dutch town Hoofddorp (see Van Dam *et al.*, 2011). The Heimanshof was initiated by Franke van der Laan, who is definitely a biocultural creative operating at the local level. Elements of the project include maintenance of the garden itself, creation of educational nature trails, establishment of a Youth Nature Club, nature management activities for healthcare clients, nature activity days for commercial enterprises and service clubs, and participation in the Nature Labour Day. The Heimanshof started in 2000 with 0.5 ha of land and a handful of volunteers and now manages 17 ha with 3,500 volunteers. Due to the growth in the number of projects and volunteers, a foundation has been established to enable the organisation to operate in a more professional way.

Volunteering for nature protection brings together both aspects discussed in the previous sections: enriching the experiences of nature and taking responsibility for nature protection.

Volunteering is not new; in fact, it has often been a part of social life for century or more. The creative moment here lies in the innovation that is needed to keep these volunteering activities alive in a modernising society. The Nature Labour Day, for example, caters well to those people who do not want to do nature management work on a regular base, but want to experience it for one day. Cultural diversity in a country like the Netherlands is not so much a reflection of the differences between people, but increasingly between different periods and activities in people's lives. Developments in volunteer work, as illustrated above, give some hint of the implications this may have for biocultural diversity in the Netherlands.

11.7 The future of biocultural diversity in the Netherlands

Biocultural creatives are not new or unknown in urban and modernised societies such as the Netherlands. Creativity is of all times and all places. Our chapter makes clear that in the Netherlands people have a highly diverse interaction with nature. The most intensive interactions take place within the domain of consumption and the least intensive within the domains of protection and decision-making. We have shown that biocultural creativity flourishes within each domain. Initiatives enrich nature with cultural values and practices and create and re-create a constant awareness of the cultural roots of nature conservation.

Although we have tried to elucidate a diversity of biocultural practices, we have left many out, especially within the activity domain of decision-making. The activities of GroenFront! (EarthFirst! NL), which takes direct action against political plans endangering the natural environment, as well as green voting initiatives and initiatives stimulating green shopping as a more passive form of nature consumption (e.g. ecolabels) have not been addressed.

Biocultural creative people do not remain within the boundaries of one particular activity domain. This is particular evident in citizens' volunteer initiatives for nature and landscape, in which many activities are not only concerned with protection, but also involve decision-making at the grass-roots level.

In our view, biocultural creativity is also an expression of public involvement with nature, which enriches our cultural connectedness with the natural environment. It should be welcomed and cherished. In its essence, creativity is about out-of-the-box thinking and its outcome in terms of biodiversity conservation is highly unpredictable. According to Turnhout *et al.* (2010) policy-makers should see this unpredictability as an opportunity for the emergence of new perspectives and ideas, which can be meaningful for nature conservation.

Thinking about citizen involvement has changed in recent years. Nature policy-makers and professionals increasingly realise that for the long-term maintenance of nature and landscape, citizens need to participate and be responsible or co-responsible for nature and landscape management. This do-it-yourself nature and landscape management requires policy-makers and planner to adopt new roles that involve facilitation, coordination and trust, and requires citizens to have a long-term commitment and perseverance (Kadijk, 2011; Van Dam *et al.*, 2011). Moreover, shared responsibilities involve the recognition that nature conservation is not only about biodiversity, but also about accessibility, multiple cultural uses and making money. Opening the doors to more involvement of citizens and entrepreneurs may also present opportunities for new biocultural creatives to engage in in nature conservation.

Freerk Wiersum has been one of the first scholars in the Netherlands to draw attention to the concept of biocultural diversity. Together with Michelle Cocks he has convincingly pointed out the social dynamic aspects of this concept. In line with his ideas, we recommend further research into biocultural diversity and its implications for nature policy and management, also in highly modernized countries like the Netherlands. In this research, we suggest, it is worthwhile to pay attention to the creative aspects of biocultural diversity.

References

Bijl, R., Boelhouwer, J., Pommer, E. and Schyns, P. (2010). The social state of the Netherlands 2009. The Netherlands Research for Social Research, The Hague, the Netherlands.

Cocks, M.L. (2006). Wild plant resources and cultural practices in rural and urban households in South Africa: implication for bio-cultural diversity conservation. PhD thesis, Wageningen University, Wageningen, the Netherlands.

Cocks, M.L., Bangay, L., Shackleton, C.M. and Wiersum, K.F. (2008). 'Rich man poor man' – inter-household and community factors influencing the use of wild plant resources amongst rural households in South Africa. International Journal of Sustainable Development and World Ecology 15: 198-210.

Cocks, M.L. and Wiersum, K.F. (2003). The significance of biodiversity to rural households in Eastern Cape Province of South Africa. Forests, Trees and Livelihoods 13: 39-58.

Cronon, W. (ed.) (1995). Uncommon ground. Toward reinventing nature. Norton, New York, NY, USA.

De Bakker, E., Van Koppen, C.S.A. and Vader, J. (2007). Het groene hart van burgers. Het maatschappelijk draagvlak voor natuur en natuurbeleid. WOt rapport 47, Wettelijke Onderzoekstaken Natuur & Milieu, Wageningen, the Netherlands.

Elands, B. and Kaae, B.C. (2010). Unique experiences. In: Pröbstl, U., Wirth, V., Elands, B. and Bell, S. (eds.) Management of recreation and nature based tourism in European forests. Springer-Verlag, Berlin, Germany, pp. 266-285.

Forestry Commission England (2007). Faith woodlands, Dec 2006 – March 2007 – The first phase. Available at: http://www.forestry.gov.uk/pdf/fce-faith-in-the-woodlands.pdf/$FILE/fce-faith-in-the-woodlands.pdf.

Goossen, C.M. (2008). Monitoring recreatiegedrag van Nederlanders in landelijke gebieden; 2006/2007. WOt werkdocument 146, Wettelijke Onderzoekstaken Natuur & Milieu, Wageningen, the Netherlands.

Hames, R (2007). The ecologically noble savage debate. Annual Review of Anthropology 36: 177-190.

Henstra, A. (2008). The child's right to nature – a call for inclusion of the right to contact with nature in the United Nations Convention on the Rights of the Child. wAarde Foundation, Beek-Ubbergen, the Netherlands.

Kadijk, J. (2011). Doe-het-zelf vergt anders denken van planners: actief burgerschap gedijt beter binnen goede systemen. NIROV, the Hague, the Netherlands. Available at: http://www.nirov.nl/Home/Nieuws/Nieuws_Items/Doe-het-zelf_vergt_anders_denken_van_planners.aspx?mId=10437&rId=477.

Posey, D.A. (1999). Cultural and spiritual values of biodiversity. A complementary contribution to the global biodiversity assessment. In: Posey, D.A. (ed.) Cultural and spiritual values of biodiversity. UNEP and Intermediate Technology Publications, London, UK, pp. 1-19.

Pretty, J., Adams, B., Berkes, F., Ferreira de Athayde, S., Dudley, N., Hunn, E., Maffi, L., Milton, K., Rapport, D., Robbins, P., Sterling, E., Stolton, S., Tsing, A., Vintinner, E. and Pilgrim, S. (2009). The intersections of biological diversity and cultural diversity: Towards integration. Conservation and Society 7(2): 100-112.

Ray, P.H. and Anderson, S.R. (2000). The cultural creatives: how 50 million people are changing the world. Three Rivers Press, New York, NY, USA.

Schama, S. (1995). Landscape and memory. HarperCollins, London, UK.

Triple, E. (2011). Landscape auction: an investment tool for nature. Available at: http://www.tripleee.nl/landscapeauction.htm.

Turnhout, E., Van Bommel, S. and Aarts, N. (2010). How participation creates citizens: participatory governance as performative practice. Ecology and Society 15(4): 26. Available at: http://www.ecologyandsociety.org/vol15/iss4/art26/

United Nations Educational, Scientific and Cultural Organization (UNESCO) (2008). Links between biological and cultural diversity-concepts, methods and experiences. Report of an International Workshop, UNESCO, France, Paris.

Van Dam, R., Salverda, I. and During, R. (2011). Effecten van burgerinitiatieven en de rol van de rijksoverheid. Deel 5 in de reeks Burgers en Landschap, Alterra, Wageningen UR, Wageningen, the Netherlands.

Van Koppen, C.S.A. (2002). Echte natuur: een sociaaltheoretisch onderzoek naar natuurwaardering en natuurbescherming in de moderne samenleving. PhD thesis, Wageningen University, Wageningen, the Netherlands.

Vreke, J. (2011). Er is geen markt voor natuur en landschap. Landwerk 2011(5): 16-18.

Wensing, D. (2008). Landscape auctions, a new financing tool for nature. Agribusiness 3(2): 32-33.

Wiersum, K.F. and Shrestha, K. (2010). Biocultural diversity in community forestry in Nepal. ETFRN News 51: 20-24.

12. 'Diversity (still) at stake': a farmers' perspective on biodiversity and conservation in Western Mexico

Peter R.W. Gerritsen
Universidad de Guadalajara, Centro Universitario de la Costa Sur, Departamento de Ecología y Recursos Naturales, Av. Independencia Nacional 151, Autlán, Jal., Mexico; prw.gerritsen@gmail.com

Abstract

Using the Sierra de Manantlán Biosphere Reserve in Western Mexico as a case study, this chapter examines the usefulness of the biodiversity concept in populated protected areas. A new concept, 'resource diversity', is presented to shed more light on the possible role of farmer participation in ecosystem management in these areas. Resource diversity is produced and reproduced in an ongoing process of co-production, which involves interactions and transformations of natural and social phenomena. This organisation of natural cycles and spatial units results in transformations of existing resource diversity. Rather than leading to a process of unilateral loss of biodiversity, it results in changing landscape mosaics with their own specific form of resource diversity (and biodiversity). I argue that conservation activities in populated protected areas should include consideration of farmers' perspectives (and resource diversity) to better address conservation issues.

Keywords: best conservation approaches, resource diversity, Sierra de Manantlán, protected areas, Mexico

12.1 Introduction

Biodiversity is important in scientific and political discussions. It is a natural scientific concept that refers to the 'variety of life' and is commonly defined as diversity at the genetic, species and ecosystem levels. It has also become a leading principle for conservation in protected areas (PAs) (Primack, 1993). The concept's limitations emerge in PAs inhabited by farmers. While biodiversity reflects values attributed by scientists – it is used as a scientific concept referring to option and intrinsic values of natural species – farmers also attribute values to biological diversity. These are mainly instrumental values (Rolston, 1994), which include not only direct use values, but also indirect use values, such as cultural and spiritual values (Posey, 1999). As farmers maintain biodiversity in return for these instrumental values, its 'human nature' is recognised.

Although many studies stress that farmed areas do contain much biodiversity (Gómez-Pompa and Kaus, 1999), the biodiversity concept is still considered to be primarily a biological phenomenon and as a consequence cannot be used to gain insight into the social processes underlying either biodiversity or its transformations (Gerritsen, 2002). This chapter argues for a critical assessment of the biodiversity concept by focusing on farmers' perspectives rather than purely on the scientific concept. Such an assessment is relevant within the debate on best conservation approaches. This is illustrated by a Mexican case study. First, I outline the relevant theoretical notions and introduce the concept of resource diversity.

12.2 Diving into theory

In general, nature is presented as an ecosystem in which the multiple components interact through a variety of biophysical processes. It can also be considered as a social construct perceived and valued by people. People value nature as a holistic unit, but may also identify and value certain components of nature. These values are context specific; in other words, they relate to prevailing socioeconomic and cultural conditions. What is meaningful to one actor can be meaningless to another, and so each actor can possess his or her own definition of nature (Gerritsen, 2010; Guyer and Richards, 1996).

Value attribution generates an appropriation of nature, which becomes part of an actor's life world. The term 'natural resource' is used to denote those parts of nature that are useful and valued, and are also recognised as being worthy of management (Slocombe, 1999). Resource management is not only determined by biophysical characteristics, but also by the social characteristics of the users of the resources and the management organisations. Traditionally, the notion of natural resources has been considered as referring to a specific subset of nature consisting of its socially valued components. In this view, natural resources are considered to be utilitarian attributes of nature only. Such an interpretation is unwarranted. People not only manage natural resources because of their utilitarian values, but also because of environmental and cultural values (Wiersum, 1997).

It is also sometimes claimed that in contrast to technologically advanced industrialised nations, which primarily attribute utilitarian values to nature, indigenous peoples perceive nature holistically. In this view, the former are portrayed as managing nature on the basis of a worldview focused on, at worst, exploitation and, at best, stewardship, while indigenous peoples are portrayed as managing nature as partners and thus living in harmony with nature. Such representations consider part of humanity as an integral part of nature. Even if this notion of the 'ecologically noble savage' has been criticised as being romantic, the discussion about different worldviews illustrates that it is not possible to distinguish clearly between 'nature' and 'natural resources' (Guyer and Richards, 1996).

As a corollary to this argument, it is possible to analytically differentiate a biological and social perspective on biological diversity. Whereas the term 'biodiversity' is used to refer to the biological perspective on nature, the term 'resource diversity' can be used to represent the social perspective on nature. Like biodiversity, resource diversity considers various system levels, from the individual species to land-use units and the landscape as a whole. As resource diversity refers to the consciously valued components of nature, it should be interpreted as a dynamic concept. Natural resource values are dynamic, and the constellation of resource diversity will change in response to transformations. In other words, natural resources can be considered to be the outcome of a process of co-production between farmers and nature, which involves ongoing interactions and mutual transformations of social and natural phenomena. As one of the effects of co-production, farmers can create diversity in natural resources (Gerritsen, 2002).

When considering tropical environments, this process of co-production has resulted in several resource management patterns: hunting and gathering, pastoralism, shifting cultivation, permanent cultivation and mixed farming (Padoch and Vayda, 1983). As a result of the use of multiple resources by many farmers, the natural environment consists of various landscape units containing their own specific resource diversity. Some of these units consist of natural ecosystems from which native species are extracted; others consist of agroecosystems that combine native and domesticated species. The distinction between natural and agroecosystems is not discrete, but gradual (Wiersum, 1997).

The coexistence of various forest system types, each with its own specific resource diversity, illustrates how farmers manipulate their natural environment in order to suit their needs. Through their selective management of natural resources, farmers induce ecological succession patterns, which result in a variety of land-use zones with different degrees of ecosystem transformation. The exact nature of these succession patterns depends on farmers' actions, but also on existing ecological conditions. Through succession management, the different resource units obtain their specific place on the farm and in the landscape. This patchiness at landscape level represents the spatial dimension of resource diversity.

Analytically, landscape patchiness management can have two contrasting impacts on diversity. In some cases, the landscape can be transformed into various units that each play a role in farming practice and that can open up new possibilities for succession management. The combination of purposeful resource enrichment combined with natural succession processes may result in landscape units in which natural and anthropogenic components are combined. These resource-enriched units may still have a high, and sometimes unique, biodiversity. In other cases, though, the diversity of resources can become impoverished. This happens when farming activities are geared to obtaining just one or a few products. Both situations can occur within a farmer-managed landscape (Wiersum, 1997).

Thus, during co-production, farmers structure their environment in time and space to meet their needs. The ensuing resource diversity then forms the starting point for a new process of co-production and can offer new possibilities and limitations for resource management. This process in turn leads to both resource diversity and biodiversity change. Some local species may diminish due to over extraction or conscious removal, while the environmental changes associated with the co-production process may result in species disappearing unnoticed due to the fact that they are meaningless in that particular land-use system. Conversely, resource systems may become enriched by locally valued species, or new species may develop spontaneously within the newly created units. Focusing on co-production and assessing its resource diversity provides insight into the social processes underlying biodiversity conservation (Gerritsen, 2002).

12.3 Research design

The research was carried out from 1993 to 2002 in the indigenous community of Cuzalapa, which is located in the Sierra de Manantlán Biosphere Reserve (Figure 12.1). Farmers' strategies and their impact on biodiversity conservation were studied. The study consisted of semi-structured interviews combined with case

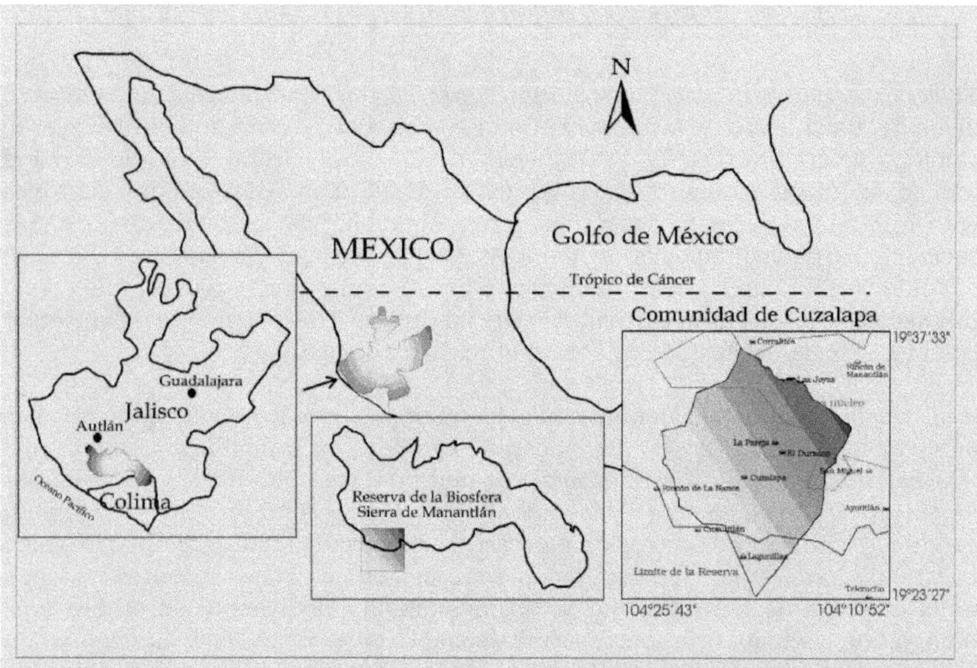

Figure 12.1. Location of the Sierra de Manantlán Biosphere Reserve and Cuzalapa (Gerritsen, 2002).

studies. Transect walks and participatory mapping was also used. Additionally, the official management plans were studied to identify the formal conservation perspective (Gerritsen, 2002).

The Sierra de Manantlán is a mountain range in Western Mexico and was declared a biosphere reserve in 1987. The reserve covers 140,000 hectares and contains a high biological diversity. It consists of three core zones of strictly protected area, covering 30% of the reserve, and buffer zone where new rules govern land use, which covers the remaining 70% of the reserve. The reserve's tenure regulations are imposed on existing properties, which consist of 32 agrarian communities holding 60% of the land and 80 private owners holding 40% of the land. The reserve´s conservation mission has expanded from one-species protection to an integrated conservation and development project. Realising this dual mission has challenged the reserve managers (Gerritsen, 2002).

Cuzalapa is located on the southern slopes of the Sierra and covers 23,963 hectares: 71% is buffer zone and 19% is core zone. The community includes 1,500 inhabitants, who live in one main village and several hamlets. Maize cultivation used to dominate, but livestock rearing has become increasingly important since the 1970s. Trees and forests are important for domestic purposes and in farming practice. There is no commercial exploitation of the forests (Gerritsen, 2002.).

12.4 Farmers' classification of the landscape

Resource diversity is described at landscape level, where farmers distinguish three land-use zones: homegardens (*huertos*), cultivation fields (*yuntas* and *coamiles*) and grazing lands (*agostadero* lands). The criteria for this classification are the potential for the use of the different parts of the landscape and the management practices within each of these areas. Within the main land-use zones, the farmers distinguish a number of subzones (Table 12.1).

Table 12.1. Land-use zones and subzones (Gerritsen, 2002).

Land-use zone	Land-use subzone
Homegardens (*huertos*)	tree gardens
	man-made forests
	coffee gardens
Cultivation fields (*yuntas* and *coamiles*)	irrigation fields
	rainfed fields
	shifting cultivation fields
Agostadero land	potential grazing land

Farmers subdivide *agostadero* lands according to field and vegetation characteristics (Table 12.2). The *agostadero* concept has a spatial and a temporal dimension. It is found not only in different landscape areas, but is also related to agricultural cycles.

The farmers do not identify forest ecosystems and secondary vegetation as a land-use zone, but as part of the grazing lands (*agostaderos*). They refer to forest resource units as *monte* and *arbolera*. These two main categories of forest resource are classified into two main classes and a number of subclasses (Table 12.3).

Farmers possess a broad ecological knowledge relating to landscape patchiness, species distribution, the relation between physical aspects and species distribution, succession processes and species growth characteristics, and other natural phenomena. This knowledge is embedded in the different farming domains (Tables 12.1, 12.2 and 12.3) and expressed through folk concepts. Farmers are also knowledgeable about insects, reptiles, mammals and birds, although they do not explicitly relate animal species to specific landscape units, which can be explained

Table 12.2. Farmers' classification of agostadero *land (Gerritsen, 2002).*

Category	Type
Fallow cultivation fields	fallow rainfed fields
	fallow irrigation fields
	abandoned shifting cultivation fields
Pastureland	natural
	newly established
Monte and Arbolera	forest and secondary vegetation

Table 12.3. Monte *and* arbolera *classes and subclasses (Gerritsen, 2002).*

Class	Type
Monte bajo (secondary vegetation)	*matorral*
	barbecho
	other *monte bajo*
Monte alto and *arbolera* (forest vegetation)	*roblera* (deciduous oak forest)
	encinera (subdeciduous oak forest)
	ocotera (pine forest)
	monte alto and *arbolera* in ravines (tropical deciduous and subdeciduous forest)
	other *monte alto* and *arbolera*

by the mobility of the animals. The farmers' knowledge is determined by the nature of the farming practice and the characteristics of the natural resources involved in the co-production, as well as other factors, such as the highly unequal land distribution and the new rules and regulations imposed by the reserve management body, which are part of the wider political-institutional context of co-production.

12.5 Succession management and landscape patchiness

The resource units are logically interrelated. Three types of relationships can be distinguished: (1) between land-use zones; (2) between land-use subzones and forest vegetation; and (3) between different types of forest vegetation. The interrelationships are of a socioeconomic and ecological nature. They are the outcome of the farmers' management of succession patterns – the active role that farmers play in managing land-use zones and forest vegetation types. The term 'succession' is therefore understood broadly; it refers to both ecological succession and succession in land-use activities. Both processes proceed simultaneously and are the temporal dimension of resource diversity. Through farming activities, farmers induce a succession pattern that leads to the creation of land-use zones and the transformation of forest vegetation. The different resource diversity units obtain a specific place on the farm and in the landscape through succession management. The resulting patchiness in the landscape is the spatial dimension of resource diversity.

Farmers manipulate succession processes in order to transform resource diversity to better suit their needs, which is also reflected in the landscape. The farmers can therefore actively influence landscape patchiness. Although the creation and transformation of landscape patchiness is one of the most visible forms of co-production, the degree of co-production varies within the different parts in the landscape due to the multiple time horizons that exist: the time horizons of crops and pasture, trees, secondary vegetation and forests are all different. Differences within forest vegetation types can be observed because certain tree species grow more quickly than others. Regarding forest vegetation, farmers state that farming practice has a bigger impact on *monte bajo* (secondary vegetation) than *monte alto* and *arbolera* (forests, see Table 12.4).

The logic that guides the creation and transformation of the different resource units is the use that can be made of each of them. However, farmers also attribute intrinsic values to the landscape and many farmers consider the community's natural resources as their natural heritage (*patrimonio natural*).

Table 12.4. Monte *and* arbolera *types and their scientific equivalents (Gerritsen, 2002).*

Common name	Common name	Scientific name
Monte bajo	*matorral*	secondary vegetation
	barbecho	secondary vegetation
	other *monte bajo*	secondary vegetation
Monte alto	*roblera*	oak forest (deciduous)
	encinera	oak forest (subdeciduous, mesophytic or with the presence of pine species)
	ocotera	pine forest or pine-oak forest
	monte alto/arbolera en las barrancas y los arroyos	tropical deciduous forest or tropical subdeciduous forest (including gallery forest)
	other *monte alto* and *arbolera*	tropical deciduous forest, tropical subdeciduous forest (including gallery forest) or cloud forest

12.6 The organisation of time and space

Succession management and landscape patchiness, the temporal and spatial dimensions of resource diversity, are logically organised through farming practice and forest resource management by farmers. Figure 12.2 schematically illustrates succession management of land-use zones and forest vegetation. It shows the succession that can take place between homegardens, cultivation fields and grazing lands, as well as between the different types of forest vegetation: *monte bajo* (divided in Figure 12.2 into *matorral* and *barbecho* vegetation), *monte alto* and *arbolera* vegetation.

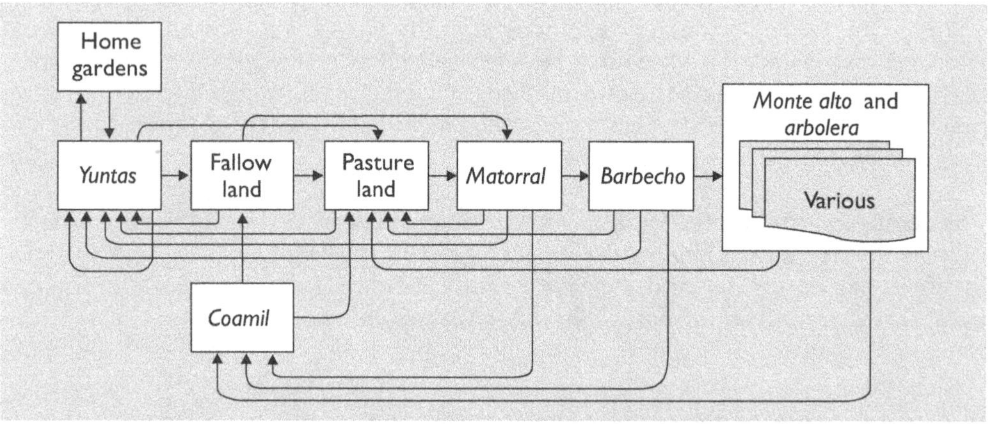

Figure 12.2. Succession management of resource diversity units (Gerritsen, 2002).

Figure 12.2 illustrates the interrelation between resource diversity units, but does not indicate the hierarchy that exists between the different succession relationships. Due to the multiple time horizons that underlie resource diversity, a diverse landscape is maintained and constructed through farming. The temporal organisation of resource diversity also implies a spatial coordination, due to the different time cycles in succession management. Figure 12.3 illustrates this by representing the spatial dimension of resource diversity, using folk concepts. It presents the patchiness of the landscape in Cuzalapa as known by its farmers.

The organisation of space depends on the time horizons underlying the different units, the specific ecological characteristics of the land and the decisions made by farmers. A number of socioeconomic factors also play a role, such as labour and financial resource availability. Traditionally, land use and land quality were related: farmers used to cultivate the best soils and graze cattle on land unsuitable for growing crops. Land quality can therefore be considered as one of the local criteria underlying the organisation of resource diversity. Nowadays, in some areas land use and land quality are disconnected; a reorganisation of time and space has taken place. For example, cows are held in forests where little fodder is available, or shifting cultivation is practised on poor soils or on very steep slopes. This is due to the current distribution of land, the increased availability of external inputs and new tools, such as herbicides and chemical fertilisers. This disconnection between land use and land quality is increasing as livestock rearing is on the increase.

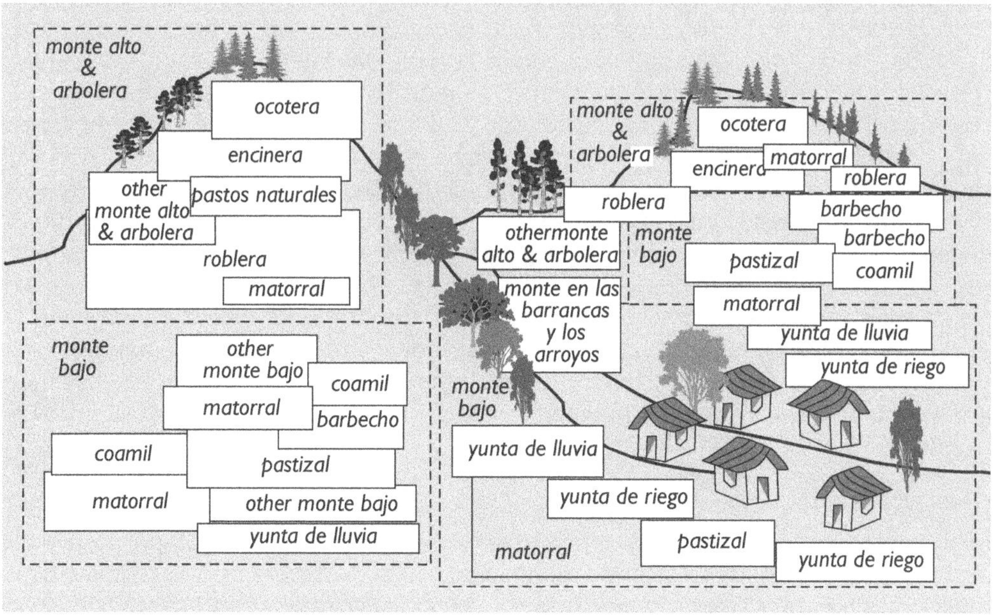

Figure 12.3. Resource diversity in the Cuzalapa landscape (Gerritsen, 2002).

It can now be considered that the high biodiversity of the Cuzalapa landscape is embedded in this resource diversity, which makes resource diversity one of the social carriers of the biodiversity.

12.7 Transformations in resource diversity

The Cuzalapa landscape has undergone various transformations over many centuries. During the last fifty years three processes have been increasingly shaping resource diversity: shifting cultivation, commercial forest exploitation and livestock rearing. Shifting cultivation played a role until the 1970s and 1980s. It was practised mostly in the lower and middle elevations of the community's land and shaped resource diversity above all in the lower-lying areas of the landscape. Its importance has since declined due to the land-distribution situation, the introduction of new technologies and the increase in livestock rearing. Forest exploitation played an important role from the 1940s to the 1980s. As a result of this activity, the majority of the forest vegetation (*monte alto* and *arbolera*) is relatively young and consists of secondary growth. Until the 1960s forest exploitation did not have any relation to indigenous farming practice as a private lumber company carried it out with external wage labourers. With the establishment of a community sawmill in the early 1980s, farmers became engaged in forest exploitation as an off-farm activity. Although this activity ceased in 1984, the forest vegetation continued to be incorporated into farming practice, as the forest plays an important, but complementary role in farming practice.

Since the 1970s, the expansion of livestock rearing has been important for shaping resource diversity. The need for pasture to feed the cattle has had an impact on resource diversity through changes in the land-use zones and the different forest vegetation types. Three types of changes can be identified. First, changes in land use resulting from the conversion of maize growing fields to pasture. Second, changes in land use as a result of the removal of vegetation cover. *Monte bajo* (secondary vegetation) and some homegardens have been transformed into permanent meadows. Third, changes in land use in combination with transformations in the forest vegetation cover have taken place. The structure and composition of much of the forest vegetation has changed as a result of the introduction of management practices aimed at increasing pasture availability. The processes of land-use change and vegetation cover removal proceeded relatively slowly until the 1970s, but have intensified from the 1980s.

Different time horizons underlie the landscape changes, including the management of the corresponding succession processes. Land-use changes, both on their own and in combination with the removal of vegetation cover, generally take place over a period of one or a few cropping seasons. The transformation of forest vegetation (*monte alto* and *arbolera*) proceeds differently from the conversion of homegardens, cultivation fields and *monte bajo*. Instead of a relatively drastic removal of the vegetation, more gradual changes can be observed here. Farmers gradually remove the understorey and open up the forest canopy to increase light penetration to accelerate pasture establishment.

12.8 Reorganisation of time and space

Figures 12.4 and 12.5 represent the transformations that have been occurring in the Cuzalapa landscape. They show resource diversity dynamics at two different moments and indicate succession management and the relative importance of the different succession relations. Figure 12.4 reconstructs the resource diversity dynamics in the 1970s and 1980s, when agriculture still was a mayor land-use activity. Figure 12.5 illustrates the resource diversity dynamics in the 1990s, when many management activities were geared towards livestock rearing and pasture availability.

Comparing these two figures reveals the transformations in resource diversity. Figure 12.5 shows the increased importance of pasture, which the farmers seek to improve through many land management activities. A number of succession relations have become more important than others. The first is the transformation of maize cultivation fields and homegardens, often the unproductive ones, into pasturelands. Overall, few homegardens have been cleared, which indicates that this does not take place on a very large scale. Furthermore, their transformation into pastureland is an indirect one, as farmers cultivate maize in the cleared homegardens before establishing pasture, due to the higher soil fertility. This allows higher maize yields to be obtained without using chemical fertiliser. The

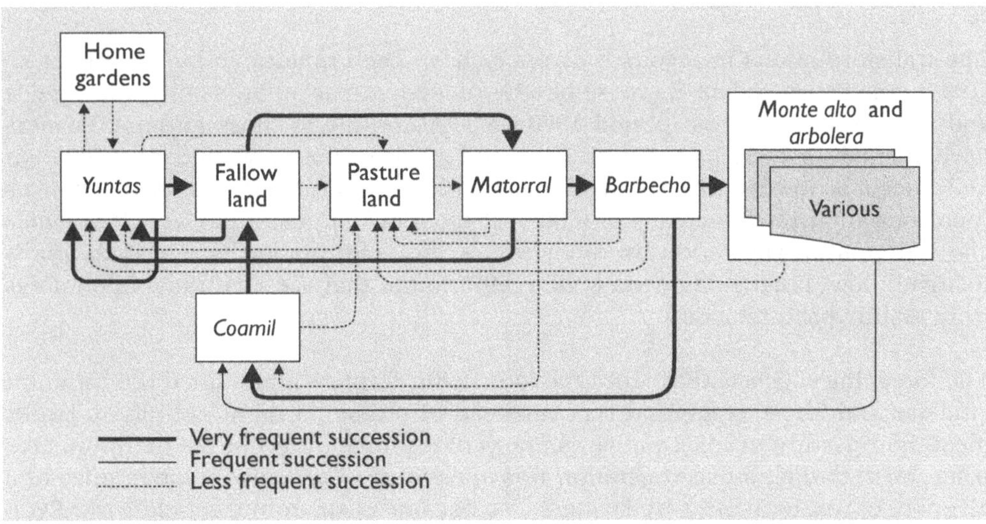

Figure 12.4. The dynamics of resource diversity in the 1970s and 1980s (Gerritsen, 2002).

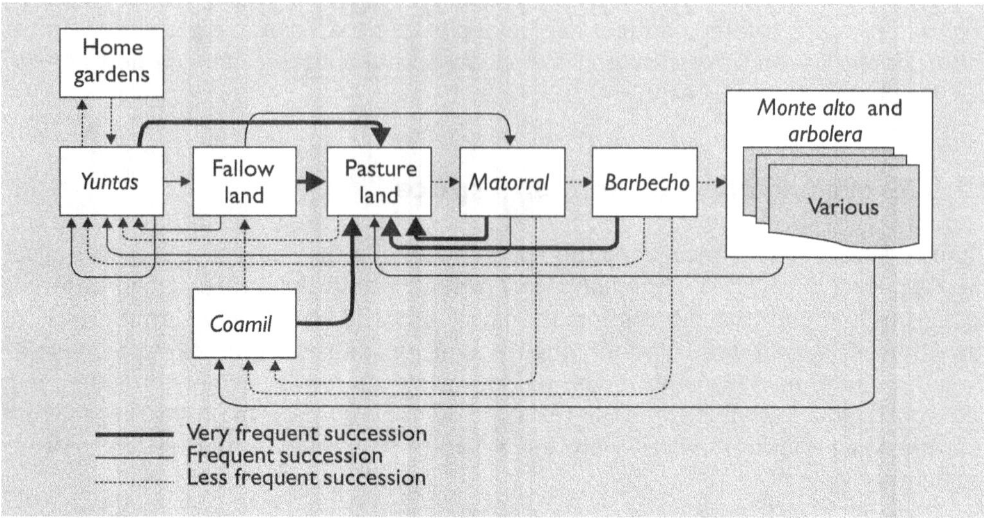

Figure 12.5. The dynamics of resource diversity in the 1990s (Gerritsen, 2002).

second relation is the clearing of secondary vegetation (*monte bajo: matorrales* and *barbechos*). Like the homegarden conversion, maize cultivation takes place before establishing pasture. Finally, the third relation refers to the conversion of *monte alto* and *arbolera* vegetation into pastureland, although it has been much less extensive than the two other processes. Again, this was combined with maize cultivation. Very little clearance of forest vegetation now takes place because of the land distribution situation and the zoning regulations in the reserve, which prohibit it. The latter can therefore be considered an additional factor in steering farmers' management practices.

The transformations in resource diversity have been quantified by Louette *et al.* (1998) who estimated an increase of 69% of 'open areas', that is cultivation fields and pasturelands, for the period 1970-1993. According to these authors, farmers have opened up 2,956 ha for cultivation and grazing, mostly secondary vegetation and forests at lower altitudes. Jardel *et al.* (2000) mentioned that these changes represent an annual decrease of 0.08% of the 'dense forest area', which is *monte alto* and *arbolera* vegetation. In other words, the deforestation rate in Cuzalapa is relatively low. Finally, Guevara *et al.* (1997) found that the additional open areas were mainly pasturelands.

The foregoing suggests that, until the late 1990s, farmers maintained the different land-use and forest vegetation types instead of removing them entirely to better meet their pasture needs. Impoverishment of resource units did not therefore take place, even though forest vegetation was opened up. The active maintenance of a diversity of resource units by farmers also became clear from the fieldwork. Even though pasture has grown in importance, farmers do not totally abandon maize cultivation, nor will they totally convert their homegardens. Besides, even though cattle graze in *monte alto* and *arbolera* vegetation, this forest vegetation continues to be important for wood, water and other non-timber forest products. In other words, farmers actively maintain the resource diversity for the specific role it plays in farming practice.

12.9 Variation in resource diversity transformations

The resource diversity transformations described above are general tendencies. In practice, these processes are more complex due to the specific social, technical and natural carriers that have influenced the characteristics, use and management of resource diversity in Cuzalapa. The land-use changes and vegetation cover transformations described above were related to altitude; with land-use changes taking place at lower altitudes and vegetation cover transformation at higher altitudes. In some cases, however, these two processes can also be observed at other altitudes or in other parts of the landscape, due to their specific biophysical and geomorphological characteristics. Social factors also play a role as the land distribution situation has contributed to the changes in farming practice. Shifting maize cultivation is hardly practised anymore because shifting fields has become almost impossible. This partly explains the emergence of *matorral* (secondary) vegetation, as it grows particularly well on intensively used soils. Other factors also play a role, such as the introduction of chemical fertiliser and herbicides and the increased importance of livestock rearing breeding.

Nowadays, much depends of the individual farmers and their strategies. Lack of land and the need for (non-timber) forest products also influences farmers' attitudes to resource diversity transformations. Many farmers no longer permit free access to their lands, which indirectly protects *monte alto* and *arbolera*. Finally, technological developments have also influenced resource diversity transformation, changing both resource diversity characteristics and farming practices.

12.10 Discussion and conclusion

This chapter has described the relation Cuzalapa farmers have with the natural environment. It became clear that they have an elaborate social definition of the natural environment and its diversity. This resource diversity contains a number of land-use zones (agroecosystems) and forest vegetation types. The boundaries between the agroecosystems and the forest vegetation types are not discrete, but gradual.

The case also illustrates that resource diversity is an outcome of co-production. This involves a specific organisation of time and space, which is purposefully created, maintained or modified by farmers in order to meet their natural resource needs.

Since the 1970s, livestock rearing has been the main motor behind transformations. This suggests an impoverishment of resource diversity, as several agroecosystems have been transformed into pasturelands and forest vegetation has been opened up to create more pasture.

There are at least two implications of this for the management of inhabited PA's. On the one hand, resource diversity gives insight into the perceptions of natural resources held by other actors than PA managers. On the other hand, it sheds light on the social dynamics underlying its constitution. Our case demonstrates that resource diversity as maintained and transformed by farmers can be regarded as a social carrier of biodiversity. However, the creation of the PA core zones has caused an artificial separation of farming practice and biodiversity. As the existing biodiversity is not just based on a pristine natural ecosystem, but on the indigenously evolved landscape pattern, such segregation will negatively influence biodiversity composition and distribution in the medium and long term unless special measures are taken. These measures will have to be based on farming practices, or will have to imitate its effects. Nevertheless, the current trends in the Cuzalapa landscape could be interpreted as indicating the need to maintain strictly protected areas in order to prevent an overall landscape homogenisation, leading to biodiversity loss.

In biosphere reserves opting for co-management, resource diversity can be a helpful tool in the discussions on natural resource management and conservation measures. While the concept of biodiversity emphasises the 'ecological side' of natural resource management, the concept of resource diversity stresses the 'social side'. Using both concepts allows the establishment of explicit interfaces for 'negotiating' sustainable land use and 'tailoring' intervention approaches to specific local contexts (Borrini-Feyerabend, 1996). Finally, all too often farmers are conceived as functional partners in conservation who have to be made 'aware' of the importance of biodiversity conservation. Co-management of PAs is possible only within a 'new' conservation professionalism (Chambers, 1997), which addresses the need to embed conservation in local notions and processes rather than just scientific ones.

References

Borrini-Feyerabend, G. (1996). Collaborative management of protected areas: tailoring the approach to the context. Issues in Social Policy, IUCN, Gland, Switzerland.

Chambers, R. (1997). Whose reality counts? Putting the first last. Intermediate Technology Publications, London, UK.

Gerritsen, P.R.W. (2002). Diversity at stake. A farmers' perspective on biodiversity and conservation in Western Mexico. Wageningen Studies on Heterogeneity and Relocalisation 4, Wageningen University, Wageningen, the Netherlands.

Gerritsen, P.R.W. (2010). Perspectivas campesinas sobre el manejo de los recursos naturales. Mundiprensa/Centro Universitario de la Costa Sur, Universidad de Guadalajara, Mexico City, Mexico.

Gómez-Pompa, A. and Kaus, A. (1992). Taming the wilderness myth. BioScience 42: 271-279.

Guevara, R.D., Carranza, J.A. and Louette, D. (1997). Levantamiento de uso actual de suelo (1991-1996), subcuenca de Cuzalapa, municipio de Cuautitlán, Jalisco, Mexico. Internal Report, Universidad de Guadalajara, Autlán, Mexico.

Guyer, J. and Richards, P. (1996). The invention of biodiversity: social perspectives on the management of biological variety in Africa. Africa 66(1): 1-13.

Jardel, E.J., Cárdenas, O.G., Ramírez, J.M., Balcazar, O.E., Chacón, J.C. and Cruz, G. (2000). Monitoreo de cambio de cobertura vegetal y uso de suelo en la región de la Sierra de Manantlán, Estados de Jalisco y Colima. Internal Report, Universidad de Guadalajara, Autlán, Mexico.

Louette, D., Martínez, L.M., Guevara, R.D., Carranza, J.A., Jímenez, R.G., Casillas, E., Esparza, J.P. and Gerritsen, P. (1998). Cambio de uso de suelo y actividad ganadera en 4 regiones de la reserva de la biosfera Sierra de Manantlán (1971-1998). Reporte preliminar. Internal Report, Universidad de Guadalajara, Autlán, Mexico.

Padoch, C. and Vayda, A.P. (1983). Patterns of resource uses and human settlement in tropical forests. In: Golley, F.B. and Lieth, H. (eds.) Tropical rainforest ecosystems, structure and function. Ecosystems of the World 14A, Elsevier, Amsterdam, the Netherlands, pp. 301-313.

Posey, D.A. (1999). Cultural and spiritual values of biodiversity. A complementary contribution to the Global Biodiversity Assessment. Intermediate Technology Publications, London, UK.

Primack, R.B. (1993). Essentials of biodiversity conservation. Sinauer Associates, Sunderland, MA, USA.

Rolston, H., (1994). Conserving natural values. Perspective in Biological Diversity Series, Columbia University Press, New York, NY, USA.

Slocombe, D.S. (1999). Natural resources. In: Alexander, D.E. and Fairbridge, W. (eds.) Encyclopedia of environmental sciences. Kluwer Academic Publishers, Dordrecht, the Netherlands, pp. 425-428.

Wiersum, K.F. (1997). Indigenous exploitation and management of tropical forest resources: an evolutionary continuum in forest-people interactions. Agriculture, Ecosystems and Environment 63: 1-16.

13. Governing biocultural diversity in mosaic landscapes

Cora J. van Oosten and Wouter Leen Hijweege
Wageningen University, Wageningen Centre for Development Innovation and Wageningen Forest and Nature Conservation Policy Group, P.O. Box 9101, 6700 HB Wageningen, the Netherlands; cora.vanoosten@wur.nl, wouter.hijweege@wur.nl

Abstract

Biocultural diversity is reflected in the existence of mosaic landscapes in which multifunctional land use offers scope for ecologically balanced and economically productive land-use patterns. Bioculturally diverse mosaic landscapes are historically shaped by people who have developed a strong sense of belonging and a deep attachment to their living environment, or place. Mosaic landscapes are therefore characterised by strong regional identities and a sense of ownership by the actors involved. However, landscapes are often cut across by administrative boundaries as they do not have a formal position in the political-administrative scaling of governance. This hampers the upscaling of current biocultural diversity management practices to higher levels of landscape governance. We therefore argue for a 'spatialisation' of governance as a means to reconnect governance to landscape and to build upon inhabitants' identities and sense of ownership to manage their biocultural diversity and govern their place. Spatial thinking and policy learning are crucial elements of landscape governance, embedded in local practice and politics of scale. Biocultural diversity can then be conserved and restored as a product of landscape governance.

Keywords: forest, landscape, place, governance, social learning

13.1 From managing biocultural diversity to governing mosaic landscapes

In the global debate on biodiversity conservation there is increasing recognition that productive land-use systems do not just reduce biodiversity of natural ecosystems, but they also create new landscape elements, offering scope for *in domo* conservation of biocultural diversity at the local level (Wiersum, 2003). Consequently, attention is gradually shifting from the conservation of single species to the conservation of the mosaic landscapes they are part of. This has led to an increased appreciation of multifunctional land-use systems in which both production and biodiversity functions are valued (Van Noordwijk *et al.*, 1997), which in turn offers scope for ecologically sound and economically productive land-use patterns (Hobbs and Morton, 1999). This also implies a gradual shift from community forest *management* approaches, which focus on the technical and operational management of forests at the local level, to higher levels of *governance* in which locally adapted forest management practices are placed within larger political and administrative systems. Such practices are not only formed by endogenous processes, but are also shaped by external influences, including political decisions driven by processes taking place at larger spatial scales.

Forest governance is also gaining ground in response to climate change. Since forests play a role as carbon sinks, they are increasingly seen as a key factor in combating climate change, making them part of the global debate on reducing greenhouse gas emissions. However, besides the shift from locally adapted forest management practice to higher levels of governance identified above, 'good' forest governance from the climate change perspective is increasingly driven by multilateral institutions, conventions and coalitions at supranational levels. This means that localised governance structures for land use, conservation and development are increasingly affected by global players with a broad and generic agenda, which does not do justice to the biocultural diversity of most forested landscapes.

Given these two opposite trends, there would appear to be a need for a mechanism which takes the biocultural diversity of mosaic landscapes into account while incorporating global agendas – a mechanism which is steered neither from below nor from above, but which functions as a *nexus* where global demands and local needs meet. Such a mechanism would not necessarily be aligned with political and administrative boundaries, but would link the multiple transboundary and multiple-scale networks in which landscape actors are involved. It would allow the inhabitants of landscapes to become more closely involved in the steering of landscape dynamics, giving them more grip on those exogenous processes affecting their landscape, strengthening their sense of belonging, and encouraging them to cross administrative boundaries and link into higher scales of political decision-making affecting their landscape. We believe that such a mechanism can be developed by connecting biocultural diversity to landscapes and their governance, thus paving the way for an emerging *landscape governance*.

13.2 Biocultural diversity, mosaic landscapes and governance defined

Before further exploring the concept of landscape governance, its three basic components of biocultural diversity, mosaic landscapes and governance have to be defined and related.

Biocultural diversity

Biocultural diversity challenges the often assumed dichotomy between people and nature. Instead of positioning people in opposition to nature, it emphasises the nature-culture continuum, ranging from more or less natural to more or less domesticated areas (Wiersum, 1997, 2003). Biocultural diversity results from the coevolution of environmental and social systems and is influenced by management practices and institutions at various levels involved (De Boo and Wiersum, 2002).

The bioculturally rich forest gardens we often find in small-scale tropical production systems are typical examples of land-use systems linking the extremes of the nature-culture continuum. Their management depends on household decisions taken in response to changing ecological conditions as well as to meet household needs and market opportunities. Forest gardens are therefore increasingly valued for contributing toward both resource production and biodiversity conservation. Nested within a larger system of management structures, the management of forest gardens forms a dynamic component of larger-scale mosaic landscape governance: management decisions at household level feed the decisions on multifunctional land use at landscape level.

Biocultural diversity as reflected in mosaic landscapes made up of patches of natural and human-influenced vegetation types are co-shaped by human interaction. Their management is rather complex, and can only be effective by involving stakeholders, their management practices, and their social and institutional networks (Wiersum, 2003). The transformation process of larger areas of natural forests into mosaic landscapes is an endogenous process. It not only reflects changes in locally applied management practices, but also changes in resource access, land-use rights and marketing arrangements.

Mosaic landscapes

The word 'landscape' is often used to describe a cultural, historical or esthetical perspective on scenery, as defined in most dictionaries. However, in the landscape literature, landscapes are conceptualised in a much broader sense. Urban *et al.* (1987) describe landscape as a mosaic of heterogeneous land forms, vegetation types and land uses. Guilmour (2008) also describes landscape as a mosaic made up of different components, but 'pieced together to form an overall landscape-level patchwork', thus emphasising the internal coherence between the various components of the mosaic. Görg (2007) adds that the concept of landscape unifies societal production and natural-spatial conditions in a particular place. According to Görg, therefore, landscape refers to the 'spatial-temporal aspects of the metabolism between nature and society', framing landscape as a realm of human-environmental interaction, tagged into place. Taylor (2008) amplifies this to a notion of a sense of identity and belonging, leading to a strong human attachment to landscapes, reflecting a collective identity in landscape and place. Shaping the landscape is 'making place', building stories and memories, and promoting a sense of local distinctiveness.

Massey's interpretation of place goes one step further, as she considers place not only as a localised meeting point, but as an intersection of partly overlapping social networks where the global and the local meet (Massey, 1994). This new interpretation of space is a reaction to the false idea that in a globalised world place no longer matters. On the contrary, Massey argues that it is because of globalisation that we see more and more geographical fragmentation and spatial disruption. And it is this fragmentation and disruption which makes people long for a renewed sense of place (Massey, 1994). Such a relational interpretation of place allows for a better understanding why most landscapes have multiple identities as products of layered sets of linkages, both locally and to the wider world. They do not always have clear boundaries, but they do have cores, which are constructed through human interaction within a particular spatial setting.

Bioculturally diverse mosaic landscapes are thus historically marked by strong regional identities, partly or completely shared by inhabitants who feel a strong sense of belonging to the area they consider to be theirs. It is this sense of 'ownership' which makes inhabitants protagonists of the biocultural diversity in the landscape. This ownership is often accompanied by a desire to be part of decision-making processes and it makes the inhabitants indispensable links in the management, conservation and restoration of landscapes. The social networks and the often informal role divisions in which actors manage their land and livelihoods often remain 'dormant' where the landscape changes gradually over longer periods of time. But sudden change, caused by environmental hazards, accelerated degradation or other external shocks, may trigger this endogenous agency, since it strengthens the shared sense of belonging and creates a common concern. This process can be actively strengthened through dialogue, storytelling, naming, mapping and using landmarks as symbols for regional identity, as shown by Buizer and Turnhout (2011) and Van Oosten (2004, 2006, 2010). Such active *place making* enhances collaboration and shared learning between actors within a landscape. Despite their diversity of interests, it mobilises actors' capabilities and intrinsic power to shape, sustain and restore biocultural diversity within mosaic landscapes from below, thus contributing to effective governance at the landscape level. However, as governance arrangements are value laden and evolve over time, their change tends to be time consuming, and the related power shifts make them contested and complex.

Governance

Whereas *management* usually refers to technical and operational functions of a system, *governance* refers to its political functions – the multiple interests and various stakeholders that give the system its direction and power – and is sometimes referred to as 'whole system management'. The current literature on governance largely comes from the political, social and administrative sciences and discusses the changing role of states, citizens and organised private actors, such as civil society organisations and private companies. Arnouts (2010) provides a good overview of current governance theory, starting with Pierre (2000), who states that 'governance denotes a conceptual or theoretical representation of co-ordination of social systems, and, for the most part, the role of the state in that process'. According to Pierre, the role of the state moves between two ends of a continuum: state centric, or 'old' governance, and society centred, or 'new' governance. Over recent decades, many countries have shifted from state-centric to society-centred governance. This does not necessarily reflect a weakening role of the state, but is due to a growing influence of non-state actors and to governments playing an enabling and (legal) frame-setting role rather than being in absolute control. It is clear that such a transition is a hard and lengthy process, marked by a blurring of boundaries between public and private realms of governance, and a redefinition of roles and responsibilities between the state, citizens and the private sector.

Society-centred governance acknowledges that society is built upon a myriad of social relations and networks that shape public life. That is why society-centred governance is increasingly referred to as multi-actor governance, collaborative governance or network governance. In this interpretation, multiple actors, all having their specific interests, mandates and claims, operate through interconnected networks and collectively decide what happens in society through bargaining, negotiation or deliberation (Arnouts, 2010). This comes fairly close to the position taken by Pierce Colfer and Pfund (2011), who define governance as 'the ways and institutions through which individual citizens and groups express their interests, exercise their rights and obligations, and mediate their differences'. This interpretation of governance gives ample space for multistakeholder dialogue and collaboration, without making a distinction between governmental and non-governmental actors. This approaches the concept of self-governance, in which non-governmental actors are considered to be able to steer their own lives, while governmental actors at most play a facilitating and frame-setting role (Kooiman 2003, quoted by Arnouts, 2010).

Multilevel governance emphasises the distinction between vertical and horizontal governance processes and the interaction between these different tiers of decision making. 'Nested' governance refers to governance systems in which roles and responsibilities between different tiers are clearly spelled out (subsidiarity principle), and to formalised shaping processes of decentralisation and devolution from higher to lower levels of control. Environmental governance is often placed under this denomination, as the governance of public goods such as forests, seas and air cuts across sectors and political-administrative scales.

13.3 The spatial dimension of governance: governance of place

As already mentioned before, politically and administratively defined governance structures usually do not coincide with the characteristics and boundaries of bioculturally diverse mosaic landscapes. This is most apparent in 'developing countries',[1] where political boundaries originated from rivalries between colonial powers, ignoring social, cultural and environmental notions of place. Current-day processes of state reform, such as decentralisation and devolution of spatial decision-making, are equally dominated by political-administrative hierarchies of scale and do not take into account the spatial dimension of landscape characteristics and regional identities. This phenomenon has disrupted the 'natural' connectedness between landscape dynamics and its inhabitants – between people and place.

The lack of such a spatial dimension in the current governance debate was recognised by Görg (2007), who stressed the importance of 'restructuring the spatial dimension of politics'. He emphasises the interconnections between socially constructed spaces and the natural conditions of place. He interprets the term 'landscape' as bridging between social and natural sciences, and *landscape governance* as a means to reintroduce the spatial dimension and the relevance of spatial scales. Such a *spatialisation* of governance could thus respond to society's renewed need for a sense of place in today's globalised world (Massey, 2005), thus confirming that, despite globalisation, place does matter.

If mosaic landscapes are considered to be spatial reflections of multiple networks cutting across ecological, geographical and political scales, then landscape governance would logically follow a network approach. In such an approach, multiple actor networks operate at different political scales, but all have an interest in or an influence on spatial decision making within a landscape. Despite differences in interests, actors are interrelated through their belonging to the place they consider to be 'theirs'; a sense of belonging which triggers agency and ownership. While not denying the existence of opposed interests and contested spaces, landscape governance could offer a basis for dialogue in an otherwise conflictive process of negotiation, marked by power imbalances and strife. Landscape governance would therefore be an unpredictable process of negotiations and trade-offs with multiple outcomes, rather than a planning process targeting at single management outcomes (Sayer *et al.*, 2008).

If landscapes are embedded in networks at various scales, then landscape governance would logically allow for upscaling of local practice. And since landscapes usually cover larger and ecologically more homogeneous areas, they provide manageable levels of operation for global actors to downscale their spatial policies. This makes landscapes suitable interfaces between the local and the global, feeding the design of international treaties and multilateral agreements at the global level and co-constructing landscapes at the local level through identities and practices rooted in place. This is what Görg means by politics of scale in which landscapes become an

[1] Although the term 'developing countries' refers to the old dichotomy between 'developed countries' and 'developing countries' which no longer exist, the term is still being used here to refer to political systems which have been strongly exogenously influenced, in this case by colonial powers.

important link between multilevel politics and natural-spatial conditions, and an appropriate realm for governance to be performed (Görg, 2007).

13.4 Enhancing landscape governance: a learning approach

Landscapes indeed are missing links in formal constellations of governance, and their use, conservation and restoration are insufficiently embedded in and enforced by formal governance institutions of law, regulation, political mandate and delegation of power. And where landscape governance does exist, it is performed through informal institutions which relate to landscape-related networks, identities, memories and shared practice, either directly or indirectly linked to politics of scale. Institutionalising landscape governance would therefore be a social process, constructed through institutions of meaning related to place.

Although not formally recognised, landscape governance *does* exist in practice. Görg (2007), for example, described a case of landscape governance in the German Südraum, where landscape governance is being practiced through active engagement of both public and private stakeholders, including policy-makers and scientists, in a process of collective action. This collective action was rooted in a deep concern over the severely degraded state of the ex-mining area and a problem-focused process of collaborative learning. It is this collaborative learning, linked to the various political scales, which has made landscape governance neither steered from above nor below, neither from the inside nor the outside, but from an integrated relational perspective embedded in politics of scale and strongly anchored in place. Despite difficulties related to the transformation of its sociopolitical context, a gradual process of linking, learning and action, anchored in the natural conditions of the landscape, has enabled a massive restoration of the Südraum and the 'shaping of a new landscape'.

Van Paassen *et al.* (2011) confirm the existence of such learning-based landscape governance. Based upon multiple case studies from all over the globe, they conclude that landscape governance entails multiple actors at multiple geographical and political scales interacting through partly overlapping networks, resulting in a web of arrangements within both the formal and the informal realms of governance. From an analysis of various case studies, they state that landscape governance is largely based on formal and informal arrangements, at multiple scales, requiring strong stakeholder interaction through transdisciplinary social learning based upon a collective sense of place.

Both Görg and Van Paassen place their notion of learning within a context of both spatial thinking and policy learning. Spatial thinking is based on a deep understanding of natural-social interactions (relations, structures, representations) within a landscape. It has the potential to empower landscape actors to better understand, explain or predict processes taking place in, or having an impact on, their landscapes. As such it contributes to better informed and consented decision-making. Policy learning is a problem-focused approach to policy-making, in which policy-makers and practitioners do not necessarily strive for 'win-win' negotiations that tend to privilege compromise over problem solving, but engage in an inter-institutional process of mobilising knowledge, identifying and sharing good practice, and developing stakeholders' capacities for their wider application. It closely follows early adopters and measures behavioural change (IUFRO, 2010). Policy learning facilitates collaboration through the creation of platforms or networks. This allows the goals or techniques of policy to be adjusted in response to past experience and new information to be generated within different sectors, which can then be upscaled and downscaled through the politics of scale. New or adapted institutional arrangements between stakeholders at various politics of scale are the envisaged outcomes of policy learning (*ibid.*).

Spatial thinking, combined with policy learning, may thus be the key to *landscape learning*, in which actors within and around mosaic landscapes learn to better govern their landscapes, respecting and enhancing the biocultural diversity they contain.

13.5 Landscape learning in practice

Social learning systems have been studied by Wenger *et al.* (2000) and Wenger (2006), who introduced the concept of *communities of practice*. Communities of practice are formed by people 'who engage in a process of collective learning in a shared domain of human endeavour. Because they share a concern or a passion for something they do and learn how to do it better as they interact regularly' (Wenger, 2006). Members are practitioners who develop a shared repertoire of resources (experiences, stories, tools and ways of addressing problems) and use these to create a shared practice. It is this experience of sharing practice that creates a sense of 'belonging', or group identity, to which members adhere. Social learning systems, be it organisations, societies or regions, thus become constellations of communities of practice, each taking care of a specific aspect of reality, a specific practice. Since most people, inhabitants or citizens take part in more than one community, they constantly move from one community to another, thus building bridges across communities, stretching their boundaries, reconfiguring relations, and creating networks of practitioners who, despite differences in professional background or specific interest, are all connected to one common ground. It is this common ground that forms the basis for social cohesion, generates collective insights, and strengthens the sense of belonging to an organisation, a society – or a landscape (Wenger, 2006).

Keen *et al.* (2005) add an environmental dimension to the concept of social learning by defining social learning as 'the collective action and reflection that occurs among different individuals and groups as they work to improve the management of human and environmental relations'. In relation to landscapes, this means that landscape learning is a social process in which improved management of human and environmental relations in a specific spatial setting requires strong collaboration between the actors involved. This is done through the linking of various communities of practice, building bridges and stretching their boundaries in order to nurture a common identity and a common concern. Within landscapes it is more likely that actors' interests are opposed and tend to compete. However, social learning not only takes place within communities of actors sharing compatible interests, thus *connecting the likewise* (Castells, 2009; Leeuwis and Aarts, 2010), but to an even greater degree connects those with conflicting interests, reconfiguring their interdependent relationships and triggering a common concern (Leeuwis and Aarts, 2010). Wals *et al.* (2009) confirm this by stating that social learning takes place when 'divergent interests, norms, values and constructions of reality meet in an environment that is conducive to learning'. Social learning can therefore be effective only if asymmetries of knowledge and power between different stakeholders are taken into account, and effectively taken care of (Giller, 2008).

Following Görg (2007), Keen *et al.* (2005), Leeuwis and Aarts (2010), Massey (2005), Van Paassen *et al.* (2011), Wals *et al.* (2009), Wenger (2000, 2006) and others, landscape learning can be an important element of landscape governance. Landscape learning is thus perceived as a fluid process of interacting communities of practice, all strongly rooted in a shared sense of place. They learn from the biocultural diversity of mosaic landscapes and learn to understand the complexity of its management and the challenges of its governance. Linking these learning processes at multiple scales of spatial decision-making helps to scale up understanding, interaction, negotiation and collective action to generate more generic lessons learned. In other words, it is a case of *stretching beyond the local fix* (Lange *et al.*, 2010) to landscape-scale solutions through politics of scale. By constructing new landscape institutions, such as multilevel networks, which are locally anchored in shared identities and common concern, biocultural diversity is managed, conserved and restored at the level of mosaic landscapes. In this sense, landscape governance may offer an opportunity to re-establish the connection between politics and place, between citizens and their environment, between the local and the global. This shaping of new mosaic landscapes based on linking and learning, interaction and action, and a renewed sense of place offers a way to create and manage the biological and cultural diversity that is sought so desperately.

13.6 Conclusion

This chapter places biocultural diversity within its wider context of mosaic landscapes. It explores the potential to embed the management of bioculturally diverse landscapes within overall systems of governance. Such landscape governance allows for a spatialisation of governance: connecting socially constructed space with its natural conditions; reconnecting citizenship to place. Embedded in politics of scale, landscape governance links local biocultural dynamics and management practice to regional and global policy networks, thus affecting decision-making processes concerning the landscape. It effectively makes use of regional identities shared by landscape actors who, despite their diversity and heterogeneity, share a common sense of responsibility and ownership. Nurturing this sense of ownership through an active process of 'place making' supports the development of agency and enables policy learning at various politics of scale. Building communities of practice within and between landscapes helps to overcome competing interests and nurtures multistakeholder dialogue by making use of endogenous agencies triggered by regional identities grounded in place. Such landscape learning helps to scale up localised understanding, negotiation and collective action to develop solutions at the landscape scale.

We argue for a shift from managing bioculturally diverse landscapes to *landscape governance*. In doing so, we add to the literature our interpretation of landscape governance as an instrument to conserve, maintain and restore landscapes through multistakeholder learning and dialogue that is anchored in place. However, we recognise that institutionalisation of landscape governance within current formal or informal institutional frameworks and politics of scale is a real challenge which needs further exploration. This should be pursued by searching for more cases of landscape governance and analysing the preconditions for good landscape governance in practice.

References

Arnouts, R. (2010). Regional nature governance in the Netherlands: four decades of governance modes and shifts in the Utrechtse Heuvelrug and Midden-Brabant. PhD thesis, Wageningen University, Wageningen, the Netherlands.

Buizer, M. and Turnhout, E. (2011). Text, talk, things, and the subpolitics of performing place. Geoforum 42(5): 530-538.

Castells, M. (2009). Communication power. Oxford University Press, Oxford, UK.

De Boo, H.L. and Wiersum, K.F. (2002). Adaptive management of forest resources: Principles and processes. Wageningen University, Wageningen, the Netherlands.

Giller, K., Leeuwis, C., Andersson, J.A., Andriesse, W., Brouwer, A., Frost, P., Hebinck, P., Heitkönig, I., Van Ittersum, M.K., Koning, N., Ruben, R., Slingerland, M., Udo, H., Veldkamp, T., Van de Vijver, C., Van Wijk, M.T. and Windmeijer, P. (2008). Competing claims on natural resources: what role for science? Ecology and Society 13(2): 34. Available at: http://www.ecologyandsociety.org/vol13/iss2/art34/

Görg, C. (2007). Landscape governance – The 'politics of scale' and the 'natural'conditions of places. Geoforum 38: 954-966.

Guilmour, D. (2008). Understanding the landscape mosaic. In: Sayer, J.A. (ed.) The forest landscape restoration handbook. Sterling, V.A., London, UK.

Hobbs, R.J. and Morton, S.R. (1999). Moving from descriptive to prescriptive ecology. Agroforestry Systems 45: 43-55.

International Organisation of Forestry Research Organisations (IUFRO) (2010). Embracing complexity: meeting the challenges of international forest governance. A global assessment report. IUFRO report presented at the 9th session of the United Nations Forum on Forests (UNFF), February 2011. IUFRO, Vienna, Austria.

Keen, M., Brown, V.A. and Dyball, R. (eds.) (2005). Social learning in environmental management – Towards a sustainable future. EarthScan, London, UK.

Kooiman, J. (2003). Modern governance: new government-society interactions. Sage Publications, London, UK.

Lange, B. and Büttner, K. (2010). Spatialization patterns of translocal knowledge networks: conceptual understandings and empirical evidences of Erlangen and Frankfurt/Oder. European Planning Studies 18(6): 989-1018.

Leeuwis, C. and Aarts, N. (2010). Rethinking communication in innovation processes: creating space for change in complex systems. Paper presented at the 9th European IFSA Symposium, 4-7 July 2010, Vienna, Austria.

Massey, D.B. (1994). A global sense of place. From Space, place and gender. University of Minnesota Press, Minneapolis, MN, USA.

Massey, D.B. (2005). Landscape as a provocation: reflections on moving mountains. Journal of Material Culture 11(1-2): 33-48.

Pierce Colfer, C.J. and Pfund, J.L. (eds.) (2011). Collaborative governance of tropical landscapes. Earthscan, London, UK.

Pierre, J. (2000). Debating governance: authority, steering and democracy. Oxford University Press, New York, NY, USA.

Sayer, J., Buck, L. and Dudley, N. (2008). Next steps in monitoring landscape approaches. In: Learning from landscapes, arborvitae special issue, Sept 2008, IUCN. Available at: http://cmsdata.iucn.org/downloads/a_avspecial_learning_from_landscapes.pdf.

Taylor, K. (2008), Landscape and memory. Paper presented at the UNESCO international workshop 'The right to landscape. Contesting landscape and human rights'. 8-12 December 2008, Cambridge, UK.

Urban, D.L., O'Neil, R.V. and Shugart, H.H. (1987). Landscape ecology: a hierarchical perspective can help scientists understand spatial patterns. BioScience 37: 117-119.

Van Noordwijk, M., Tomich, T.P., De Foresta, H. and Michon, G. (1997). To segregate or to integrate? The question of balance between production and biodiversity conservation in complex agroforestry systems. Agroforestry Today 9: 6-9.

Van Oosten, C. (2004). Fading frontiers: local development and cross-border partnerships in Southwest Amazonia. Geographical Studies of Development and Resource Use, Royal Dutch Geographical Society (KNAG), Utrecht, the Netherlands.

Van Oosten, C. (2006). Through the looking glass: understanding regional integration from below. In: Van Lindert P., De Jong, A., Nijenhuis, G. and Van Westen, G. (eds.) Development matters, geographical studies on development processes and policies. Utrecht University, Utrecht, the Netherlands.

Van Oosten, C. (2010). Constructing regional integration from below: cross-border partnerhsips and local development in Southwest Amazonia. In: Van Lindert, P. and Verkoren, O. (eds.) Local development and governance in Latin America, geographical Perspectives. Springer Verlag, Berlin, Germany.

Van Paassen, A., Van den Berg, J., Steingröver, E., Werkman, R. and Pedroli, B. (2011). Knowledge in action – The search for collaborative research in sustainable landscape development. Mansholt Publication Series no. 11, Wageningen Academic Publishers, Wageningen, the Netherlands.

Wals, A.E.J., Van der Hoeven, N. and Blanken, H. (2009). The acoustics of social learning: designing learning processes that contribute to a more sustainable world. SenterNovem, Utrecht, the Netherlands.

Wenger, E.C. (2000). Communities of practice and social learning systems. Organisation 7(2): 225-246.

Wenger, E.C. (2006). Communities of practice: a brief introduction. Available at: http://www.ewenger.com/theory/index.htm.

Wenger, E.C. and Snyder, W.M. (2000). Communities of practice: the organizational frontier. Harvard Business Review, January-February: 139-145.

Wiersum, K.F. (1997). Indigenous exploitation and management of tropical forest resources: an evolutionary continuum in forest-people interactions. Agriculture, Ecosystems and Environment 63: 1-16.

Wiersum, K.F. (2003). Use and conservation of biodiversity in East African forested landscapes. In: Tropical forests in multi-functional landscapes, Seminar proceedings. Prince Bernhard Centre, Utrecht University, Utrecht, the Netherlands, pp.33-39. Available at: http://www.bio.uu.nl/pbc/publications/proceedings/Wiersum.pdf.

Wiersum, K.F. (2004). Forest gardens as an 'intermediate' land-use system in the nature-culture continuum: characteristics and future potential. Agroforestry Systems 61: 123-134.

14. The (onto)politics of classifying biocultural diversity: a tale of chaos, order and control

Severine van Bommel[1] *and Esther Turnhout*[2]
[1]*Wageningen University, Department of Social Sciences, Communication and Innovation Studies Group, Hollandseweg 1, 6706 KN Wageningen, the Netherlands;*
[2]*Wageningen University, Department of Environmental Sciences, Forest and Nature Conservation Policy Group, Droevendaalsesteeg 3, 6708 PB Wagenignen, the Netherlands; severine.vanbommel@wur.nl; esther.turnhout@wur.nl*

Abstract

The issue of classification plays a central role in Wiersum's work on biocultural diversity. The design of classification systems has enabled Wiersum to classify landscapes into natural, cultural and various intermediate categories. These classification systems do not merely mirror the world, but can only be understood in the light of the social and political values and desires they highlight and seek recognition for. In this chapter we employ a performative perspective of classification by analysing the social work that classification systems do in practice: how they influence not only how the world is known, but also how it is acted upon, and how social and material relationships are remade in the process. We conclude that by performing a world that consists of various natural, cultural and mixed categories, Wiersum's work (1) privileges local/indigenous communities to manage the nature-culture mixtures; (2) creates a nature-culture continuum to allow for coordination across the nature-human divide; and (3) creates a network of scientists and practitioners from diverse disciplines who can arrive at a division of labour in the research into and management of the biological, human and cultural categories that are distinguished.

Keywords: biocultural diversity, classification systems, categories, boundary work, performativity, onto-politics

14.1 Introduction: biocultural diversity as an ordering device

'Now imagine that the forest is a huge information space, and each of the trees and bushes are classification systems ... You may write a basic manual of forestry ... What will your product look like? Who will use it?' (Bowker and Starr, 2000: 31-32). When Bowker and Starr (2000) introduced the forest metaphor to describe classification systems, they seemed to have the work of Freerk Wiersum on biocultural diversity in mind. Indeed, Wiersum's work is very much concerned with classifications designed to introduce a certain order to chaotic things. As Scott (1998) reminds us, forests, nature and the values associated with them need to be made legible in order to enable management and control. Biocultural diversity is a concept which does exactly that. It is defined as the diversity of life in all its manifestations – biological, cultural and linguistic – which are interrelated (and likely coevolved) within a complex socioecological adaptive system (Newing, 2010). In the 1990s, biocultural diversity emerged as a new field in both academia and policy-making as a response to the growing concern over the 'converging extinction crisis' (Harmon, 1996), which was thought to be affecting both human and non-human nature alike (Mathez-Stiefel *et al.*, 2007). The concept of biocultural diversity conveys the idea that many of the planet's areas of highest biological diversity are inhabited by indigenous and traditional people, providing what the Declaration of Belem (issued by the International Society of Ethnobiology in 1988) calls an 'inextricable link' between biological and cultural diversity. As such, biocultural diversity provides a way to put humans, plants, animals and other forms of life in the same category, thereby enhancing order, increasing possibilities for management and reducing chaos. Although in the academic world early observations of the link between biodiversity and cultural diversity were made by a number of researchers working in the field, including Nietschmann (1992), Nabhan (1997) and Stevens (1997), Wiersum in particular pioneered this area. Drawing on a variety of sources in the natural, social, applied and policy sciences, his work argues that humans have maintained, enhanced and even created biodiversity through culturally diverse practices of managing 'wild' resources and raising domesticated species (such as the protection, dissemination and use of culturally important 'wild' species, agroforestry, forest gardens, homegardens, etc.), leading to landscapes with a mosaic of vegetation types. Emphasising how humans and nature have always influenced and shaped each other therefore makes a lot of sense. However, what stands out most in the work of Wiersum are his numerous classifications. To draw on Bowker and Starrrs' forest metaphor, Wiersum's forest manual consists of many trees and bushes, each of which offers a classification system to turn the chaotic field of biocultural diversity into a more ordered and manageable entity.

In this chapter we draw on insights from the field of science and technology studies (STS) to analyse Wiersum's work on biocultural diversity, specifically focusing on the classifications he has designed, and discuss the political project that these classifications pursue, their role in network building, and their social and material implications.

14.2 Classification systems, boundaries and performativity

The issue of classification may seem boring and insignificant at first. Why does it matter how many different types of forest are distinguished? What is the relevance of adding new categories to existing classifications? The answer is simple: classifications have an impact on the items that get classified. Different types of forest come with different prescriptions for management, different income levels come with different tax percentages, different types of publications come with different credits. When tied to these practical effects, the number of categories and the definition of their boundaries become crucially important.

Classification systems become problematic when confronted with things or people that do not fit into the categories. Nature has proven to be a particularly rich source of things that cannot easily be fitted into established categories. Victorian naturalists, for example, did not know what to do when the first platypus arrived from Australia (Ritvo, 1997). At first, they thought it was an artificial fabrication using different parts of mammals and birds. Later, when it had been decided that it must be a mammal, they refused to believe that it laid eggs. It is therefore not surprising to find that scientists involved in the study of nature are deeply concerned with issues of classification and only seem to agree on them after considerable negotiation. Their ambition is to develop the perfect classification system that adequately reflects the order of nature. For example, Stephen Jay Gould, the American palaeontologist, evolutionary biologist and historian of science, said 'taxonomy (the science of classification) is often undervalued as a glorified form of filing – with each species in its prescribed place in an album ... classifications are theories about ... natural order, not dull catalogues' (Gould, 1990: 98). This view explains why classifications of nature and biodiversity are so hotly debated; if the aim is to discover and represent the true order of nature, it is important to get it right.

STS researchers have criticised this view of classifications as mirroring nature, pointing out that it is humans that impose these systems onto a nature that does not tell us which categories it consists of. Furthermore, STS scholars have drawn attention to the values they reflect, the work they do and the purposes they serve (Bowker and Starr, 2000). According to Bowker and Starr (2000: 10), 'A "classification system" is a set of boxes, metaphorical or not, into which things can be put in order to then do some kind of work, for example knowledge production.' Classification systems produce order and enable understanding: 'Deciding what is to count as likeness and distinguishing like from unlike is the way in which our understanding becomes an orderly affair' (Keshet, 2011: 144). This order is a crucial requirement for not only understanding, but also coordination and control. For central governments to be able to create and implement rules about issues as diverse as planning, forest management or taxation, their territories and their human and non-human inhabitants need to be made legible (Scott, 1998). Classification systems are vital to the creation of this legibility. They reduce complexity and chaos by creating a well-ordered architecture that accommodates both similarities and differences.

The creation of difference by defining categories and setting their boundaries also enables coordination across these differences (Halffman, 2003; Shapin and Schaffer, 1985). Classification systems allow those that belong in different categories to cooperate and come to productive divisions of labour. Even though scientists involved in the study of nature might lack consensus about crucial issues, their common commitment to classification ties them together: 'biologists do not agree on the nature of species; sociologists bicker about the nature of society; literary critics diverge on notions of genre and style. What endures, however, are debates about the categories' (Starr, 2002: 11). As such, classification systems function as 'boundary objects' (Starr and Griesemer, 1989; Turnhout, 2009): they are flexible and open enough to stimulate debate among different social and scientific domains and allow for different interpretations and modifications; at the same time, these boundary objects are stable enough to serve as a common anchor point that ties these different domains together.

Developing categories and classifying items is a messy affair riddled with bootstrapping problems (Bowker, 2000). The categories of classification systems are not empty boxes with fixed boundaries that existed before the things that go into them (Jones, 2009). Instead, the two are mutually constitutive: the development of categories and the definition of their boundaries cannot be separated from the process of naming, labelling and identifying the items that are made to belong to that category. Thus, the development of a classification system requires preconceived ideas about what items need to be put into the categories, and vice versa. In order to know which items are out there, you first need a classification system (Bowker, 2000; Turnhout and Boonman-Berson, 2011).

The bootstrapping problems involved in classification systems imply that the latter can in principle always be overthrown when non-fitting (or hybrid) items emerge. However, this does not always happen, because these hybrids are not always counted or taken into account. According to Latour (1993), scientists are obsessed with categories and dichotomies, for example between nature and culture or between natural and social science. And science is thoroughly intolerant of hybrids (Latour, 1993), because these disturb order and threaten control. If hybrids cannot be silenced, ignored or force-fitted into existing categories, the classification system is modified; existing categories are redefined or new categories are added. The playtypus provides an excellent example of this. At first, it was ignored and its existence denied. Later, its existence was silenced by refusing to accept some of its hybrid characteristics. Only after repeated shipments of specimens, including one that was killed in the process of hatching, was the category of mammals adapted to allow for spoon-billed, egg-laying animals (Ritvo, 1997). This restored order and stabilised animal taxonomy again. However, this order is provisional and temporal. Any process of setting boundaries creates its own boundary problems: hybrids are an artefact of the classification system in place.

It is now clear that classification systems are designed to reflect not just any differences, but only those differences that are considered to make a difference – and conversely, they hide those differences that are not reflected in the categories. Rather than representing – by mirroring – nature, they make it to be a certain way, casting some things into the shadow of invisibility, while drawing attention to others (Bowker and Starr, 2000). This in turn, influences not only how the world is known, but also how it can be acted upon. It privileges certain items and courses of action over others that are not reflected in the classification system. Therefore, classifying is not only about knowing and describing certain realities, but it is also about creating these realities by describing them (Callon, 1998; Mol, 1999). 'Since knowing is also about performing, it is therefore about accepting the responsibilities that go with knowing. This, then, is a new kind of located and situated critical project, one that is profoundly political, but not foundational' (Law, 2008: 634). Classifying can therefore be seen as a political strategy. Mol (1999) has coined the term 'ontological politics' (or onto-politics) for this particular strategy. The term onto-politics focuses attention on the relation between the making of an ontological claim (something exists) and the political project that the claim serves (this something matters). Classification systems do just that. When valued items are put into categories, they have to be counted and taken into account in policy and management. Classification systems are not passive instruments for these practices of control, but active forces in their constitution.

In the remainder of this chapter, we use the various effects and implications of classification systems outlined in this section to analyse Wiersum's work on biocultural diversity and the classifications he develops and uses.

14.3 Biocultural diversity in Wiersum's research

Wiersum's interest in biocultural diversity started in the early 1990s and emerged from the assessment of the success of social forestry projects in developing countries (see, for example, Wiersum, 1990, 1991, 1993). When international agencies such as the UN Food and Agriculture Organization (FAO), the World Bank and the World Wide Fund for Nature first started these projects in the late 1980s and early 1990s, it was thought that local communities needed to be made aware of the values of forests and they needed to be educated about appropriate forest management practices. His research thus focused mostly on obtaining a better understanding of the various options for stimulating reforestation and tree growing by local communities. Gradually this led to the understanding that many local communities already employed indigenous forest management systems and that these played an important role in maintaining forest cover. Wiersum was therefore one of the first to challenge the dominant view that local people are the major cause of environmental degradation and that nature needs to be protected from their influence. As Wiersum (2003: 34) put it: 'ecological scientists and conservation authorities mostly refer to the need to maintain biodiversity for use by future populations and they focus mainly on the need of conserving rare and endangered species occurring in wilderness areas ... It is often considered that local people are one of the main driving forces between deforestation and loss of biodiversity. Consequently, it is proposed that forest and biodiversity conservation is best served by professional conservation agencies protecting natural forests.' He later argued (Wiersum, 2003: 35) that 'this ... view cannot be maintained and a more complex picture emerges'.

Rather than a deficit of knowledge and understanding on the side of communities, Wiersum deduced that it might be foresters and other professionals who required knowledge and training about the complex realities of indigenous forest management. Consequently, Wiersum became interested in understanding indigenous forest management practices and the extent to which these shaped and maintained habitats and ecosystems around the world. At first, his studies aimed at understanding forest use, but gradually he broadened his focus to include the practices by which local people purposefully maintained forests, or even augment valuable forest resources. Wiersum argued that human creativity in maintaining and increasing forest resources had in fact resulted in a wide variety of forest types, including protected natural forests, transformed and modified forests, and various types of tree plantations. He argued that 'various types of community forest management have been in existence for centuries ... local people living in or near forests should not a priori be considered as mere gatherers of forest products and/or as people who are transforming forests into agriculture. In many cases they are active forest managers who are involved in purposeful activities to safeguard the continuous availability of the valuable forest resources' (Wiersum, 1997: 2). This implies a positive outlook on human use and management of forest resources: 'the protection and purposeful regeneration of useful species does not necessarily involve the transformation into an agricultural landscape, but may result in a domesticated landscape, characterised by a mosaic of managed forests and agroforestry systems' (Wiersum, 1997: 2-3); and 'Human-induced changes in native forests and introduction of cultivation does not always involve ... disappearance of a forest cover and loss of biodiversity ... In some cases it has even been shown that ... instead of depleting forests local people were rather stimulating a forest extension' (Wiersum, 2003: 33).

Based on these insights, Wiersum came to conceive forest-human interactions as an evolutionary continuum of reciprocal relationships; these interactions started to become the key not only to understanding the forests themselves, but also the (indigenous) people who lived in and off these forests. Thus, rather than a dichotomy between natural and domesticated areas, Wiersum proposed a nature-culture continuum which allowed for hybrid categories ranging from more or less natural to more or less completely domesticated areas. Wiersum thereby challenged the image of 'pristine' environments, unaffected by humans, that can be brought back to their 'original' state. This profoundly problematised management strategies that focus on fencing off areas because, if human as well as biological processes have formed the forest, it becomes impossible to find a natural benchmark against which to measure and evaluate human impacts.

Wiersum demonstrates his commitment to contributing to improving governance and management of forests by arguing that effective conservation should focus on hybrid landscapes and should involve local stakeholders: 'forested landscapes in which a mosaic of wild and human-influenced vegetation types co-exist ... Such a forested landscape approach offers options for involvement of various stakeholder groups in biodiversity conservation' (Wiersum, 2003: 38). A similar argument can be found in Wiersum *et al.* (2007: 1) in which it is argued that 'Efforts should be made to combine forest conservation and livelihood improvement strategies ... this can be accomplished by building on local forest cultures and management arrangements, as well as on local understanding of the relations between increased production and forest maintenance, coupled with locally-trusted arrangements for fair and equitable access to NTFPs.' Thus, Wiersum is arguing for new types of governance and management strategies that involved local stakeholders. This desired contribution to governance and management is facilitated by the particular shape that Wiersum's contributions take: classification systems.

14.4 Classifications in research on biocultural diversity

The works of Wiersum discussed in the previous section focus on the extent to which biocultural diversity is shaped by a multitude of factors, including institutions, livelihood and forest management practices, and cultural values and beliefs. Biocultural diversity thus emerges as a complex affair, never stable and continuously changing. Contingencies and chaos abounds, so it seems. Although such accounts enrich and complement existing, overly simplistic accounts of nature and biodiversity as purely biological affairs, they do pose a problem because such diversity and complexity is not easy to manage. Order must therefore be brought to the existing chaos. And this is where Wiersum's specific contribution to our understanding of human-non-human relationships comes in.

A considerable portion of Wiersum's oeuvre is dedicated to discussing existing classifications and designing new ones. For example, Wiersum (1997) discussed the different classifications of indigenous forest management that can be identified in the scientific literature and found each of them to be flawed, because they assume rigid dichotomies between natural and cultural or anthropogenic areas. He proposed his own classification of indigenous forest management, which consists of four types: protected native forests, resource-enriched native forests, reconstructed native forests and mixed arboriculture. He argued that the classification offered 'provides a first approximation of the varied and often complex ways in which different local communities have actively engaged in maintaining the forest vegetation and manipulating it to suit their need' (Wiersum, 1997: 5). Wiersum also introduced a diagram which represents different stages in forest management and includes the relationships between the classifications and their different elements (Table 14.1). These are meant to communicate the diversity, dynamics and complexities of forest areas, including forest gardens, in an orderly fashion. Clearly, the objective of this endeavour is to reduce complexity and bring order to chaos.

However, the classification proves insufficient to accommodate the full variety of the nature-culture continuum. In a later article on biocultural diversity (Wiersum,

Table 14.1. Different stages in forest management, including the relationships between the classifications and their different elements (Wiersum, 1997).

forest conservation

modification of forest

forest transformation

increasing manipulation of the forest ecosystem

Uncontrolled procurement of forest products	**forest/tree crop management**		
	protection of forest resources	**cultivation of wild trees**	**cultivation of domesticated trees**
casual/ opportunistic gathering/ collection of wild forest products	protection of patches of forest against collection of products		
	controlled gathering/collection of forest products	enhancement of (re) procuctive potential of valued species through ecosystem manipulation purposeful dispersal of seeds/seedlings of wild species in natural forests purposeful incorporation of valuable species in fallow fields	cultivation of exotic or local cultivars in forest environment
		purposeful regeneration of wild species on cleared fields: • homegardens • agricultural fields	plantations of selected & improved cultivars: • timber plantations • fruit orchards • tree crop plantations • growing of domesticated trees on croplands

2004), he came back to this classification and elaborated it further. He discussed the characteristics and future potentials of forest gardens as an 'intermediate' land-use system in the nature-culture continuum. The article distinguishes five types and fourteen subtypes of indigenous forest management practices (Table 14.2).

The article also presents several other classifications, including different types of forest: 'a differentiation is made between old-growth, or virgin forests, and secondary forests, the latter being those that develop after clearing of the original forest. In addition, a third type of forests may be distinguished, i.e. altered forests, in which man has exercised an appreciable influence on forest composition, but many elements of the original forest are maintained' (Wiersum, 2004: 124). He

Table 14.2. Different classifications of indigenous forest management practices (Wiersum, 2004: 125).

1. Control of access to resources by definition of legitimate user groups
2. Maintenance of the resource through controlled utilisation and protection
- only certain species harvested according to stand composition, e.g. size, stand structure, etc.
- rotational harvesting regimes
- using harvesting techniques that do not cause tree mortality, e.g. limited harvesting, coppicing/pollarding/lopping
- control of pests and diseases, e.g. sanitary pruning
- fire control practices
3. Stimulation of the production of required products within existing vegetation
- selecting coppice shoots, rejuvenation pruning, ringing trees to stimulate fruiting
- decreasing water/nutrient/light competition for trees by weeding and thinning non-valuable species
- optimisation of soil conditions (e.g. mulching) to favour desired species
4. Stimulating regeneration of valued species
- protection of natural regeneration
- stimulating root sprouting
- planting of cuttings
- transplanting of seedlings obtained from natural forests or plantations
- incidental or purposeful seeding
5. Cultivation of genetically improved varieties
- selection of (high-yielding) cultivars

then identified two types and five subtypes of intermediate forests (Table 14.3). The types of intermediate forest are distinguished on the basis of two variables, the change in structure and composition of forest vegetation.

In his more recent work, classifications continue to feature prominently. Wiersum and Shrestha (2010), for example, offer a classification of different types of communities according to their dependence on the forest, ranging from communities that are totally dependent on the forest to communities that recently started to practise agriculture and communities that fully engage in agriculture. Subsequently, they identified two kinds of forest needs these communities can have: local forest-related needs and the production of forest products for national and international markets. This overview of the diversity in forest needs then enabled them to argue that the criteria and indicators used in forest certification schemes need to adequately reflect this diversity.

The categorisation of biodiversity in the work of Wiersum produced order in the chaos of the continuously changing and never stable patchwork of human-forest interactions. This, in turn, enabled the development of appropriate governance and management strategies.

Table 14.3. Major types of 'intermediate' forest types (Wiersum, 2004: 126).

A. (Modified) forests with prevalent tolerant forest management practices
Natural forests modified by gathering of forest products and selected protection of specific resources: specific areas in natural forests which are favoured and protected because of their value for extraction of useful materials. Examples: forest patches with individually claimed trees
Resource-enriched natural forests: natural forests, either old-growth or fallow vegetations, whose composition has been altered by selective protection and incidental or purposeful propagule dispersion of food and/or commercial species. Examples: enriched natural forests, enriched fallows
B. Transformed forests with prevalent intrusive forest management practices
Reconstructed natural forests: (semi-)cultivated forest stands with several planted useful species, tolerated or encouraged wild species of lesser value and non-tree plants (herbs, lianas) composed of mainly wild species. Example: forest gardens
Mixed arboriculture: cultivated mixed stands, almost exclusively of planted, and often domesticated, tree species. Examples: homegardens, mixed smallholder plantations
Interstitial trees on croplands: either naturally regenerated and protected trees, or planted and sometimes domesticated trees scattered over agricultural fields. Examples: protected trees on swidden fields, scattered fruit trees cultivation on/along crop fields

14.5 The (onto)politics of classifying biocultural diversity

There is a particular logic and strategy to Wiersum's work on biocultural diversity. First, he sets out to further complicate the notion of biodiversity by adding human and cultural dimensions and proposing a nature-culture continuum rather than a dichotomy. In doing so, Wiersum seeks to gain attention and appreciation for complexity and the important role of humans and cultural values. However, his commitments reach far beyond mere recognition of existence, but extend to questions of governance, management, coordination and control. Not only do these human and cultural values need to be recognised, they also need to be incorporated into policy and management in order to protect biodiversity and secure livelihoods. Thus, his framing of forests as (partly) cultural artefacts, underpinned by the notion of biocultural diversity, legitimises human use and promotes the role of local people in management activities.

His focus on classifications fits in with this strategy in two ways. First, adding human and cultural categories into existing classifications of biodiversity reinforces the existence of the former as entities in the world. Second, making these ontological claims in the form of categories and classifications enables human and cultural dimensions not only to be counted, but also to be taken into account. By offering classifications, Wiersum imposes order onto otherwise chaotic human-non-human relationships. He reduces the complexities of biocultural diversity and makes its biological, human and cultural categories amenable to management and control. This, in a nutshell, is the onto-politics that Wiersum practices.

Wiersum's work on biocultural diversity is used in many fields. In environmental sciences, Wiersum's work is cited, for example, by Obiri *et al.* (2002) in an article on the ecological status of the standing crop of high-value tree species in coastal scarp forest of the Eastern Cape Province of South Africa. It is also cited by Correia *et al.* (2010) in an article that compares the tree structure and diversity of agroforests with neighbouring natural forest to assess the role they play in biodiversity conservation. In the social sciences, Wiersum's work is cited, for example, by Pretty *et al.* (2009) in a discussion on how nature and culture intersect. It is also cited by Ros-Tonen and Dietz (2005) in a book that discusses the relationship between sustainable forest management and local livelihoods. But Wiersum's work is not only cited by scientists, it is also used by professionals working in the field of forest and landscape conservation and restoration. Wiersum has cooperated closely with organisations such as Tropenbos, the European Tropical Forest Research Network (ETFRN), The Consultative Group on International Agricultural Research (CGIAR) and the Center for International Forestry Research (CIFOR). This is illustrated by his presentations at their workshops, his contributions to their publications and his participation in their projects. Some of his work can be downloaded from the websites of CGIAR or CIFOR. Furthermore, his work is regularly mentioned in the 'publications of interest' section in FAO information bulletins, for example in the March 2001 FAO information bulletin on Non-Wood Forest Products and the July 2007 issue of FAO Non-Wood News. This demonstrates Wiersum's commitment to speak to a variety of scientific and non-scientific audiences.

When looking more closely at what these different communities do with Wiersum's work, it seems that the classification schemes in Wiersum's work are more or less accepted. No-one seems to question whether or not forest types and/or indigenous forest management practices can or should be put into categories to make biocultural diversity legible. Instead, particular categories or interpretations (rather than the classification scheme as a whole) have become targets for debate (see, for example, De Jong *et al.*, 2006; Kull *et al.*, 2011; Lewis, 2005). So, even people who disagree on theoretical and practical matters use his work in a similar fashion, which shows the value of classifications as boundary objects (Starr and Griesemer, 1989; Turnhout, 2009). By using classification systems, Wiersum spans the boundaries of natural scientists, social scientists and professional practitioners to create a network of people with different perspectives who might debate the biological, human and cultural categories of the classification systems, but who nonetheless share a general commitment to group them together. The classifications in effect form relationships and allow durable cooperation across different social worlds. They provide an infrastructure for ongoing network construction.

As successful as this strategy might be, we do see some of the problematic bootstrapping effects of classification systems as well. Inevitably, classification systems and the categories they include produce boundary problems when classifiers find themselves confronted with things that do not fit. Wiersum's response was not to silence these hybrids by forcing them into existing categories or denying their existence or relevance. Instead, he replaced his 1997 classification of four categories with three new ones, one of which consists of no less than 14 subcategories (Wiersum, 2004). Obviously, this cannot go on endlessly. The increasing complexity that comes with ever more sophisticated classifications will inevitably stand in the way of orderly management and control, thereby creating the need to rethink the categories and reclassify them in order to the reflect the differences that – in a certain context at a certain time – make a difference. Classification systems do not mirror reality, nor do they intend to. They only make sense in the light of the social and political values and desires they intend to highlight and seek recognition for. These political aspects of Wiersum's work do not form a threat to the scientific value of his classificatory systems; on the contrary, without them, their meaning and scientific value is lost as well.

14.6 Performativity: the social and material implications of classifying biocultural diversity

The recognition of the politics of classification changes the debate about classification systems and categories. Instead of focusing on problems of representation – and the endless questions of whether the classification system adequately reflects reality, whether the right categories are distinguished, and whether the right boundaries are drawn – we need to focus on the performativity of classification systems. This means taking a closer look at what these classifications do in practice; how they influence not only how the world is known, but also how it is acted upon, and how social and material relationships are remade in the process.

In Wiersum's case, we can deduce three main interrelated, performative effects of classifications of biocultural diversity. First, they promote specific conservation strategies. By adding human and cultural dimensions to our understanding of biodiversity, the notion of biocultural diversity highlights the role of local/indigenous communities, knowledge and practices. Second, they generate and reinforce difference. The concept of biocultural diversity promotes a symmetrical approach to human-non-human relationships by ordering biological, human and cultural phenomena under a single umbrella. However, it does not collapse the differences between nature and culture. The nature-culture continuum positions ideal typical natural and cultural landscapes at its two extremes and classifies the various hybrids according to their degree of naturalness or 'culturalness'. By creating distance between pure nature and pure culture, the continuum provides the necessary space for the various nature/culture hybrids in the middle and allows for coordination across the nature-culture divide. Indeed, the creation of difference goes hand in hand with the creation of possibilities for coordination. Third, they create a network of natural and social scientists and practitioners who can arrive at a division of labour (along traditional, professional and disciplinary lines) in the research into and management of the biological, cultural and hybrid categories that are distinguished.

To conclude, by performing a world that consists of various natural, cultural and mixed categories, Wiersum's work privileges local/indigenous communities to manage the nature/culture mixtures, creates a nature-culture continuum to allow for coordination across the nature-human divide, and creates a network of scientists and practitioners from diverse disciplines that can arrive at a division of labour in the research into and management of the various biological, human and cultural categories.

It needs to be noted that although Wiersum's criticisms of the dominant paradigm within current conservation and natural resource management, which focus on fencing off areas – and particularly the dichotomy between humans and nature it is based on – are quite radical, the strategy he pursues is thoroughly pragmatic. By designing new classification systems that include hybrid categories, he tinkers with, but does not fundamentally problematise or overthrow, the dominant paradigm, and leaves professional and disciplinary boundaries more or less intact. This pragmatic strategy has proven essential in pursuing interdisciplinary and professional cooperation and advocacy for the role of local/indigenous people in conservation and management practices.

As we have demonstrated in this chapter, classification systems, although often portrayed as neutral tools of description and representation, reflect specific values and desires and can only be understood in relation to these values and desires. However, when classification systems become naturalised and taken for granted, the political project they pursue disappears from view, particularly if presented in scientific publications. Critical engagement with the performative effects of these tools of representation – including not only classification systems but also models, scientific methods and public participation techniques (Callon, 1998; Law, 2009; Turnhout and Boonman-Berson, 2010; Van Egmond and Zeiss, 2010) – is important for highlighting the politics they practice. For the case of biocultural diversity, this involves asking questions such as: What do these tools make biocultural diversity to be? What is being left out? Whose interests do they serve? Which actors and perspectives do they subjugate? Subsequently, if these tools are recognised as ways to practice politics, attention should also be focused on ensuring relations of accountability between those who are being represented and those who are doing the representing.

References

Bowker, G.C. (2000). Biodiversity datadiversity. Social Studies of Science 30(5): 643-6831.

Bowker, G.C. and Starr, S.L. (2000). Sorting things out: classification and its consequences. The MIT Press, Cambridge, MA, USA.

Callon, M. (1998). Introduction. The embeddedness of economic markets in economics. In: Callon, M. (ed.) The laws of the markets. Blackwell, Oxford, UK, pp. 1-57.

Correia, M., Diabate, M., Béavogui, P., Guilavogui, K., Lamanda, N. and De Foresta, H. (2010). Conserving forest tree diversity in Guinée Forestière (Guinea, West Africa): the role of coffee-based agroforests. Biodiversity and Conservation 19(6): 1725-17471.

De Jong, W., Ruiz, S. and Becker, M. (2006). Conflicts and communal forest management in northern Bolivia. Forest Policy Economics 8: 447-4571.

Gould, S.J. (1990). Wonderful life: the Burgess Shale and the nature of history. Hutchinson Radius, London, UK.

Halffman, W. (2003). Boundaries of regulatory science. Albatros, Boechout, the Netherlands.

Harmon, D. (1996). Losing species, losing languages: connections between biological and linguistic diversity. Southwest Journal of Linguistics 15: 89-108.

Jones, R. (2009). Categories, borders and boundaries. Progress in Human Geography 33: 174-189.

Keshet, Y. (2011). Classification systems in the light of sociology of knowledge. Journal of Documentation 67(1): 144-158.

Kull, C.A., Shackleton, C.M., Cunningham, P.S., Ducatillon, C., Dufour-Dror, J.-M., Esler, K.J., Friday, J.B., Gouveia, A.C., Griffin, A.R., Marchante, E.M., Midgley, S.J., Pauchard, A., Rangan, H., Richardson, D.M., Rinaudo, T., Tassin, J., Urgenson, L.S., Von Maltitz, G.P., Zenni, R.D. and Zylstra, M.J. (2011). Adoption, use, and perception of Australian acacias around the world. Diversity and Distributions 17: 822-836.

Latour, B. (1993). We have never been modern. Harvard University Press, Cambridge, MA, USA.

Law, J. (2008). On sociology and STS. The Sociological Review 56: 623-649.

Lewis, J. (2005). Challenges of interdisciplinarity for forest management and landscape perception research. In: Tress, B., Tress, G. and Fry, G. (eds.) From landscape research to landscape planning: aspects of integration, education, and application. Springer, Dordrecht, the Netherlands, pp. 83-94.

Mathez-Stiefel, S.L., Boillat, S. and Rist, S. (2007). Promoting the diversity of worldviews: An ontological approach to biocultural diversity. In: Haverkort, B. and Rist, S. (eds.) Endogenous development and biocultural diversity: the interplay of worldviews, globalization and locality. COMPAS series on Worldviews and Sciences Nr. 6, COMPAS and CDE, Leusden, the Netherlands, pp. 67-81.

Mol, A. (1999). Ontological politics: a word and some questions. In: Law, J. and Hassard, J. (eds.) Actor network theory and after. Blackwell, Oxford, UK, pp. 74-89.

Nabhan, G.P. (1997). Cultures of habitat: on culture, nature, and story. Counterpoint, Washington, DC, USA.

Newing, H. (2010). Bridging the gap: interdisciplinarity, biocultural diversity and conservation. In: Pilgrim, S. and Pretty, J. (eds.) Nature and culture: rebuilding lost connections. Earthscan, London, UK, pp. 23-40.

Nietschmann, B.Q. (1992). The interdependence of biological and cultural diversity. Occasional Paper 21, Center for World Indigenous Studies, Olympia, WA, USA.

Obiri, J., Lawes, M. and Mukolwe, M. (2002). The dynamics and sustainable use of high-value tree species of the coastal Pondoland forests of the Eastern Cape Province, South Africa. Forest Ecology and Management 166: 131-148.

Pretty, J., Adams, B., Berkes, F., Ferreira de Athayde, S., Dudley, N., Hunn, E., Maffi, L., Milton, K., Rapport, D., Robbins, P., Sterling, E., Stolton, S., Tsing, A., Vintinnerk, E. and Pilgrim, S. (2009). The intersection of biological diversity and cultural diversity: Towards integration. Conservation and Society 7: 100-112.

Ritvo, H. (1997). The platypus and the mermaid, and other figments of the classifying imagination. Harvard University Press, Cambridge, MA, USA.

Ros-Tonen, M.A.F. and Dietz, T. (2005). African forests between nature and livelihood resources: interdisciplinary studies in conservation and forest management. The Edwin Mellen Press, Lewiston, NY, USA.

Scott, J. (1998). Seeing like a state: how certain schemes to improve the human condition have failed. Yale University Press, New Haven, NY, USA.

Shapin, S. and Schaffer, S. (1985). Leviathan and the air pump: Hobbes, Boyle and the experimental life. Princeton University Press, Princeton, NJ, USA.

Starr, S. (2002). Got infrastructure? How standards, categories and other aspects of infrastructure influence communication. The 2nd Social Study of IT, LSE workshop on ICT and Globalization, 22-23 April. Available at: http://www.lse.ac.uk/collections/informationSystems/newsAndEvents/2002events/ICTAndGlobalisation.htm.

Starr S.L. and Griesemer, J.R. (1989). Institutional ecology, 'translations' and boundary objects: amateurs and professionals in Berkeley's Museum of Vertebrate Zoology, 1907-39. Social Studies of Science 19(3): 387-420.

Stevens, S. (1997). Conservation through cultural survival: indigenous peoples and protected areas. Island Press, Washington, DC, USA.

Turnhout, E. (2009). The effectiveness of boundary objects: the case of ecological indicators. Science and Public Policy 36: 403-412.

Turnhout, E. and Boonman-Berson, S. (2011). Databases, scaling practices, and the globalization of biodiversity. Ecology and Society 16(1): 35. Available at: http://www.ecologyandsociety.org/vol16/iss1/art35/.

Van Egmond, S. and Zeiss, R. (2010). Modeling for Policy – Science-based models as performative boundary objects for Dutch policy making. Science Studies 23(1): 58-78.

Wiersum, K.F. (1990). International experiences in social forestry. Social forestry in Indonesia. FAO/RAPA, Bangkok, Thailand, pp. 135-143.

Wiersum, K.F. (1991). A comparison of project approaches. In: Savenije, H. and Huijsman, A. (eds.) Making haste slowly. Strengthening local environmental management in agricultural development. Development oriented research in agriculture Vol. 2, Royal Tropical Institute, Amsterdam, the Netherlands, pp. 103-114.

Wiersum, K.F. (1993). La foresterie rurale au Sénégal. Participation villageoise et gestion locale. In: Van den Breemer, J.P.M. (ed.) Leiden development studies 12, pp. 135-159.

Wiersum, K.F. (1997). Indigenous exploitation and management of tropical forest resources: an evolutionary continuum in forest-people interactions. Agriculture, Ecosystems and Environment 63: 1-16.

Wiersum, K.F. (2003). Use and conservation of biodiversity in East African forested landscapes. In: Tropical forests in multi-functional landscapes. Seminar proceedings, Utrecht University and Prince Bernhard Centre, Utrecht, the Netherlands, pp.33-39. Available at: http://www.bio.uu.nl/pbc/publications/proceedings/Wiersum.pdf.

Wiersum, K.F. (2004) Forest gardens as an 'intermediate' land-use system in the nature-culture continuum: characteristics and future potential. Agroforestry Systems 61(1): 123-134.

Wiersum, K.F. and Shrestha, K. (2010). Biocultural diversity in community forestry in Nepal. ETFRN News 51: 20-24.

Wiersum, F., Bognetteau, E. and Haile, A. (2007). Integrating forest conservation and livelihood improvement in South-West Ethiopia. Policy Briefing Note 2, NTFP South-West Ethiopia Research & Development Project, Mizan Teferi, Ethiopia.

Part IV. Forest governance

15. Forest governance: a state of the art review

Bas Arts and Ingrid Visseren-Hamakers
Wageningen University, Forest and Nature Conservation Policy Group, Droevendaalsesteeg 3, 6708 PB Wageningen, the Netherlands; bas.arts@wur.nl

Abstract

Forest governance refers to new modes of governing that go beyond the confines of the state. These modes are believed to be more (cost-)effective and legitimate than conventional steering by governments. Over the years, a large body of knowledge has emerged on this topic, which makes a review of the state of the art in forest governance both timely and useful. In parallel, a critical literature has developed that warns against too optimist, naïve and technocratic accounts of forest governance. Examples are critical governance studies, critical institutionalism, political ecology and governmentality studies. This chapter addresses these critical studies and proposes the 'Triple G' framework – government, governance and governmentality – as a way of analysing the forest policy field in greater depth.

Keywords: forest governance, forest government, forest governmentality, Triple G

15.1 Introduction

Forest governance has become a new 'growth industry' in forest sciences and practices. In November 2011 a Google search for 'forest governance' produced about 565,000 hits, Google Scholar about 3,800 hits, Scopus about 130 hits and ISI Web of Sciences about 100 hits. These numbers have doubled in only five years or so (Breeman *et al.*, 2009). In our view, this rapid growth in the literature justifies a review of the state of the art on this new topic. This review will be 'positive, yet critical', but far from complete given the size of the literature and the need to focus. Generally, the governance concept refers to new modes of forest governance that go beyond the confines of the state, such as policy networks, certification schemes, social corporate responsibility, public participation, community forestry, markets for ecosystem services, and (public-)private partnerships. These are believed to be more capable of managing public 'goods and bads' related to forests than conventional government. However, over the years a critical literature on forest governance has developed that puts this optimism into perspective. Some criticise the institutional bias of the governance literature, others the naïve belief in managerialism and technocracy, and still others the lack of addressing power and domination. As alternatives, theories like critical governance, critical institutionalism, political ecology and governmentality have been applied to the forest policy field.

This chapter briefly introduces these accounts and then examines governmentality in more depth, analysing the forest policy field from the perspective of the 'Triple G' framework: government, governance and governmentality. The structure of the chapter also follows this logic, with sections on governance in general (which also covers the relationship between government and governance), forest governance, forest governmentality and Triple G.

15.2 Governance

The key buzzword in political sciences, public administration and management sciences for the last two decades has been *governance* (Bevir, 2010; Held and McGrew, 2002; Hooghe and Marks, 2001; Pierre, 2000; Pierre and Peters, 2000). Etymologically, the concept can be traced back to the Greek verb *kubernan*, which means 'to pilot' or 'to steer' (Kjaer, 2004). In its broadest interpretation, governance is defined by us as:

> The many ways in which public and private actors from the state, market and/or civil society govern public issues at multiple scales, autonomously or in mutual interaction.

This definition includes both conventional government and innovative governance arrangements beyond the state. In this broad interpretation, governance is about governing *by*, governing *with* and governing *without* the state. Consequently, the concept acknowledges the actual and potential role of both public and private actors in providing public 'goods' – such as welfare, health, environment – and in managing or solving public 'bads' – such as poverty, disease, pollution – across multiple scales of governance. Generally, the administrative scale – from the municipality to the nation state to the United Nations (UN) – is of the utmost importance for governing public issues. Recently though, various new governing mechanisms have developed beyond this scale, from local to global political spaces. Examples are 'glocalised' networks of NGOs around the world (Arts, 2004) as well as many partnerships among public and private actors that operate between the local and the global level (Visseren-Hamakers, 2009). To 'govern' can of course mean many things, including disciplining others, persuading others or exchanging resources with others to let them do things in accordance with certain policy objectives (Dunn, 2003). All are implied in modern or postmodern governance, often in hybrid combinations.

The broad definition of governance as detailed above is often confused with a stricter interpretation, which is dominant in the literature and which will be followed in the section on forest governance below. Here, the term refers to a 'paradigm shift' in the way in which we govern modern/postmodern societies and organisations today (Arnouts, 2010; Breeman *et al.*, 2009; Pierre and Peters, 2000). According to this interpretation, the old paradigm of top-down, state-led, command

and control ways of steering has lost its legitimacy. The 'big' government of the welfare state lost credibility during the economic crisis of the 1980s, while its bureaucracy and elitism raised questions about efficiency and democracy (Pierre, 2000). As a consequence, public administrations became subject to intense reform programmes, like those of the new public management movement, and new modes of governance emerged as alternatives for managing public affairs (Kjaer, 2004). Examples are network-like arrangements, self-regulation by businesses, (public-) private partnerships, emission trading schemes and certification programmes (Agrawal *et al.*, 2008; Kickert *et al.*, 1997; Kolk). Some refer to this development as a 'shift from government to governance' (Rosenau and Czempiel, 1992), implying that authority and competencies have moved away from the state to other bodies, like international organisations, NGOs and businesses (Pierre and Peters, 2000). This 'shift' literature started the debate on whether the state is 'losing' power while others are gaining it, as if we are dealing with a zero-sum game (Pierre, 2000; Skjaer, 2004). In the 1990s some theorists even claimed the 'retreat' or the 'death' of the state (Albrow, 1996; Strange, 1996). Today, however, most analysts agree that the state has remained strong, given its legally-based authority and its numerous political-economic and military resources, although it has been transformed under influence of the governance shifts and the various rounds of public administration reforms since the 1980s (Arts *et al.*, 2009; Kjaer, 2004; Pierre and Peters, 2000).

Table 15.1 gives an overview of the various conceptualisations of governance in the literature. 'Old' or *state*-centric governance refers to traditional state steering, 'new' or *society*-centric governance to innovative modes of governance, and 'all' governance refers to both these categories (Knill and Lehmkuhl, 2002; Pierre, 2000). One could claim that these three conceptualisations are all analytical in nature; they just

Table 15.1. Four conceptualisations of governance.

Conceptualisation	General definition	Global dimension	Local dimension
'Old' governance	state steering (top-down, command and control)	intergovernmental arrangements (international diplomacy, organisation and law)	local authorities (provinces, municipalities)
'New' governance	new modes of governance (from partnerships to self-regulation)	new modes of global governance (from issue networks to private standardisation)	new modes of local governance (e.g. public participation)
'All' governance	all mechanisms to govern public issues	all mechanisms to govern global issues	all mechanisms to govern local issues
'Good' governance	reform of public administration (new public management)	reform of third world government (World Bank, IMF)	reform of local politics (e.g. decentralisation)

try to characterise the various forms and histories of conventional and innovative governing mechanisms in modern/postmodern societies. Normatively, however, one may add a fourth category, namely 'good governance', which is the advocacy of reform of the public sector and/or of corporate management in accordance with a number of good governance criteria, such as cost-effectiveness, transparency, accountability and participation, among others advocated by the European Union (EU), the International Monetary Fund (IMF) and World Bank (Kjaer, 2004; Woods, 2000). Examples of 'good governance' programmes are new public management (NPM) which applies business principles to public administration for improved cost-effectiveness, and good corporate governance (GCG), which applies principles of government to business practices for improved accountability.

Table 15.1 also identifies the global dimension of governance, since this concept has become an important topic in the study of international relations (Held and McGrew, 2002; Nye and Donahue, 2000). In this literature it is generally referred to as 'global governance', but it has a slightly different meaning in this domain and has provoked some other debates, simply because the international system lacks a central authority (Rosenau and Czempiel, 1992). Yet some theorists make similar distinctions as in the above, because we observe comparable innovations in policymaking and public steering at the global level as domestically. Therefore, Heritier (2001) distinguishes between two meanings of global governance: a 'restricted' meaning, referring to only *new* governance mechanisms (such as global issue networks of public and private actors) and a 'broad' conceptualisation that refers to *all* modes of public and private coordination in the global arena. The international relations literature also discusses 'good governance'. Under this label, the World Bank and the IMF, for example, operate programmes to promote democratic, effective and transparent governments in developing countries (Woods, 2000). Finally, another part of the global governance literature discusses reforming the governance structure of the UN system, focused on improving the intergovernmental system (Biermann *et al.*, 2009).

Besides global governance, Table 15.1 also includes local governance, the subnational dimension of the governance spectrum. Here again we can identify the four types of governance (old, new, all, good). Local governance often implies two phenomena: decentralisation on the one hand and public participation on the other (Van den Arend and Behagel, 2011; Ribot *et al.*, 2006). The former refers to the transfer of policy-making authority from central to lower administrative tiers, the latter to empowerment of the public in local policy and decision-making beyond regular democratic elections. Both trends are believed to increase the effectiveness and legitimacy of local policies as well as to reduce the assumed gap between politics and citizens.

If we take Table 15.1 as a whole, we can also speak of multilevel governance (MLG). Originally, this concept was introduced in the realm of European studies to refer to the multilevel character of EU policy-making, consisting of European institutions on the one hand and Member States on the other (Hooghe and Marks, 2001). Later, the MLG concept was also used in other fields of subnational, national and global policy analysis to argue that the old distinction between domestic and international politics has become blurred, and hence outdated (Held and McGrew, 2002). Also, the original government-centred approach to MLG by Hooghe and Marks has been broadened to include non-state actors as well. After all, local authorities and NGOs are believed to affect global and European politics, whereas global agreements and European directives are conceptualised as having direct impacts on local practices (Arts *et al.*, 2009).

Of course, the shift from government to (multilevel) governance does not take place in a vacuum; it is rooted in a number of trends. Pierre and Peters (2000) distinguish three types of displacement of state power and control: *upward* to international organisations, *downward* to subnational authorities and *outward* to semi-public and private bodies. In other words, governance is rooted in the globalisation, decentralisation and privatisation of public issues and political spaces (Van Tatenhove *et al.*, 2000). These trends are both planned and autonomous. To take the example of globalisation: international organisations gain more authority, not only because the economy is globalising or because many environmental issues are transboundary in nature, which necessitates regulation beyond the boundaries of the nation state, but also because some politicians and bureaucrats actively advocate the strengthening of international organisations, as in the case of the EU. Hence, political globalisation is both planned (advocated) and autonomous (following socioeconomic and environmental trends).

15.3 Forest governance

Forests have been considered public goods – and their related problems 'public bads' – for a long time. As early as the Middle Ages, in many regions in Europe deforestation and forest degradation led to the introduction of game reserves for the nobility, the protection of forests on steep slopes to combat erosion, and active management of timber resources for the construction of naval ships (Umans, 1993). Later, after the onset of industrialisation in Europe, many national forest laws were introduced in the 19th century as a response to the rapid depletion and degradation of forest resources and as a means for gaining state revenues from forests through taxation and public land ownership (Scott, 1998). These laws required the use of (elements of) scientific forestry, with its sustainable yield principle and silvicultural methods and tools (monocultures of commercial species, forest zoning, rotational harvesting, replanting, afforestation, etc.) (Jeanrenaud, 2001; Wiersum, 1995). For example, Belgium introduced a forest law in 1847 which provided for government subsidies to forest owners – both communities and individuals – in order to stimulate tree planting. The Netherlands introduced its first forest law in 1922, which in principle required that each tree that was felled had to be replaced by the planting of a new one. Also, many of these laws were meant to regulate forest ownership and forest-related conflicts (Krott, 2005). Often, ownership resided with the state, although many countries accepted private ownership of forestlands as well. However, the old 'forest commons', with their customary rights and responsibilities for rural communities, which had existed for centuries in many European regions, started – or were forced – to disappear due to the modernisation process, although many farmers openly resisted the new laws and policies that denied their traditional access and use rights of the forest resources (Jeanrenaud, 2001). While scientific forestry brought enormous gains in terms of increased timber production and employment, both ecological and social issues related to forests were marginalised for too long in this paradigm, which forced the forest sector to change (Wiersum, 1995). Part of this change has been a shift from forest government to forest governance (Agrawal *et al.*, 2008).

The European system of forestry and forest policy was exported to most of the European colonies (Peluso, 1994; Scott, 1998), where a strong symbiosis of state forestry and scientific forestry emerged, for example in Indonesia, Ghana and India. Forest reserves were issued and declared state property, particularly where the rich forests and valuable timber were located, and plantation economies and concession systems were introduced. These systems, which generally (and often violently) excluded local people from their lands and suppressed traditional forest institutions, have remained in place in many countries in the post-colonial era. However, although built upon modern ideas of scientific forestry, most of these plantation and concession systems have brought neither sustainable forestry nor justice for forest-dependent people (Bose *et al.*, 2011). Deforestation and forest degradation have continued in many regions around the world (up to 13 million hectares on an annual basis; FAO, 2010), particularly in the global South, while forest-related land and resource conflicts have continued (Marfo, 2006). In many cases, colonial and post-colonial states proved to be bad managers of the forests, in various ways: (1) by exploiting the resource itself to the extreme, often in conflict with local livelihoods and their own conservation objectives; (2) by issuing concessions to private companies or public enterprises without any effective monitoring mechanism in place; and (3) by being absent as a manager, leaving the forests open to, often illegal, local use. This situation led to protests by NGOs, which claimed the need for forest conservation, opposition by grass-roots movements, which fought for local land and forest rights, and to pressure by international donors, who advocated sustainable forest management (Agrawal *et al.*, 2008). For all these reasons, many developing countries around the world have recently reformed public forest policy and law, a process which is called 'forest governance'.

One dominant feature of forest governance in the tropics has been participatory forest management, or PFM (an umbrella term for initiatives like community forestry, community-based forest management, collaborative forest management and joint forest management). The central idea behind PFM is that local management of forests, either by communities themselves or jointly with regional forest departments, can be as – or even more – (cost)effective in conserving and using forest resources than the central state institutions. India, Nepal, Mexico, Bolivia, Kenya and Tanzania have pioneered different forms of PFM from the early 1990s onwards. Many countries, from Ethiopia to Albania, followed later. However, the results of PFM have so far been generally reported as 'mixed' (Blomley *et al.*, 2008; Charnley and Poe, 2007; Dressler *et al.*, in press; Mustalahti and Lund, 2010). Where success is reported, this usually relates to the forest condition rather than to enhancing local livelihoods or empowering local people, including women. Also, PFM has been subject to serious power struggles. Elite capture in village forest committees has been frequently reported, as well as conflicts between forest officials and communities over valuable timber resources, land rights, monitoring and the like. Moreover, there has been little 'real' transfer of management and use rights and decision-making power from the state to local communities. Finally, PFM initially focused too much on small-scale, local development and therefore depended heavily on the support of NGOs and donor money. Mechanisms such as community enterprises, access to national and global markets, and certification schemes for guaranteeing quality and sustainability have only recently been introduced in the context of PFM (Wiersum, 2009).

Although PFM has been one of the dominant dimensions of forest governance today, it is not the only one. Agrawal *et al.* (2008), for example, refer to the broader term *decentralisation* – of which PFM is part – and to *certification* of forest lands and products. (We exclude the third form of forest governance distinguished by Agrawal *et al.* (2008), namely 'forest concessions'; we do not see how this arrangement can be considered a *new* mode of forest governance; see the short description of colonial forestry in the above). Besides PFM, decentralisation involves the de-concentration of administrative competencies and/or the transfer of political authority from the central state to subnational administrations (Ribot *et al.*, 2006). The local administration gains technical competencies and/or formal authority from the central state and is to be held accountable by local communities in executing these. Decentralisation is therefore believed to bring politics closer to the people, to increase policy effectiveness and to enhance democratic checks and balances at the regional and community levels. This trend has become very influential worldwide in the forest sector too, although some question both the intentions and performance of such decentralisation programmes in forestry. Overall, decentralisation is an example of forest governance *with* or *without* the state, because it may include local forest departments, local communities, or both.

Forest certification has also become a dominant stream in current forest governance (Cashore *et al.*, 2004; Visseren-Hamakers and Glasbergen, 2007). This entails a market-based mechanism of independent labelling and monitoring that guarantees to both consumers and producers that timber products originate from sustainably managed forests. One of the first organisations in this field was the Forest Stewardship Council (FSC), established in 1993. Because this was an NGO-led initiative, with stringent requirements on sustainability, other industry initiatives followed later, the Programme for the Endorsement of Forest Certification (PEFC) being the largest today. Together, these two initiatives now cover more than 300 million hectares forests around the world and thousands of companies and products, although most of these are located and traded in the global North. Certification is a good example of forest governance *with or without* the state, since FSC is a private initiative and PEFC is a public-private one. A public aim (sustainability) is pursued through (public-)private means (market transactions). Also, governments can enhance certification, for example by subsidising certified companies, through tax exemptions for certified forest owners, or through public procurement policies. However, governments are hesitant to prefer one certification standard over another, arguing that this would generally oppose WTO rules.

The overview by Agrawal *et al.* (2008), however, does miss some relevant current forest governance initiatives. A crucial one is *payment for ecosystem services*, or PES (Constanza *et al.*, 1997; Farber *et al.*, 2002). The core idea is that forest and other ecosystems provide services to society, such as water regulation, soil protection and climate change regulation, that are currently not accounted for in the economic system or in policy. Giving the provision of these services a (shadow) price allows them to compete more equally with services that are already accounted for, such as timber production. However, critics argue that such approaches contribute to the ongoing commodification, privatisation and marketisation of nature, 'sacrificing' its intrinsic, social and cultural values 'on the altar' of the economy (Liverman, 2004). An application of PES in forest governance is REDD+ (Levin *et al.*, 2008). This acronym stands for 'Reducing emissions from deforestation and forest degradation in developing countries', while the '+' stands for the sequestration of carbon through forest conservation, sustainable management of forests and enhancement of forest carbon stocks in developing countries (UNFCCC, 2010). Since deforestation and forest degradation are said to contribute to about 15 to 20% of the climate change problem, avoiding these phenomena helps to mitigate climate change. The main idea of REDD+ is that developing countries are paid for their forest conservation and management practices by earning carbon credits and trading these on international carbon markets, while developed countries can buy these credits to implement their international commitments under the post-Kyoto Agreement, which is under negotiation. Currently, REDD+ is being piloted in a number of projects through bilateral initiatives (e.g. between Norway and Indonesia) and multilateral initiatives (through the UN and World Bank). REDD+ is a good example of market governance *with* the state. The ultimate aim is to have self-governing markets for trading credits on ecosystem services, but governments are essential for their design and functioning (being both rule makers and buyers of credits). The broader PES concept also includes examples of market governance *without* the state.

The shift from forest government to forest governance, which we have so far mainly dealt with in empirical terms, also has its origin in theoretical debates. Since Hardin's (1968) seminal work *The tragedy of the commons,* scientific discussion has continued on how to properly manage 'common pool resources' (CPRs). These resources are characterised as being *rival* (consumption of the resource by A renders consumption by B impossible) but *non-exclusive* (both A and B have access to the resource) (Holcombe, 1997). Examples are 'open access regimes' like fishery grounds, village forests or common grasslands, the latter being the example Hardin uses. He assumes that each herdsman will be inclined to optimise private gains, to continuously add livestock to the grassland, thus undermining its carrying capacity as a collective outcome in the long term - hence a tragedy of the commons is born. To overcome this tragedy, access to the CPR should be restricted, either through state regulation or through private ownership. Although Hardin refers to both options, in the 1970s and 1980s state regulation was generally considered the most promising mechanism to stop a tragedy of the commons in practice, probably because it matched the tradition of state forestry so well (see Scott, 1998).

Since then, however, Hardin's theory and assumptions have been strongly criticised. One of the most-well known critics is Nobel laureate Elinor Ostrom, who in her book *Governing the Commons* (1990) showed that local community institutions can be very successful in managing CPRs. She used many examples around the world to substantiate this claim, from grazing institutions in the Swiss Alps to the institutions of Zanjera irrigation systems in the Philippines. However, whether such local, communal institutions are robust depends on a number of characteristics, which Ostrom derived from a comparison of these successful cases, such as clearly defined boundaries of the resource, participation of resource users in decision-making, rules that fit the local conditions, monitoring of rule compliance, sanctions against non-compliance and conflict resolution mechanisms. Also, Ostrom criticised Hardin's simplistic idea of rationalism in his tragedy of the commons thesis, questioning whether choices by individuals are only based on cost-benefit calculation and utility maximisation, and whether remote herders on a common grassland, who are hardly or only indirectly affected by Western culture, would follow this rational path of decision-making. Instead, Ostrom builds upon the notions of *bounded* rationality and *institutional* choice (Ostrom, 2011). Not only is rationality bounded, because people often lack crucial information to make good decisions, it is also mediated by the rules of the game in a specific social setting. This implies that the choices made by individuals are deeply influenced by institutions, hence by rules, norms, beliefs and values that are valid for a specific group of people. Individuals do not only choose the option that maximises their own gains, but the one that is also 'appropriate' in a given community.

Ostrom's work has been very influential in embedding most of the (forest) governance literature in neo-institutionalism (Bevir, 2010; Kjaer, 2004). As a consequence, a strong belief in institutions in general ('rules do and can guide behaviour') and in the 'right' institution to solve CPR problems in particular (the state, the market or the community) has dominated the literature on (forest) governance. Although neo-institutionalism is indeed inspiring and does offer interesting perspectives on CPR problems and solutions, it is also limited. Therefore, in the next section, we examine some of the more critical approaches to (forest) governance.

15.4 Forest governmentality

The concept of governance and its neo-institutional basis are contested. Criticism is aimed particularly at its managerial and instrumental approach, its lack of theorising about politics and power, and its optimism on institutional design and rule following (Arts *et al.*, 2009; Bevir, 2010; Hewson and Sinclair, 1999; Van den Arend and Behagel, 2011). Too easily, governance analysts believe that 'apolitical' networks and markets can do the public job, based on cooperation, trust, expertise and exchange. Also, relationships between public and private actors in different arenas and at different levels of policy-making are often considered 'free of power games'. If politics are to be located on a continuum between 'rationality' and 'power', which academics often do (Dunn, 2003), the governance literature is definitely at the 'rational side'. Nonetheless, the governance of (controversial) issues often includes interest-driven bargaining, social conflicts and power games. In addition, the question of democracy becomes utterly relevant here, because those who govern through the new governance arrangements do not necessarily have a democratic licence to do so. Finally, due to their institutionalist background, many governance theorists and practitioners have a lot of confidence in institutional design and rule following. Particularly in a context of weak and absent states, or of weak societies, this position is overly optimistic about institutional capacity (building).

As a consequence of the above criticisms, other more critical approaches have been developed. Cutler *et al.* (1999), for example, criticised the shift from government to governance literature, viewing private steering mechanisms as an erosion of public authority, or as the private capture of what should be public. Important issues for these *critical governance* authors include the lack of transparency and accountability of these new forms of governance. Other authors criticised the neo-institutional bias of much of the literature. Cleaver (2002), for example, criticises Ostrom's neo-institutionalism, showing that communities do not simply adopt externally crafted resource institutions, but mix them with – or even reject them on the basis of – socially-embedded rules, norms and beliefs (De Koning, 2011). Cleaver calls this process *institutional bricolage* (and this approach 'critical institutionalism'; see De Koning and Cleaver in Chapter 17 of this volume). This does not necessarily exclude institutional design, but criticises its unquestioned status and adds a sociocultural logic to institution building. With that, the perspective shifts from 'intentions, rules and effects' to 'intentions, rules-in-action and unintended consequences'.

A third, more radical, perspective on forest governance is *political ecology*, although the topic is discussed less in this literature than might be expected on the basis of the popularity of the governance concept. This approach – a marriage between human ecology and neo-Marxist political economy – particularly focuses on power, inequality and injustice in late capitalist and neoliberal resource management and governance (Bryant and Baily, 1997; Peluso, 1994). Unequal access to resources, elite capture in decision-making, state dominance over land and resources, exploitation of people and forests by (multinational) corporations, neoliberal reform of natural resource sectors, etc. are recurrent themes in this literature. These authors do not believe that a shift to governance will contribute to poverty alleviation or to environmental sustainability, because for them governance represents not a shift away from, but a continuation of, late capitalism and neoliberalism.

A fourth critical, but post-Marxist, approach builds upon the notion of 'governmentality' (Dean, 2010; Foucault, 2000). This concept literally refers to 'modes of thought on government': to the way we think, talk and produce knowledge about governing society and its subjects, and to the way we act upon those ideas. Questions like 'How are subjects and identities produced?', 'How are we governed?' and 'How is power exercised' are key to governmentality studies (Dean, 2010). Crucial in this approach is that government by the state and self-government by individuals are intrinsically intertwined. Through discourses, practices and technologies of governance in the family, the school, the hospital, the prison, the university, the parliament, etc., a society produces certain subjects, but these identities are self-confirmed, embraced and strategically used by citizens at the same time. So government is not so much about governing others, but about letting others govern *themselves* by creating 'responsible' subjects ('conduct of conduct'). Therefore, power is to be conceived differently than in mainstream political analysis: not as something embodied in a state, a president or a king, who give orders to docile citizens, but as disciplinary practices that are shaped both by (hegemonic) regulatory logics and (in principle) free subjects.

Two branches of governmentality studies are particularly relevant for forest governance, one on political reform and neoliberalism (Rose and Miller, 2010; Triantafillou, in press) and one on the environment (Agrawal, 2005; Bose *et al.*, in press; Peluso and Vandergeest, 2006). The first branch theorises about governance (reform) as 'control at a distance' by the state. Governance (reform) is not meant to seriously transfer authority to lower administrative tiers, but to maintain control in a different manner. This can be realised, for example, through performance measurement and reporting obligations at the level of de-central authorities vis-à-vis the central state. Governance and administrative reform also offers ample opportunities for states to decentralise their unresolved policy problems and financial deficits, while maintaining central control over valuable assets and resources. Hence, what looks like empowerment of and increased autonomy for de-central authorities at first glance, might actually turn out to be a way for the central state to dispose of its wicked problems. The second branch of literature focuses on the creation and self-confirmation of environmental subjects and of forest-related identities by states and NGOs. Agrawal (2005) speaks of 'environmentality' and Bose *et al.* (in press) of 'forest governmentality' to capture these processes. Whereas in the past traditional forest dwellers were often identified as encroachers and backward people, today they are suddenly applauded as noble savages and guardians of the forests, holding valuable indigenous knowledge on how to manage the forests more sustainably. And whereas the former legitimised the violent exclusion of such people from their forest lands, their new identities define them as interesting vehicles for implementing forest policy programmes, both by governments and NGOs. Here power is not only directly and visibly exercised by excluding people from their lands by the state or by enforcing certain norms upon them by NGOs, but by producing certain subjects and identities through societal discourses, state institutions and social practices that work indirectly, invisibly, collectively and in the *long durée*.

If we take the above critical perspectives on board, the following two propositions on forest governance seem justified:

1. Decentralisation of forest management is not so much the transfer of authority from the government to local authorities, but 'control at a distance' to solve wicked policy problems locally, which tends to reaffirm power structures instead of changing them.
2. Community participation in forest governance is not so much an increase of decision-making power for ordinary people, but the shaping of environmentally responsible subjects and the creation of mutual consent around forestry problems and objectives.

These two propositions turn many of the governance claims upside down, namely that governance implies (more) local participation, ownership and empowerment than the conventional regulatory approaches. However, this optimism remains to be seen, or worse, it may even be misplaced, because governance may work exactly the other way around than officially advocated. This should make us cautious about accepting too optimistic or too naïve governance claims.

15.5 Triple G

A governmentality perspective on forest governance differs radically from the mainstream governance literature or from the more conventional governmental policy literature. Different questions are asked, different concepts are used, different empirics are investigated and different answers are found. This probably raises the question: 'Who is right?'. However, we consider this to be the wrong question. In our view, these three perspectives may complement each other if they are considered to be different *knowledge claims* (Buizer *et al.*, 2011). The three perspectives of government, governance and governmentality may analyse similar phenomenon, like the forest policy field, but they do so in different ways, probably producing different, maybe even contradictory conclusions. A government analysis may claim that the forest law is effective in enhancing environmental sustainability in a certain region, whereas a governance perspective may claim that local stakeholders are empowered by the social networks related to the forests. A governmentality study may claim that these networks are still controlled by the state and NGOs, for example through external audits and donor money. Instead of evaluating whether or not A is right and B is wrong, we prefer the various findings to be mutually confronted as different knowledge claims. These claims may (partly) converge, strengthening each side of the story; they may overlap, pointing at a core of mutually confirmed knowledge; or they may exclude each other, each being credible in its own right because each takes a different perspective – reality is, after all, multifaceted. In any case, a much richer picture of a certain phenomenon is now produced by interpreting it from different angles and by mirroring these.

Thus, we opt for a combination of perspectives when studying the forest policy field, which we would like to refer to as 'Triple G': (forest) government, governance and governmentality. The objective is not so much to integrate these perspectives – this we consider impossible – but to see them as different knowledge claims that can converge or diverge. Of course, some scholars might object to this approach, because the three Gs originate from different and incompatible philosophies of science (positivist, interpretative, critical; see Crotty, 1998). However, according to a pragmatist approach to science, such different perspectives *can* be applied together (Maxcy, 2003). The question in pragmatism is not so much whether perspectives fully match, ontologically, epistemologically and methodologically, but whether they work or not, in isolation or in combination. We think – in line with this view – that the Triple G framework offers a more in-depth and reflexive analysis of the forest policy field than focusing on the concept of forest governance alone.

15.6 Conclusion

The immense and rapidly growing forest governance literature has stimulated us to write this state of the art review. It shows that the term has various meanings, from steering in general to new modes of governance that go beyond the confines of the state, which can be local, national, global and multilevel in nature. The most important new modes of 'forest governance' are: (1) participatory forest management; (2) decentralisation of forest administration; (3) forest certification; and (4) payment for environmental services (particularly REDD+). Theoretically, the literature has been strongly embedded in neo-institutionalism, not in the least because of the contributions to the field by Nobel laureate Elinor Ostrom. Although her work is very inspiring, it has also attracted criticism. In particular, her faith in 'optimal' institutional design and in 'automatic' rule following by actors has provoked alternatives theories in response. Examples are (1) the critical governance literature, which questions the democratic content of the governance shift; (2) critical institutionalism, which shows the many unexpected ways people can respond to newly introduced institutions; (3) political ecology, which addresses issues of power, inequality and injustice reproduced in new forest institutions; and (4) governmentality studies, which claim – amongst others – that governance should be considered as 'old wine in new bottles' namely 'control at a distance' by the central state. We were mainly inspired by the last perspective, which prevents us from taking too narrow a view of forest governance. We therefore conclude that forest governance is best addressed in accordance with the so-called Triple

G framework, referring to government, governance *and* governmentality. By mirroring the different knowledge claims of these three perspectives, too optimist and naïve accounts of forest governance can be prevented. This is essential now that the faith in the new modes of forest governance is so strong.

Acknowledgement

The authors would like to thank Mirjam Ros-Tonen and Jelle Behagel for their very valuable comments and suggestions on reading the first version of this chapter.

References

Albrow, D. (1996). The global age. State and society beyond modernity. Polity Press, Oxford, UK.

Agrawal, A. (2005). Environmentality: technologies of government and the making of subjects. Duke University Press, Durham, NC, USA.

Agrawal, A., Chhatre, A. and Hardin, R. (2008). Changing governance of the world's forests. Science 320: 1460-1462.

Arnouts, R. (2010). Regional nature governance in the Netherlands: four decades of governance modes and shifts in the Utrechtse Heuvelrug and Midden-Brabant. PhD Thesis. Wageningen University, Wageningen, the Netherlands.

Arts, B. (2004). The local-global nexus: NGOs and the articulation of scale. Journal of Economic and Social Geography 95(5): 498-511.

Arts, B., Lagendijk, A. and Van Houtum, H. (eds.), 2009. The disoriented state: shifts in governmentality, territoriality and governance. Springer, Dordrecht, the Netherlands.

Bevir, M. (2010). Democratig governance. Princeton University Press, Princeton, NJ, USA.

Biermann, F., Davies, O. and Van der Grijp, N. (2009), Environmental policy integration and the architecture of global environmental governance. International Environmental Agreements: Politics, Law and Economics 9(4): 351-369.

Blomley T., Pfliegner, K., Isango, j., Zahabu, E., Ahrends, A. and Burgess, N. (2008). Seeing the wood for the trees: an assessment of the impact of participatory forest management on forest condition in Tanzania. Oryx 42(3): 380-391

Bose, P., Arts, B. and Van Dijk, H. (in press). 'Forest governmentality': a genealogy of subject-making of forest-dependent 'scheduled tribes' in India. Land Use Policy, in press. DOI: http://dx.doi.org/10.1016/j.landusepol.2011.11.002.

Breeman, G., Goverde, H. and Termeer, K. (eds.) (2009). Governance in de groen-blauwe ruimte. Handelingsperspectieven voor landbouw, ruimte en water. Van Gorcum, Assen, the Netherlands.

Bryant, R. and Baily, S. (1997). Third world political ecology. Routledge, London, UK.

Buizer, M., Arts, B. and Kok, K. (2011). Governance, scale and the environment: the importance of recognizing knowledge claims in transdisciplinary arenas. Ecology and Society 16(1): 21. Available at: http://www.ecologyandsociety.org/vol16/iss1/art21/.

Cashore, B., Auld, G. and Newson, D. (2004). Governance through markets – forest certification and the emergence of non-state authority. Yale University Press, New Haven, NY, USA.

Charnley, S. and Poe, M. (2007). Community forestry in theory and practice: where are we now? Annual Review of Anthropology. 2007. 36: 301-336.

Cleaver, F. (2002). Reinventing institutions: bricolage and the social embeddedness of natural resource management. European Journal of Development Research 14(2): 11-30.

Constanza R., d'Arge, R., De Groot, R., Farber, S., Grasso, M., Hannon, B., Limburg, K., Naeem, S., O'Neill, R.V., Paruelo, J., Raskin, R.G., Sutton, P. and Van den Belt, M. (1987). The value of the world's ecosystem services and natural capital. Nature 387: 253-260.

Crotty, M. (1998). The foundations of social research. Meaning and perspective in the research process. SAGE, London, UK.

Cutler, A.C., Haufler, V. and Porter, T. (eds.) (1999). Private authority and international affairs. State University Press, New York, NY, USA.

De Koning, J. (2011). Reshaping institutions. Forest policy in Bolivia and Ecuador. PhD Thesis Wageningen University, Wageningen, the Netherlands.

Dean, M. (2010). Governmentality. Power and rule in modern society. Sage, London, UK.

Dressler W., Büscher, B., Schoon, M., Brockington, D., Hayes, T., Kull, C.A., McCarthy, J. and Shrestha, K. (in press). From hope to crisis and back again? A critical history of the global CBNRM narrative. Environmental Conservation, in press. DOI: http://dx.doi.org/10.1017/S0376892910000044.

Dunn, W. (2003). Public policy analysis. An introduction. Pearson Prentice Hall, Upper Saddle River, NJ, USA.

Food and Agriculture Organisation of the United nations (FAO) (2010). Global forest resources assessment 2010. FAO, Rome, Italy.

Farber, S., Constanza, R. and Wilson, M. (2002). Economic and ecological concepts for valuing ecosystem services. Ecological Economics 41: 375-392.

Foucault, M. (2000). Power. Essential work of Foucault 1954-1984. Edited by J. Faubion. Penguin Books, London, UK.

Hardin, G. (1968). The tragedy of the commons. Science 162(3859): 1243-1248.

Held, D. and McGrew, A. (2002). Governing globalization. power, authority and global governance. Polity Press, Cambridge, UK.

Héritier, A. (2001). New modes of governance in Europe: policy-making without legislating? Max-Planck-Projektgruppe Recht der Gemeinschaftsgüter, MPP Preprint 01/14, Bonn, Germany.

Hewson, M. and Sinclair, T. (eds.) (1999). Approaches to global governance theory. State University of New York Press, Albany, NY, USA.

Holcombe, R.G. (1997). A theory of the theory of public goods. Review of Austrian Economics 10(1): 1-22.

Hooghe, L. and Marks, G. (2001). Multi-level governance and European integration. Rowman and Littlefield, Lanham, MD, USA.

Jeanrenaud, S. (2001). Communities and forest management in Europe. WG-CIFM Report. Available at: http://www.iucn.org/dbtw-wpd/edocs/2001-061.pdf.

Kickert, W., Klijn, E. and Koppejan, J. (eds. (1997). Managing complex networks. Strategies for the public sector. Sage, London, UK.

Kjaer, A. (2004). Governance. Polity Press, Cambridge, UK.

Knill, C. and Lehmkuhl, D. (2002). Private actors and the state: internationalization and changing patterns of governance. Governance 15(1): 41-63.

Kolk, A. (2000). Economics of environmental management. Pearson Education Ltd, Edinburg, UK.

Kooiman, J. (1993). Modern governance. new government-society interactions. Sage, London, UK.
Krott, M. (2005). Forest policy analysis. Springer, Dordrecht, the Netherlands.
Leverman, D. (2004). Who governs, at what scale and at what price? Geography, environmental governance, and the commodification of nature. Annals of the Association of American Geographers 94(4): 734-738.
Levin, K., McDermott, C. and Cashore, B. (2008). The climate regime as global forest governance: can reduced emissions from deforestation and forest degradation (REDD) initiatives pass a 'dual effectiveness' test? International Forestry Review 10(3): 538-549.
Marfo, E. (2006). Powerful relations: the role of actor-empowerment in the management of natural resource conflicts: a case of forest conflicts in Ghana. PhD Thesis. Wageningen University, Wageningen, the Netherlands.
Maxcy, S. (2003). Pragmatic threats in mixed method research in the social sciences. The search for multiple modes of inquiry and the end of the philosophy of formalism. In: Tashakori, A. and Teddle, V. (eds.) Handbook of mixed methods in behavorial and social sciences. Sage, London, UK, pp. 51-89.
Msutalahti, I. and Lund, J. (2010). Where and how can participatory forest management succeed? Learning from Tanzania, Mozambique, and Laos. Society and Natural Resources 23: 31-44.
Nye, J. and Donahue, J. (2000). Governance in a globalizing world. Brookings Institution Press, Washington, DC, USA.
Ostrom, E. (1990). Governing the commons. The evolution of institutions for collective action. Cambridge University Press, Cambridge, UK.
Ostrom, E. (2011). Background on the institutional analysis and development framework. The Policy Studies Journal 39(1): 7-27.
Peluso, N. (1994). Rich forests, poor people. Resource access and resistance in Java. University of California Press, Berkeley, CA, USA.
Peluso, N. and Vandergeest, P. (2006). Empires of forestry: professional forestry and state power in Southeast Asia, part 1. Environment and History 12: 31-64.
Pierre, J. (ed.) (2000). Debating governance. authority, steering and democracy. Oxford University Press, Oxford, UK.
Pierre, J. and Peters, G. (2000). Governance, politics and the state. Macmillan, London, UK.
Ribot, J., Agrawal, A. and Larson, A.M. (2006). Recentralizing while decentralizing: how national governments reappropriate forest resources. World Development 34(11): 1864-1886.
Rose, N. and Miller, P. (2010). Political power beyond the state: problematics of government. British Journal of Sociology 61: 271-303.
Rosenau, J. and Czempiel, E. (eds.) (1992). Governance without government: order and change in world politics. Cambridge University Press, Cambridge, UK.
Scott, J. (1998). Seeing like a state: how certain schemes to improve the human condition have failed. Yale University Press, Cambridge, MA, USA.
Strange, S. (1996), The retreat of the state. The diffusion of power in the world economy. Cambridge University Press, Cambridge, UK.
Triantafillou, P. (in press). Decentralization as control at a distance. Journal of Political Power 5(1).
Umans, L. (1993). A discourse on forestry science. Agriculture and Human Values, Fall 1993: 26-40.
United Nations Framework Convention on Climate Change (UNFCCC) (2010). Report of the Conference of the Parties on its 16th session, held in Cancun from 29 November to 10 December 2010. Available at: http://unfccc.int/bodies/body/6383/php/view/reports.php.

Van den Arend, S. and Behagel, J. (2011). What participants do. A practice based approach to public participation in two policy fields. Critical Policy Studies 5(2): 169-186.

Van Tatenhove, J., Arts, B. and Leroy, P. (eds.) (2000), Political modernization and the environment. the renewal of policy arrangements. Kluwer Academic Publishers, Dordrecht, the Netherlands.

Visseren-Hamakers, I. and Glasbergen, P. (2007). Partnerships in forest governance. Global Environmental Change 17: 408-419.

Wiersum, F. (1995). 200 years of sustainability in forestry: lessons from history. Environmental Management 19(3): 321-329.

Wiersum, F. (2009). Community forestry between local autonomy and global encapsulation: *quo vadis* with environmental and climate change payments? Paper presented at First Community Forestry International Workshop, Pokhara, Nepal, September 15-18, 2009.

Woods, N. (2000). The challenge of good governance for the IMF and the World Bank themselves. World Development 28(5): 823-841.

16. Exploring forest governance in Tanzania

Josiah Z. Katani[1] *and Innocent H.Babili*[2]
[1]*Sokoine University of Agriculture, Faculty of Forestry and Nature Conservation, Department of Forest Mensuration and Management, P.O. Box 3013, Morogoro, Tanzania;* [2]*Sokoine University of Agriculture, Institute of Continuing Education, P.O. Box 3044, Morogoro, Tanzania; josiahkatani@yahoo.com; babhili@yahoo.com*

Abstract

This chapter explores forest governance in Tanzania. It is based on a literature review of the governance concept in general and an analysis of selected case studies - Suledo, Duru-Haitemba and Bereku forest reserves - in particular. We identify various types of forest governance and show how these are related to the decentralisation process in Tanzania. Our findings indicate that decentralisation through participatory forest management (PFM) in Tanzania integrates various governance types, including good governance, multilevel governance, corporate governance, network governance and global governance. We also show that community-based forest management (CBFM) and joint forest management (JFM), the two dominant forms of participatory forest management in Tanzania, have generally improved the condition of forest resources in the selected case studies. However, the contribution of CBFM and JFM to improved livelihoods of local communities is still rather low. Both CBFM and JFM still face a number of challenges, including inadequate additional income for local communities and insufficient support from external stakeholders.

Keywords: participatory forest management, decentralisation, livelihoods, forest condition, Tanzania

16.1 Introduction

This chapter explores how forest management in Tanzania has shifted from government to governance during the last 25 years and what its implication are for forest resources and local people's livelihoods. The chapter addresses the following research questions:

1. How has forest governance evolved over time during different administrations and periods of decentralisation in Tanzania, from the colonial era to the present?
2. How is participatory forest management (PFM) related to the forest governance debates and what is its relevance for Tanzania?
3. To what extent has PFM contributed to the improvement of the forest condition and to the livelihoods of local communities?
4. What are the future challenges for effective forest governance in Tanzania?

In the first section we introduce the subject matter by providing information on forest cover in Tanzania, the types of forest lands and the history of forest management in Tanzania. The second section covers the organisation of forest governance in Tanzania, including the role of formal and informal institutions and the implications of different forest governance types for forest management. The third section presents different case studies that provide information on the effects of forest governance on the resource base and on the livelihoods of rural communities. The last section formulates some conclusions that synthesise key issues related to forest governance in Tanzania. It also identifies some future challenges for effective forest governance in the country.

16.2 Background information

Tanzania is estimated to have about 35.2 million ha of forests and woodlands (FAO, 2006). The current Forest Act of 2002 defines different categories of forests to allow easy monitoring and management of the resources. According to the Government of Tanzania (URT, 2002) there are four types of forests in Tanzania. The first is national forest, which consist of forest reserves and forest on general lands owned and managed by the central government. Second, forests under local authority management, which consist of forest reserves and forest on general lands owned and managed by local authorities, including city, municipal, town and district councils. Generally speaking, forest reserves are under a strict protection regime, whereas forests on general lands are characterised by an open access regime. The third type are village forests, which consist of village forest reserves, community forest reserves created by groups of local people outside village forests reserves, and forests which are not reserved. Management of non-reserved forests on village land is vested in the village council. The last type are private forests, which consist of forests on village land held by one or more individuals under customary rule, and forests on general or village land for which the rights of occupancy or lease have been granted to individuals, groups, corporate bodies or any other organisation.

Forest management in Tanzania has undergone several changes, which include shifts from management by indigenous people to management by the state and subsequently to PFM. This historical trend in forest management can be traced back to five different periods (Ylhäisi, 2003): the pre-colonial period; the German colonial period; the British colonial period; the independence period; and the current period. During the pre-colonial era, forest management was based on customary laws whereby different ethnic groups had different rules for owning, accessing and managing forests. The customary laws protected some forest areas and tree species for different purposes, such as conservation of water sources, top hills conservation, ritual forests and forests for fodder production. The German colonial period lasted from 1885 to 1919. The German colonial government appropriated all the important forest resources because it considered that the local people surrounding the forests posed a threat to the environment. The Forest

Conservation Ordinance issued in 1904 created forest reserves from the Crown Land and it excluded local people from using the forest for farming, grazing and other uses (Ylhäisi, 2003). The exclusion of local people from control over natural resources started to erode the role of indigenous people in managing the resources because it undermined the authority of traditional institutions that used to control the resources. The British colonial government, which ruled from 1919 to 1961, had the same interest as the Germans in forest protection (Katani, 2010). During this period more forest lands were converted into forest reserves. At the same time, large tracts of forest were cleared to establish large-scale export crop plantations, including coffee, tea, sisal, rubber, cotton and tobacco. This resulted in displacement of local communities from their areas (Mbonile, 2005; Ylhäisi, 2003). After independence in 1961, the government adopted the reserved area policy and more lands were converted to forest reserves. The role of customary laws continued to be weakened, especially when the chieftainship was formally abolished by the independent Government of Tanganyika in 1963. The current PFM is the result of major changes in forest governance that began in the 1990s. Their purpose is to rescue forests threatened by high rates of deforestation caused by the failure of the previous policies and institutions, which aimed to bring more forests under state tenure and exclude local communities from forest management (Zahabu, 2008).

16.3 Contemporary forest governance in Tanzania

The concept of governance is used in various fields with different meanings. This chapter does not discuss which definition or meaning is relevant to forest management, but rather attempts to explore and present a clear overview of the usage of the concept of governance in forest management in Tanzania. Generally, the concept of governance is interpreted as a shift from government to governance (Arts and Leroy, 2006; Rhodes, 1997; Van Kersbergen and Van Waarden, 2004). This implies that forest governance is not restricted to government activities related to forests alone, but that it involves all kind of activities by public and private actors that guide, steer, control or manage society in the use and/or conservation of forest resources. In other words, the governance concept embraces the idea that the state is no longer the only dominant actor in forest management, but involves both governmental and non-governmental actors.

Current forest governance in Tanzania is conceptualised as decentralisation (Mniwasa and Shauri, 2001), involving the transfer of legal and political authority from the central government and its agencies to local organisations and institutions. In Tanzania, decentralisation has shifted ownership and forest management from the government to include other actors as well. The decentralisation process, which started in 1982, was considered to be the best option for managing natural resources. The Local Government (District) Authorities Act of 1982 gave power to the village councils to propose bylaws related to natural resources management, while the district councils were given the power to approve these bylaws. This process of decentralisation was also reflected in the development of the National Land Policy of 1995 and the National Forest Policy of 1998. These two policies recognise and emphasise the active participation of local communities and local government in the management of land and forest resources. This decentralisation process was set in motion by the central government to gradually distance itself from the direct management of the resources (URT, 1998).

Under the umbrella of decentralisation, the new forest governance has resulted in the adoption of the PFM approach, also known as collaborative forest management, which effectively started in the 1990s. The PFM approach has developed into two approaches to forest management: joint forest management (JFM) and community-based forest management (CBFM). Both approaches are based on three dimensions of social relations: benefit sharing, right holding and duty bearing. Under the JFM arrangement, which is applied mainly in national or local authority forest reserves, the right holder is usually the government (central or local government) and the communities are duty bearers. However, the guidelines for PFM also provide opportunities for JFM between a community and a private owner. JFM is formalised and legalised through the signing of a Joint Management Agreement (JMA) between village representatives and either the district council or the Director of Forestry and Beekeeping. Under the CBFM arrangement, on the other hand, local people are both owners and managers (Kajembe and Kessy, 2000). It applies to 'village land', land that is surveyed and registered under the provisions of the Village Land Act (1999).

The main objective of PFM is to improve forest governance through the establishment of effective and accountable local institutions, which should lead (1) to better livelihood opportunities for forest dependent communities, and (2) to improvement of the local forest condition through sustainable management and protection (Blomley and Ramadhani, 2004). The JFM and CBFM approaches have spread rapidly throughout the country and about 13% of all the forest in Tanzania is currently under PFM arrangements (see Table 16.1). However, visions on PFM differ. Wily (1998) argued against the conventional form of JFM that 'allows' communities living next to forests to use certain areas or products. According to Wily, this tends to increase rather than decrease the policing and administrative burdens on the state. Wily (1998) and Misana (2001) made it clear that the proper mechanism of CBFM should not be 'benefit sharing', as is the case with most JFM ventures, but 'power sharing'. CBFM calls for a transfer of 'power' to the community and not just the right to use certain forest products (Kajembe and Nduwamungu, 2004).

The decentralisation of forest management in Tanzania through PFM has contributed to the emergence of different types of forest governance. Some have been promoted by donors, while others have emerged spontaneously as a result of different actors becoming involved in the management of forest resources. The emerging types of governance in forest management include good governance, multilevel governance, corporate governance, network governance and global governance.

Table 16.1. Current coverage of community-based forest management and joint forest management across mainland Tanzania (MNRT, 2008).

Joint forest management (JFM)		**Community-based forest management (CBFM)**	
Area of forest under JFM	1.77 million ha	Area of forest under CBFM	2.35 million ha
Number of forest reserves under JFM	246	Number of declared village land forest reserves	331
Number of villages engaged in JFM	863	Number of villages engaged in CBFM	1,457
Number of villages with signed joined management agreements	155	Number of gazetted village land forest reserves	64
Number of districts engaged in JFM	58	Number of districts engaged in CBFM	63
Most common forest type under this management regime	montane forest and mangroves	Most common forest type under this management regime	miombo, acacia and coastal woodlands
% of forest reserved by central or local government under JFM	13%	% of unreserved forests now under CBFM	12%

Good governance

Good governance is promoted by the World Bank and has become prominent in international aid from around 1989-90 as a general guiding principle for recipient governments to properly handle the development assistance and to make changes in the institutional environment to enhance policy effectiveness (Doornbos, 2001). The central themes of good governance include accountability, democracy, human rights, rule of law, anti-corruption, transparency, empowerment and participation (World Bank, 2007). However, the concept of good governance in the academic community has been concerned with developing the understanding of different ways in which power and authority relations are structured in different contexts (Doornbos, 2001). The introduction of PFM as one of the reforms in forest management in Tanzania was supported by international organisations such as the World Bank, DANIDA, NORAD and FINNIDA.

Multilevel governance

Multilevel governance refers to different levels of government, private actors and international relations. The key issues in this type of governance are the sets of implicit or explicit principles, norms, rules and decision-making procedures guiding the actors, the power relations resulting from these rules and the substance of policies (Van Kersbergen and Van Waarden, 2004). Multilevel governance seeks to analyse the transfer or relationship of competences at different levels (supranational, national states, local). In the Tanzanian context, decisions related to forest management under PFM have been taking place at different hierarchal levels, including village environmental committees, village councils, village assemblies, district councils, the ministry and international organisations. Under PFM, actors at the lowest level, the village community, have sometimes gained more power over the management of forests than actors at higher levels of administration, especially in CBFM. For example, village environmental committees can some times bypass the village government and appeal to the district authority (Mahonge, 2010), or to the ministry when the next highest tier of authority is reluctant to solve a natural resources management problem. Also, through funding the establishment of JFM, international organisations have influenced decisions related to forest management, including imposing the priority of forest conservation.

Corporate governance

Corporate governance as defined by Lubatkin *et al.* (2007) refers to those administrative, monitoring and incentive mechanisms that are intended to reduce conflicts among private actors due to differences in incentives. It also includes the relationships between the many parties involved and the goal they want to achieve. The parties involved in corporate governance include the chief executive officer, the board of directors, management, shareholders and stakeholders. Van Kersbergen and Van Waarden (2004) refer to corporate governance as 'good governance' in the private sector. Hirst (2000) argues that corporate governance improves the accountability and transparency of the management of a company. Firms or companies are situated within a given society and political context and so the decisions made by individuals within the firm will be influenced by the context, but at the same time society will be influenced by the decisions made by the firm. Recognising the relationship between the firm and its social context, corporate social responsibility is becoming an important element in corporate governance (Aguilera *et al.* 2006).

The establishment of private forests in Tanzania has been encouraged under PFM (Blomley and Ramadhani, 2004) and is supported by Policy Statement 3 (*ibid.*: 17): 'To enable participation of all stakeholders in forest management and conservation, joint management agreements, with appropriate user rights and benefits, will be established. The agreement will be between the central government, specialised executive agencies, private sector or local governments as appropriate in each case and organised local communities or other organisations of people living adjacent to the forest'. The involvement of the private sector in forest management brings new challenges in meeting the interests of local communities under situations in which power struggles among the actors cannot be avoided. This becomes more apparent when big companies or firms are involved and no effective institutions are in place to manage conflicts and to take care of local community interests.

Network governance

Network governance is conceptualised as pluricentric forms of governance. It is characterised by exchange of resources and negotiation, and is considered to be self-organising (Van Kersbergen and Van Waarden, 2004). The networks can involve public policy organisations, private organisations and combinations of public and private organisations. Therefore, the networks can be seen as patterned relationships between individuals, groups and organisations and they involve exchanges that are not derived from authority structures or from legal contracts. This kind of governance takes into consideration the fact that public services are not exclusively provided by government, but by a network of several government, private and voluntary actors. It is based on the idea of the interdependence of actors and how they combine resources to achieve a certain goal. The analysis of network governance in this chapter does not seek to explain how the network should be established, but rather to explain how network governance in forest management has benefited local communities under PFM. Klijn (2005) argues that to solve societal problems, a public actor needs information and resources from other actors. However, the network perspective recognises the fact that the power of an actor in influencing decisions depends on the resources available to it and the way it depends on the resource of other actors. PFM in Tanzania has involved a network of actors, including local communities, district councils, private sector actors, researchers, community-based organisations and donors. To meet the interests of every actor in forest management, the decision-making process has become complex.

Global governance

Global governance is a broad term because it can be defined by referring to the economic, political and environmental dimensions of globalisation. Here we do not focus on the definition, but discuss how global governance, in terms of the economic, political or environmental contexts, has influenced the management of forest resources. Generally, global governance refers to several types of coordination mechanisms in the global domain and can be reflected in terms of international institutions like the UN Security Council, the World Bank, the International Monetary Fund, conventions and treaties. Global governance has been perceived differently by different authors. Arts (2006) sees a positive development in global forest policies and Held (1995) claims that globalisation has led to a cosmopolitan democracy. Cerny (1999) claims globalisation has caused policies to become less democratic.

Global governance is becoming popular in forest management and exerts widespread influence on policy formulation at the national level. This is demonstrated in various treaties and conventions signed by different countries with respect to natural resource management and conservation. Tanzania has signed a number of conventions, including the Ramsar Convention, CITES, The Convention on Biological Diversity (CBD) and the Kyoto Protocol. These conventions do not replace national laws; rather, they provide a framework within which national governments adopt their own legislation to implement them. Currently, the Tanzanian national forest policy is under review and issues related to climate change, including Reduced Emission from Deforestation and Forest Degradation (REDD), payments for environmental services and carbon credits have been included. New reforms in forest policy aim to address the challenge of a lack of proper incentive mechanisms in PFM for local communities to effectively participate in forest management. The REDD is considered to be a new potential incentive mechanism for PFM (Zahabu, 2008). However, Babili and Wiersum (personal communication) note that REDD can act as a disincentive due to its emphasis on the protection of forests rather than using them for enhancing the livelihoods of local people. Also, a danger of global governance is that signing international conventions or treaties can strengthen national government and weaken the power local people have over the management of their natural resources. This is a common phenomenon, especially for forests designated as world heritage nature reserves, where development projects such as construction of tarmac roads through forests and national parks are not permitted. Kessy (1998) argues that the heavy orientation of the management of areas of global conservation value (due to their high biological diversity) towards conservation can conflict with local values that emphasise resource use, which in turn could be a disincentive to local people in PFM.

16.4 Impact of new forest governance on forest resources and livelihoods of local people

We use three case studies, two cases of community-based forest management (CBFM) and one case of joint forest management (JFM), to discuss new forest governance and its impact on forest resources and livelihoods of local people. The cases are Suledo village forest reserve, Duru-Haitemba village forest reserve and Bereku forest reserve.

Suledo village forest reserve, Kiteto District

Suledo was gazetted as a Village Land Forest Reserve in 2007 and is under a CBFM regime. The reserve is situated in the southern dry lands of the Maasai steppe in central Tanzania and covers 167,416 ha. Suledo village forest reserve is legally owned by nine villages (Blomley and Iddi, 2009; ZEC, 2009): Sunya, Asmatwa, Loltopes, Olgira, Laiseri, Lengatei, Lesoit, Olkitikiti and Eng'ongongare. The forest has been divided into management zones corresponding to the administrative boundaries of the villages. Each village owns and manages its respective area.

Community participation in managing forests began in 1994 when two villages demarcated their forest areas to solve conflicts between farmers and pastoralists. In the period to 2007 more villages became involved and the District Land Use Management Programme (LAMP) supported by SIDA (Swedish Development Agency) assisted with land surveys, land use plans, village bylaws and the demarcation of the forest borders. The management plan, which guides the use and protection of forest resources, has been in place since 2000 and is reviewed every five years. The main issues covered in the management plan include the details of the forest, management objectives, implementation of plan, the reporting and follow-up system, record keeping, financial management, forest borders, protection of forests, legal issues, handling of trespasses or illegal activities, forest improvement and rehabilitation, harvesting of forest products and dividends payable to the district council. Suledo forest is managed under the overall responsibility of the Zonal Environmental Committee (ZEC), a joint management committee representing all nine villages. This committee can enter into harvesting contracts with individuals or companies. Despite its power in managing the forest reserve, ZEC does not fit into the village government structure and its legal status has been challenged by other actors at village and district levels. Below ZEC there are other committees, including the village environmental committees responsible for forest management at village level and sub-village environmental committees responsible for forest management at sub-village level. ZEC also works in collaboration with other stakeholders through general meetings in which it presents its annual report on the management and harvesting of the Suledo forest. Stakeholders include Suledo community, the Prime Minister's Office of Regional Administration and the local government (PMORALG), the Forestry and Beekeeping Division of the Ministry of Natural Resources and Tourism, NGOs, forest research institutions, donors and other relevant institutions.

Participatory forest management has led to an improvement in the condition of Suledo forest in terms of forest cover and species composition and has contributed to high levels of commercially viable species, such as mpingo (*Dalbergia Melaloxylon*), mninga (*Pterocarpus angolensis*) and mkalakala (*Brachystegia spp.*) (Blomley and Ramadhani, 2004). In 2010, communities started harvesting the forest selectively in a 500 ha pilot coupe and the Ministry of Natural Resources and Tourism gave them a hammer for marking logs and timber as a means of controlling illegal harvest. A number of successes have been reported in Suledo (Blomley and Iddi, 2009; Blomley and Ramadhani, 2004). In 2002, Suledo community was awarded the UNDPs Equator Initiative Prize of USD 30,000 for 'extraordinary efforts to reduce poverty through the conservation and sustainable use of biodiversity'. The money was used for village development. The first round of harvesting has created new jobs for villagers, village projects have been started, households have realised income and profits have been shared between the villages and the district. The governance issues attributed to the success of Suledo include the change in the legal situation (the Forest Act allows community ownership); transparent cooperation among actors (communities, village governments, the joint management committee, district authorities, the ministry); capacity building in financial management, legal issues and tender processes; equity in profit sharing among actors (individuals, the communities, the district, government and the harvester); and only minor problems with corruption.

Duru-Haitemba village forest reserve, Babati District

Duru-Haitembe village forest reserve was established in 1994 and is under a CBFM regime (Zahabu, 2008). Information from Babati District indicated that Duru-Haitemba village forest reserves were previously earmarked for gazettement as a district council owned forest (Babili and Wiersum, personal communication), but opposition to this by local people resulted in the establishment of CBFM in which a SIDA-funded Regional Forestry Programme supported the process of engaging local communities in the conservation and management of the surrounding forests (Kajembe *et al.*, 2003). However, other literature sources indicate that Duru-Haitemba was initially earmarked for reservation as a state forest reserve (Blomely and Iddi, 2009). The forest is typical dry miombo woodlands covering an area of 9,000 ha. Duru-Haitemba is located about 20 km south of Babati township and is adjacent to eight registered villages: Duru, Riroda, Endagwe, Hoshan, Endanachan, Gidas, Bubu and Ayasanda. Duru-Haitemba forest contains about 145 species of trees and shrubs, including climbers and creepers. The dominant species are *Brachystegia microphylla, Brachystegia spiciformis, Julbernardia globiflora, Albizia versicolor, Brachystegia boehmii, Combretum collinum, Acacia nigrescens* and *Tamarindus indica* (Kajembe *et al.*, 2003).

In 1990/1991 the government proposed the reservation of Duru-Haitembe as state forest reserve but the local community opposed this, which led to gazettement of the forest as a village reserve (Blomley and Iddi, 2009; Kajembe *et al.*, 2003; Zahabu, 2008). The forest has been divided into management zones corresponding to the administrative boundaries of the adjacent villages. The eight villages have obtained title deeds and the law allows villages to work as formal government structures and legal corporate entities that can sue and be sued and are allowed to fine offenders (Blomley and Iddi, 2009; Kajembe *et al.*, 2003). The management of the forest is governed by village bylaws, which were formally approved by Babati District Council in mid 1995. Each village has a Village Forest Committee (VFC) which is responsible for the management of their own area of the village forest reserve. Members of the VFC are selected from villagers, taking account of gender aspects. For example, in Ayasanda village the current committee has 10 members, 6 of whom are men and 4 are women (Zahabu, 2008).

Before the Duru-Haitemba forest was handed over to the community it was highly degraded, but after gazetting the prevailing trend of forest degradation was reversed (Blomley and Iddi, 2009). The forest stocks of Duru-Haitemba increased year on year from 2005 to 2008, indicating the effectiveness of the current management in protecting the forest (Zahabu, 2008). However, it is acknowledged that the Duru-Haitemba project has contributed little to local livelihoods (Babili and Arts, personal communication) and that the value of the forest for commercial harvesting is low. The land under CBFM was degraded and it will take a long time before the forest becomes commercially viable again (Blomley and Ramadhani, 2004). The successes of Duru-Haitemba are attributed to community participation and a marked degree of power sharing that give the community full responsibility and control of the resource, while the government has become a technical advisor, facilitator and watchdog (Kajembe *et al.*, 2003).

Bereku forest reserve, Babati District

Bereku forest reserve is under JFM with Babati District (Babili and Wiersum, personal communication). Bereku forest was gazetted as a government forest reserve during British colonial government rule, with declaration number 294 of 5 September 1941. Bereku forest reserve is located about 20 km south of Babati District. It is a hilly miombo woodland covering an area of about 5,370 ha and is situated on mountains with an elevation ranging between 1,280 to 1,810 m. The forest is largely composed of species of *Brachystegia microphylla*, *Brachystegia speciformis* and *Julbernardia globiflora* (Lupala, 2009). The forest is adjacent to five villages – Ayasanda, Haraa, Tsamasi, Boay and Gijedaboshka – which collaborate with the government in joint management. Each of these villages manages its own portion of Bereku forest adjacent to it. The areas managed by each village are as follows: Ayasanda, 1,025 ha; Haraa 469.6 ha; Tsamas, 1,846.2 ha; Boay, 1,491.0 ha; Gijedaboshka, 540.5 ha (Babili and Kajembe, personal communication).

Community participation in managing the forests effectively started in 2002, although the process for establishing JFM started earlier, in 2001. The establishment of JFM was consistent with new forest policy of 1998, which promotes collaboration between various stakeholders in the management of forest resources, including government forest reserves. The Forest Act of 2002 gives additional support to the new approach, which requires the involvement of more actors than the government alone in the management of the resources. The process for establishment JFM in Bereku forest was facilitated by catchment forest officers in the district. However, meetings for engaging village actors and activities for demarcation of borders for respective villages was supported financially by the district council's LAMP programme, which was supported by SIDA. Key activities in the initiation of the JFM in the village included persuading local people to accept collaboration with the government in managing the forest, and the formulation of village rules and management plans for managing forest areas adjacent to the respective villages. By 2006, documents for forest management rules and plans had gone through different stages of endorsement, including village assemblies, ward development committees and the district council, before they were signed by the district council and village leaders of the respective villages. The main issues covered in the management plans include information about the composition of the forest, management objectives, implementation of the plans, the reporting and follow-up system, record keeping, financial management, forest borders, protection of forests, legal issues, dealing with trespass and illegal activities, and forest improvement and rehabilitation. The management plans and rules specify the role of the village environmental committees (VECs) as managers of the forest on behalf of village communities. The VECs (also known as VFCs) are assisted by local forest guards, who are responsible for conducting regular patrols and reporting to the committee. The village community members are also managers and are encouraged to report non-compliance with the rules. The rules guiding management of the forest are divided into four categories: free use, permitted uses that are not paid for but need notification to the VEC, permitted activities that need a paid permit, and totally prohibited activities. Most of the issues regarding the management of forest resources are settled at the village level. However, offenders can be taken to court if they repeat offences several times. Forest management disputes can also be taken to court if an offender challenges the decision of the VEC. The JFM stakeholders in Bereku forest include the village communities, village leaders, the district council, donors, and the catchment forest officer, who represents the government. The success of JFM therefore depends on the extent to which these actors support each other in the management of the forest.

Before the introduction of JFM, the Bereku forest was highly degraded and illegal timber harvesting was common, despite being prohibited in the catchment forest. Both the government and local people acknowledge that following the introduction of JFM the forest cover has improved because the trees have regenerated. Over 80% of the villages practising JFM recognise that forest cover has increased. Local people also benefit more from the forest, including permission to graze cattle at affordable fees as opposed to illegal fees, which local people used to pay to the forest officer before JFM. Community members are also free to collect firewood. The VECs also collect fees from people who collect firewood for commercial purposes and use this money to support forest management and village government activities. Occasionally, villages are allowed to harvest timber from dry trees to support public services. However, over 80% of the people practising JFM in Bereku forest assert that JFM has not improved their household income. Furthermore, local people experience some challenges in managing the forest, including inadequate support from the government in resolving management conflicts and motivating VEC and forest guards to carry out their forest management tasks.

Comparison

The three cases presented in this chapter show some similarities and differences in governance issues. The similarities between the three cases are based on the fact that the establishment of CBFM in the Duru-Haitemba and Suledo village forest reserves and JFM in the Bereku forest reserve have been supported by external donors. Their support underpins a common feature in all three cases: the formation of committees by democratic election as a way of ensuring good governance. These committees have been embedded in the village government system, making them accountable to the village general assemblies, which are comprised of all villagers aged 18 years and above. These local level structures are also linked to the district council and the Ministry of Natural Resources and Tourism, a typical multilevel governance structure. The loci and focus of decision making in forest management have shifted from the central government – with an emphasis on protecting the resource by excluding local communities from access – to addressing both livelihood and resource sustainability considerations. The donors have also promoted the conservation of biodiversity within these forests, which is the main agenda in global forest governance. This is demonstrated by the award of the UNDP's Equator Initiative Prize to Suledo for 'extraordinary efforts to reduce poverty through the conservation and sustainable use of biodiversity'. Moreover, researchers from both Tanzania and abroad are interested in all three cases and are investigating how PFM has improved the forest resource and the livelihoods of adjacent villages. The researchers have become part of the network governance of these PFM cases because they provide the information needed by other actors to make informed decisions about the management of these forests. Klijn (2005) argues that to solve societal problems public actors need information and resources from other actors. Therefore, the successes achieved in these cases, in terms of enhanced local people's livelihoods and sustained forest resources, are the result of the concerted efforts of the PFM network.

Corporate governance is more prominent in the Suledo case than the Duru-Haitemba and Bereku cases because the Suledo village forest reserve has a high proportion of commercially viable species, such as mpingo (*Dalbergia melaloxylon*), mninga (*Pterocarpus angolensis*) and mkalakala (*Brachystegia spp.*), and commercial selective harvesting has started. This has attracted big timber harvesters who are supposed to enter into contracts with the Zonal Environmental Committee (ZEC), the joint management committee representing all nine villages. The villages managing Suledo forest have become business entities/firms with the main objective of maximising financial gains, while at the same time sustaining the resource base. The local people, who are the principals, have delegated responsibility to ZEC as an agent to serve their interests. Each year ZEC presents its annual report on the management of and harvesting in the Suledo forest in a general meeting. These two governance structures (ZEC and the general meetings) are meant to safeguard both the interests of local communities and the sustainability of the forest resources in a situation where power struggles are likely. This is in line with the concept of corporate governance, which aims to create administrative monitoring and incentive mechanisms that should reduce conflicts among organisational actors (in this case the nine villages adjacent to Suledo forest) due to differences in incentives (Lubatkin *et al.*, 2007). The other two cases, the Bereku and Duru-Haitemba forest reserves, have committees at the village level only, without a higher-level committee representing all the villages adjacent to that particular forest, as is the case with ZEC in Suledo. The Duru-Haitemba village forest reserve was established on degraded land and has not attained high-level commercial timber value. The Bereku forest reserve is a catchment forest and commercial timber harvesting is not allowed, which is the common feature of JFM. Because there is no commercial harvesting in these two forests, power struggles are less common.

The development of governance types and structures under PFM seems to be determined by the types of actors involved and the level of financial gains expected from the forest. Businesses are interested in making a financial profit from the resource, conservationists are interested in protecting the resource and local communities are interested in their livelihoods. This means that governance structures in forest reserves which lack commercial timber harvesting value have to be designed in ways that create financial incentives that avoid free riding. However, forests in which commercial harvesting takes place are at great risk from emerging new power struggles. These struggles may take place within management bodies like ZEC, within villages, between villages and between other actors. If they are not well managed, they could also result in free riding. In Suledo, some actors have challenged the legality of ZEC, which they see as detached from the local government structure. The governance structures should therefore not be thought of as static, but should be dynamic arrangements that are able to address the emerging challenges in managing these forest reserves.

16.5 Conclusion

Four key conclusions can be drawn from this chapter. First, PFM is a new phenomenon in Tanzania as a result of the shift from centralisation to decentralisation in forest management. PFM is characterised by the involvement of a wide range of actors in forest governance for benefit sharing, right holding and duty bearing. These actors include public bodies such as the Division of Forestry and Beekeeping (Ministry of Natural Resources and Tourism), district councils and village councils, and private actors such as NGOs, individual local villagers and companies. Other actors are donors and international agencies.

Second, PFM is divided into two forms, namely CBFM and JFM. Each of the two forms has distinctive characteristics. Under CBFM, village forest reserves are established and local people are involved as both owners and managers. In contrast, JFM is mainly associated with national or local authority forest reserves in which the right holder is usually the central or local government, while communities are duty bearers. However, the guidelines for PFM also provide opportunities for JFM between a community and a private owner. The JFM arrangement is formalised and legalised through the signing of a JMA between village representatives and either the district council or the Director of the Division of Forestry and Beekeeping.

Third, the three cases studies demonstrate PFM's strengths and weaknesses. Among the strengths, for example in the Suledo case, are the efforts made by local communities to improve the forest resource condition, which have started to receive recognition not only from stakeholders within the country, but also from abroad. However, the legal status of Suledo's ZEC is still contested, because it does not fall within the official village government structure. One of the strengths of the Duru-Haitemba case has been the gradual recovery of the forest from a highly degraded condition and the trend towards restoration. Additionally, local opposition to the initial gazettement of Duru-Haitemba as a government forest reserve, which led to the establishment of CBFM, shows that local people have an important role to play in shaping their own destiny in forest governance. One of the weaknesses of the Duru-Haitemba case is its low contribution to the livelihoods of local people. Similarly, while the condition of forest resources in the Bereku forest reserve has improved, there has been little contribution to the livelihoods of local people.

Fourth, manifestation of the different types of governance in PFM, including good governance, multilevel governance, cooperate governance, network governance and global governance, very much depends on who the actors involved are and what benefits can be expected from the forest. For example, forests with high timber value and due for commercial exploitation attract more actors, including companies. These kinds of forest need to be monitored by more coordinated and effective governance structures to avoid power struggles and forms of economic and social injustice. However, a complicated governance structure may lead to more bureaucracy and less legitimacy, which in turn may lead to ineffectiveness of the governance system that jeopardises management of forest resources.

A number of challenges for the future governance of the forest resources in Tanzania still remain. One of those challenges concerns how to balance the conservation of forest resources on the one hand and human use and management on the other. Even though the management of forest resources has shifted from government to governance, forest institutions still tend to emphasise the conservation of resources over use by local people. This is because the central government is still the most powerful actor in influencing the formulation of CBFM and JFM rules. Another challenge is how the government and international organisations can sustain local support for the management of forest resources through the provision of appropriate incentives to local communities. Finally, despite its overall success in Tanzania, PFM cannot be applied to forests which are not close to village communities. Bringing such forests under PFM is another challenging and as yet unresolved issue.

References

Aguilera, R.V., Williams, C.A., Conley, J.M. and Rupp, D.E. (2006). Corporate governance and social responsibility: a comparative analysis of the UK and US. Journal of Compilation 14(3): 147-158.

Arts, B. (2006). Forests, institutions, discourses: a discursive-institutional analysis of global forest politics. Inaugural address, Wageningen University, Wageningen, the Netherlands.

Arts, B. and Leroy, P. (2006). Institutional dynamics in environmental governance. Springer, Dordrecht, the Netherlands.

Blomley, T. and Iddi, S. (2009). Participatory forest management in Tanzania: lessons learned and experiences to date. Ministry of Natural Resources and Tourism, Division of Forestry and Beekeeping, Dodoma, Tanzania, 69 pp.

Blomley, T. and Ramadhani, A. (2004). Going to scale with Participatory Forest Management: early lessons from Tanzania. TNRF Occasional Paper No. 4, 10 pp. Available at: http://www.tnrf.org/node/5609.

Cerny, P. (1999). Globalization and the erosion of democracy. European Journal of Political Research 36: 1-26.

Doornbos, M. (2001). 'Good governance': the rise and decline of a policy metaphor? Journal of Development Studies 37(6): 93-108.

Food and Agriculture Organization (FAO) (2006). Global forest resource assessment 2005. FAO Forestry Paper 147, Rome, Italy, 320 pp.

Held, D. (1995). Democracy and the global order: from the modern state to cosmopolitan governance, Polity Press.

Hirst, P. (2000). Democracy and governance. In: Pierre, J. (ed.) Debating governance: Authority, steering and democracy. Oxford University Press, Oxford, UK.

Kajembe, G.C. and Kessy, J.F. (2000). Joint forest management in Urumwa Forest Reserve, Tabora, Tanzania. A process in the making. In: Virtanem, P. and Numerin, M. (eds.) Forests, chiefs and peasants in Africa: local management of natural resources in Tanzania, Zimbabwe and Mozambique. Silva-Carelica 34. University of Joensuu Joensuu, Finland, pp. 141-158.

Kajembe, G.C. and Nduwamungu, J. (2004). Participatory forest management in Tanzania. A paradigm shift. Paper presented at the Earth Day on April 22, 2004 at the Auditorium of the American Embassy, Dar es Salaam, Tanzania.

Kajembe, G.C., Monela, G.C. and Mvena, Z.S.K. (2003). Making community-based forest management work: a case study from Duru-Haitemba village forest reserve, Babati, Tanzania. In: Kowero, G., Campbell, B.M. and Sumaila U.R. (eds.) Policies and governance structures in woodlands of southern Africa. Center for International Forestry Research, Jakarta, Indonesia, pp. 16-27.

Katani, J.Z. (2010). The role of multiple institutions in the management of micro spring forests in Ukerewe, Tanzania. PhD Thesis, Wageningen University, Wageningen, the Netherlands, 202 pp.

Kessy, J.F. (1998). Conservation and utilization of natural resources in the East Usambara forest reserves: conventional views and local perspectives. Tropical Resource Management Papers Series No. 18. Wageningen Agricultural University, Wageningen, the Netherlands, 168 pp.

Klijn, E.H. (2005). Networks and inter-organizational management: challenging, steering, evaluation and the role of public actors in public management. In: Ferlie, E., Lynn, L.E. and Pollitt, C. (eds.), The Oxford handbook of public management. Oxford University Press, Oxford, UK, pp. 257-281.

Lubatkin, M., Lane, P.J., Collin, S., Very, P. (2007). An embeddedness framing of governance and opportunism: towards a crossnationally accommodating theory of agency. Journal of Organizational Behaviour 28: 43-58.

Lupala, Z.J. (2009). The impact of participatory forest management on miombo woodland tree species diversity and local livelihoods. A case study of Bereku miombo woodland, Babati, Tanzania. Master Degree Thesis, CBM Swedish Biodiversity Centre, Uppsala, Sweden, 71 pp.

Mahonge, C.P.I. (2010). Co-managing complex social-ecological systems in Tanzania. The case of Lake Jipe wetland. PhD Thesis, Wageningen University, Wageningen, the Netherlands.

Mbonile, M. (2005). Migration and intensification of water conflicts in the Pangani Basin, Tanzania. Habitat International 29(1): 41-67.

Misana, S. (2001). Empowering local communities to manage resources. Paper presented at the World Mountain symposium 2001, Moshi, Tanzania.

Mniwasa, E. and Shauri, V. (2001). Review of the decentralization process and its impact on environmental and natural resources management in Tanzania. Available at: www.leat.or.tz/publications/decentralization/decentralization.pdf.

Ministry of Natural Resources and Tourism (MNRT) (2008). Participatory forest management in Tanzania. Facts and figures. Forestry and Beekeeping Division, Ministry of Natural Resources and Tourism, Dodoma, Tanzania.

Rhodes, R. (1997). Understanding governance: policy networks, governance, reflexivity and accountability. Open University Press, Buckingham, UK.

United Republic of Tanzania (URT) (1998). National forest policy. Government Printer, Dar es Salaam, Tanzania, 59 pp.

United Republic of Tanzania (URT) (2002). The Forest Act No. 14 of 2002. Dar es Salaam, Tanzania, 127 pp.

Van Kersbergen, K., and Van Waarden, F. (2004). 'Governance' as a bridge between disciplines: cross-disciplinary inspiration regarding shifts in governance and problems of governability, accountability and legitimacy. European Journal of Political Research 43: 143-171.

Wily, L. (1998). Devolution: the critical institutional change in future resource management. A case from the forestry sector of Tanzania. The World Bank/WBI's CBNRM Initiative, Washington, DC, USA.

World Bank (2007). Worldwide governance indicators: 1996-2005. Available at: http://www.worldbank.org.

Ylhäisi, J. (2003). Forest privatisation and the role of community in forests and nature protection in Tanzania. Environmental Science and Policy 6: 279-290.

Zahabu, E. (2008). Sink and source. A strategy to involve forest communities in Tanzania in global climatic policy. PhD Thesis, University of Twente, Enschede, the Netherlands, 235 pp.

Zonal Environmental Committee (ZEC) (2009). The SULEDO forest conservation and sustainable harvesting plan. Prepared by the Zonal Environmental Committee (ZEC), 57 pp.

17. Institutional bricolage in community forestry: an agenda for future research

Jessica De Koning[1] and Frances Cleaver[2]
[1]University of Wageningen, Forest and Nature Conservation Policy Group, Droevendaalsesteeg 3, 6708 PB Wagenignen, the Netherlands; [2]University of Bradford, School of Social and International Studies, Bradford, BD7 1DP West Yorkshire, United Kingdom; Jessica.deKoning@wur.nl; f.d.cleaver@bradford.ac.uk

Abstract

This chapter outlines the concept of institutional bricolage as a tool for understanding how community forest arrangements actually work. We characterise two contrasting schools of institutional thinking and show how bricolage belongs to a 'critical institutionalist' rather than a 'mainstream institutionalist' perspective. The key elements of bricolage are outlined to elaborate the concept. These are further explored through an examination of the different practices adopted by local actors in shaping institutional arrangements. Illustrations are drawn from studies of community forestry in Bolivia and Ecuador and areas for further work are identified.

Keywords: institutional bricolage, community forestry management, critical institutionalism, articulation, aggregation, alteration

17.1 Introduction

Community forestry within a forest governance system in developing countries is often perceived as a promising way of managing natural resources, as it draws on both legislative rules introduced by governments and local knowledge of resources and local conditions. However, the formal community forestry institutions introduced by the government often work differently in practice. This has led to a critical rethinking of the faith placed in such formally designed institutions in community-based natural resource management. One of the critical perspectives proposes that the concept of institutional bricolage helps us to understand how institutions really work. Institutional bricolage is the creative piecing together of different arrangements, styles of thinking and sanctioned social relationships to produce new or adapted institutions (Cleaver, 2001). Institutional bricolage is a relatively new concept and although it has been applied in different contexts, it is in need of further development. This chapter furthers such conceptualisation by synthesising insights from the work of De Koning (2011) and Cleaver (2001, 2005, 2012) and by illustrating these with examples from community forestry in Latin America. In so doing it aims to set out an agenda for future comparative research on institutional bricolage in natural resource management.

17.2 Institutional thinking in community-based forest management

In forestry, ideas about the role of institutions in regulating forest use are often influenced by institutional theories, such as 'new institutional economics' and 'common pool resources management'. Although proceeding from different insights, both theories highly value the rational design of new and appropriate institutions for doing the job (North, 1990; Ostrom, 1990, 1991). This applies not only to forestry alone, but also to natural resource management in general. Mainstream institutional approaches see institutions as mechanisms for furthering 'good governance' of natural resources at the local level. Well-designed institutions are thought to play a key role in regulating property rights, preventing degradation and depletion of resources, promoting robust management and creating sustainable livelihoods (Cleaver, 2001; De Koning, 2011). Such approaches suggest that local natural resource management can be strengthened through policy reform, capacity building and particularly through redesigning community level institutions (Wiersum, 1997).

Communities are thought to have an advantage in resource management, being able to draw on their local knowledge of resources, environmental conditions and technology. It is argued that the interplay of communities and institutions may help to mitigate poverty and powerlessness by specifying a place for marginalised people in decision making, formulating rules of access to resources for the benefit of all and organising labour-intensive management activities which generate economic as well as environmental benefits (Wiersum, 1997). Forest governance, as the overarching structure, is a popular concept believed not only to provide 'better' regulations for forest use in general, but also to improve local development, sustainable resource use and social justice through the design of public institutions (Leach *et al.*, 1999; Paavola, 2007). It is suggested that by participating in these public institutions, local communities can hold state officials and service providers accountable, thus contributing to good governance and mitigating state failure. From this perspective, local traditions of cooperation are often seen as providing the building blocks of good resource management, but need strengthening and formalising, for example through the codification of property rights (Wiersum, 1997).

Such approaches draw heavily on 'mainstream institutionalist' (MI) thinking. From this perspective, epitomised by the work of Elinor Ostrom (see, for example, Ostrom, 1990, 2005), the role of institutions is to provide information and assurance about the behaviour of others, to offer incentives to behave in accordance with the collective good and to monitor opportunistic behaviour. All these factors aid individual decision making and choice, and reduce the transaction costs of cooperation. In this model, institutions need to be formal and public, and designed for a certain purpose. Preferably, they are 'crafted' through public decision-making processes. 'Design principles' for robust and long-enduring institutions are identified (Ostrom, 1990), based on desirable attributes of the resource managing community and of governance mechanisms. The scale above the community level is dealt with by concepts of the desired 'nesting' of local resource management arrangements within larger units of organisation. This body of ideas is popular with policy makers because it blends neoliberal economics with the desirability of decentralised local management and offers clear guidance for building robust institutions.

These ideas have heavily influenced approaches to community forest management. Many policy initiatives are derived from the basic idea that new formal institutions need to be designed in order to achieve sustainable forestry. This was visible with the publication of the Brundtland report in 1987, which focused on the standardisation of sustainable forest management (Brundtland Commission, 1987). The approach has further evolved in the development of many forestry-related institutional instruments and mechanisms, such as community forest management, certification schemes like the Forest Stewardship Council (FSC), formally established partnerships for forest products, and payments for reducing carbon emissions, like REDD (Karky and Skutsch, 2010; Rametsteiner and Simula, 2003; Ros-Tonen *et al.*, 2008).

However, the results of all these different schemes for forest management are, to say the least, varied. It appears that despite many policy regulations and interventions designed to manage forests better, the actual practices of sustainable forest management remain nebulous, subject to interpretation and dynamic (De Koning, 2011; Kant and Berry, 2005; Leach *et al.*, 1999; Ribot *et al.*, 2010; Schanz, 2004; Wiersum, 1995). Drawing on such evidence, new perspectives have evolved which are critical of the dominant focus on formally designed institutions in community-based management. In addition to exploring complexity, these perspectives scrutinise the uneven costs and benefits of public participation and the ways in which power works through local institutions (Campbell *et al.*, 2001; Cleaver, 2005; Matose, 2008; Ribot *et al.*, 2010; Wong, 2010).

Such approaches contribute to 'critical institutionalist' (CI) thinking. This perspective, though drawn from disparate insights, has some common features. It rejects the rational choice and functional assumptions of mainstream institutionalism and emphasises instead the complexity of institutions entwined in everyday social life, their historical formation and the interplay between traditional/modern and formalised/informal arrangements. In this view, institutions for natural resource management are not necessarily designed for a specific purpose, but borrowed or adapted from other arrangements (such as multipurpose village assemblies, women's groups or savings clubs). People's rationalities are 'emotional', 'moral' or 'social' as well as economic. Actions and preferences are also shaped by non-conscious factors – by taken-for-granted everyday practices and ideologies. From a CI perspective, the working of power and the patterning of inequality means that some people are more able to shape collective rule making and benefit from outcomes than others. Table 17.1 summarises the key features of institutionalist thinking, both mainstream and critical.

Table 17.1. Key features of mainstream and critical institutionalist thinking.

Mainstream institutionalism	Critical institutionalism
Formal/public institutions in nested layers	Networks of arrangements, blurred boundaries, intersecting domains/scales
Institutions formed through design and crafting	Institutions pieced together through processes of bricolage
Decision making and negotiations conducted in public fora	Decision making and negotiations embedded in everyday life, formal and informal institutions
'Bounded rationality' models of agency – individuals as strategic resource appropriators	Agency shaped consciously and non-consciously; individuals with complex identities
Behaviour shaped by incentives, rules, sanctions	Behaviour shaped by blended social rationalities
Efficiency of resource management through crafted institutions	Social inclusion/exclusion reproduced through institutions

Both schools of thought face a number of challenges in explaining institutional formation and functioning. These include the need to balance policy legibility with the recognition of complexity as well as the need to avoid both too much abstraction (which overlooks local specificities) *and* too much detail (which may obscure the detection of patterns). Institutions must also be understood in local and non-local contexts. They are animated by people, and analyses should offer explanations of how and why this occurs and with what effects. Institutions are also authoritative arrangements, so tracking the ways in which power is visibly and invisibly channelled through them is a key challenge.

One way of addressing these challenges is through the theory of institutional bricolage. This is a critical institutionalist approach as it has evolved from questioning the faith placed in designing robust institutions to fight resource degradation as well as the assumptions about the direct relationship between policy and local practices (Cleaver, 2001; Leach *et al.*, 1999; Mehta *et al.*, 1999). Institutional bricolage implies the piecing together of 'different institutions, styles of thinking, social relationships.' (Cleaver, 2001: 16). In recent years the concept of bricolage has been deployed to frame research in different fields of natural resources management, such as water, forest, land and wildlife management, and in different geographical regions, such as Africa, Latin America and Europe (Cleaver, 2002; De Koning, 2011; Osei-Kufuor, 2010; Sandstrom, 2008; Sehring, 2009). Despite its increasing popularity, and some commonalities of approach, institutional bricolage has been a somewhat loosely defined concept so far. Here we aim to redress this deficiency by outlining some of the key elements of bricolage and identifying the practices through which institutions are formed and adapted at the community level. In doing so, we aim to provide a conceptual tool for the better understanding of people-natural resource interactions.

17.3 Institutional bricolage

Bricolage is a French word. It means 'to make creative and resourceful use of whatever materials at hand, regardless of their original purpose'. We use the concept to explain institutional formation and functioning. Institutional bricolage is a process through which actors consciously and unconsciously reshape or piece together different arrangements at hand (Cleaver, 2001, 2002; De Koning, 2011). Bricolage consists of adaptive processes through which people inscribe configurations of rules, traditions, norms and relationships with meaning and authority. In so doing they modify old arrangements and invent new ones, but innovations are always authoritatively linked to acceptable ways of doing things. In this process of institutional bricolage, institutional components from different origins are continuously reused, reworked or refashioned to perform new functions (Cleaver, 2001; De Koning, 2011; Galvan, 1997; Lanzara, 1999; Sehring, 2009). These refurbished arrangements are everyday responses to changing circumstances. Therefore, the concept of bricolage helps us to understand how such processes of innovation, adaptation and legitimisation occur.

Rather than being designed, or even deliberately crafted, institutions are patched together, consciously and non-consciously, from the social, cultural and political resources available to people, based on the logic of dynamic adaptation (Chase Smith *et al.*, 2001: 42). As such, institutions are not 'things', but the results of what people do; they must be continually reproduced or re-enacted by people to exist (Lund, 2006). No single factor (or group of factors) is sufficient to explain their success. Rather, institutional processes are dynamic, played out through very different forms in varying contexts and, to this extent, elude design.

17.4 Key elements of bricolage

The emphasis on complexity, diversity and improvised adaptation should not mislead us into thinking that any patched together arrangement will suffice. Bricolage is more than an opportunistic short cut to institutional design. Rather, we see it as an articulation of the dynamic relationship between social structure and individual agency. Here we identify five key characteristics of the formation of institutions through bricolage – processes which shape and are shaped by individual action and by broader societal relationships and patterns of resource distribution (Cleaver, 2011).

1. *Everyday practice, 'necessary improvisation' and innovation.* Institutions are formed through the necessary improvisation of daily practice. People piece together institutional arrangements to address their everyday challenges and to respond to changes in their social fields. In the reworking of the existing institutional arrangements, actors innovate. Institutions so formed are therefore a patchwork of the new and the second-hand. They include:

- taken-for-granted ways of doing things;
- well-worn and accepted practices adapted to new conditions;
- organisational arrangements invented or 'borrowed' from elsewhere;
- new or adapted devices to ensure social applicability.

2. *Multipurpose institutions, ebb and flow.* Such pieced together institutions often have multipurpose functions. They are rarely organised only according to single purposes, sectoral divisions or particular projects. Even if they begin that way, they often evolve to encompass other purposes. Bricolage is a fundamentally dynamic process characterised by unevenness and temporal intermittence.

3. *Naturalisation, leakage of meaning, invention of tradition.* Bricolage involves the piecing together of old and new to make something different. However, a key aspect of this process is that the 'something different' must appear familiar, it must work through a routinely accepted logic, it must socially fit. There are a number of ways in which new configurations are made to seem familiar and legitimate:
   - tradition is called upon both unconsciously and consciously and can be reinvented;
   - meaning in the form of legitimising symbols, discourses and power relationships leaks from one institutional setting to another;
   - institutional arrangements constructed through bricolage are often 'naturalised' by analogy with existing 'right ways of doing things' – categorisations, hierarchies, notions of proper order derived from the social, natural or spiritual worlds.

4. *Conscious and non-conscious action, moral rationalities.* Bricoleurs shape institutions both consciously and non-consciously. People devise arrangements of roles and rules for natural resource management and these are influenced by habituated everyday practices and conventions (such as the ways of collecting water from the well), by overlapping social identities (for example, the forest user is also a mother, wife, member of a particular caste or class or ethnic group), by moral world views (such as the desirability of living in peace together) and by conscious and non-conscious psychological motivations (including emotions of belonging, the desire for recognition, the preference for companionship, dislike of conflict).

5. *Authoritative processes and unequal outcomes.* Bricolage is an authoritative process shaped by relations of power. Individual bricoleurs are able to exercise different levels of influence over the formation and functioning of institutions as a result of their social positions. Authority, reputation and assets all matter when it comes to making and breaking rules.

   Power shapes institutional functioning, but is often invisible, embedded in the taken-for-granted nature of the social order. In processes of institutional bricolage, inequalities can be challenged through public negotiation and in the daily practical enacting of resource access, where endless variations on bending the collective rules are possible. However, neither public spaces nor 'everyday spaces' are neutral, but are sites where power is exercised. The costs to poor or socially marginal individuals of challenging the rules are often high in terms of loss of reputation, goodwill and patronage, payment of fines, time spent resolving disputes and restricted resource access.

17.5 Practices of institutional bricolage

Delineating broad characteristics of bricolage still leaves us with the question of how these processes are combined or blended to produce hybrid institutions in particular contexts. Building on the broad elements of bricolage outlined above, De Koning (2011) tracks three types of practice adopted by local actors which shape the way institutions are formed and altered at the community level. These practices are aggregation, alteration and articulation (the three A's of institutional bricolage). The practices show the different ways in which local communities respond to institutions and help to explain the varied evolution of institutions. Drawing on examples from community forestry in Bolivia and Ecuador, De Koning shows that these local practices are interwoven articulations of the key elements of institutional bricolage outlined above.

The process of institutional bricolage that happens when bureaucratic institutions are introduced into pre-existing socially embedded settings can be described metaphorically as throwing a rock into a pond (Figure 17.1). As the rock enters the water, it creates a ripple on the surface that widens out and becomes smaller. In this case, the rock is the formally designed institution that has been 'thrown into the pond' of local institutions, knowledge, technologies, practices and conditions. However, the comparison with institutional bricolage stops there. As it turns out, the rock does not necessarily sink to the bottom of the pond, nor is it as solid as it would appear.

There are at least three possible outcomes of this rock-like item being thrown into the water that relate directly to the identified practices. First, the rock enters the water 'normally', but then dissolves in the water, like sugar. Second, the water resembles soft ice and the rock leaves a mark or a dent in it. In this situation, the rock transforms into a film on the surface, like oil. Third, the rock bounces off the water, as if thrown onto thick ice. In this case, the rock does not enter the water and is forced to go in another direction. These three scenarios resemble three possible consequences or outcomes of the different processes of bricolage. Each of these outcomes relates to one of the three practices: aggregation, alteration and articulation.

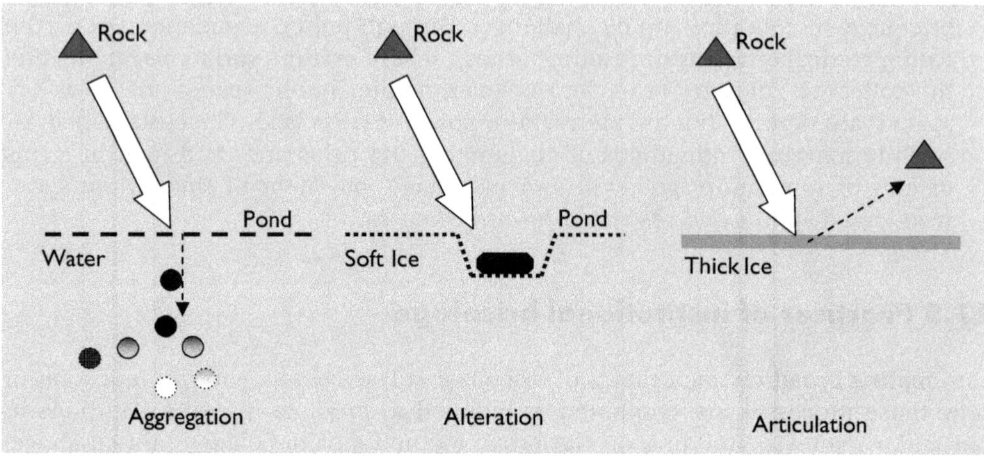

Figure 17.1. 'Rock in pond' metaphor of institutional bricolage.

Aggregation

Aggregation, or the recombination of different types of institutions and social-cultural elements, often involves embedded institutions and social characteristics, such as culture, routines, traditions, social norms, needs, expectancies and experience. These are recombined with formal institutions that are introduced into a community, such as forest regulations and standards or technical norms on logging. In this recombination, the formal institutions are given either a new meaning or a new purpose. The outcome of this process can be described as a more or less balanced situation in which both types of institution correspond, or are even in harmony. This process results in the necessary correlation between the formal institutional frameworks and locally embedded institutional, social and cultural elements (Cleaver, 2002).

A community forest management plan was implemented in a Bolivian community. This involved introducing government regulations aiming at sustainably managing the forests by the community. Through creative recombining, the community quickly added other elements to the forest management plan. Through *innovation* and some *necessary improvisation*, the forest management plan not only became a plan to regulate timber extraction, but also a land title document. By establishing a forest management plan for a certain area of forest, the community resolved a long-lasting problem of ambiguous informal land titles. By bringing the elements of formal regulations and the need for secure land titles together, the community ensured a *social applicability* of the formal regulations.

Another consequence of aggregation is the creation of multipurpose institutions. Although community forestry is geared solely to improving forest use by installing community level institutions, these institutions almost naturally, or *unconsciously*, become *multipurpose*. Local associations for forest management also become a social security system. Members of a local Bolivian forest association chipped in when one of its members became seriously ill and she needed to be hospitalised. In another case, this local association also functioned as a neighbourhood watch to protect some cattle from theft. These externally introduced community institutions therefore become much more than just single-purpose forest institutions; they were also given more symbolic purposes when community members recombine their identity with these institutions. In the example of the forest association, community members used this formal structure to differentiate themselves from other communities without such a local organisation, or even from members of the same community who were not part of the association. In this way, local identities, livelihoods, aspirations and status were strengthened by linking them to the reshaped formal regulations.

However, the functioning of these collective multipurpose institutions is also shaped by the exercise of power. In collective community institutions in which a leader is formally appointed, this leader can have a very big say in the way the aggregation practices develop. For example, one of the leaders of an Ecuadorian association for forest management was able to combine this association with his personal political ambitions, enabling him to influence certain collective decisions and create a platform for his own political messages. This resulted in a change in the original collective and democratic character of the association as his position in the organisation became more and more dominant.

Alteration

Alteration, the adaptation of institutions, can be related to both imposed institutions and locally embedded institutions, such as cultural beliefs or social norms. Local actors are often found tweaking and tinkering with different institutions to make them better fit their livelihoods or identity. This practice can vary from making small changes to complete reinterpretations of certain institutions in which it is hard to predict the actual outcome.

Necessary improvisation to ensure *social applicability* is an important element in practices of alteration. In one case, forest users in a small community in Ecuador 'invented' certain conditions that allowed them to continue their daily practices in the forest. These conditions were used to ignore or avoid those formal regulations that were negatively perceived by them, so negating the introduced arrangement. For example, farmers constructed the image of themselves as cattle farmers in order to avoid the stringent regulations for forest users. By claiming this image, thus *inventing a tradition* of cattle farming, they felt justified in continuing their business as usual. In their eyes, forest management was meant for forest users and not for cattle farmers. Alteration, however, also often entails the *adaptation of well-worn practices* to new circumstances. In the same example, farmers decreased the amount of timber they wanted to extract. By portraying timber extraction as more of an exception in which only very little timber was extracted, they felt justified in continuing these practices without a forest management plan. These inventions and reinterpretations, however, could easily change on a day to day basis. Alteration practices depend very much on the time, the location and the situation.

Alteration is also a practice that involves authoritative individuals. Forest users with certain political or economic resources have many more possibilities to alter institutions than others. In the Ecuadorian community already referred to, the circumstances to make a living were very poor. The constant rain resulted in high numbers of people with serious diseases or arthritis. The poor Amazon soils had become exhausted and the price of cattle had dropped. Only a few better-off farmers were still able to make a living, but the poorer members of the community were no longer able to do so, which made it also virtually impossible for them to participate in institutional bricolage practices of any kind.

Articulation

Articulation involves the asserting of traditional identities and culture to resist introduced institutional arrangements. The introduced formal institution bounces off a shield of local perceptions of traditions and identity. Articulation is as visible in the actual practices as it is in discursive practices. In the cases investigated, articulation occurred when the forest regulations were directly in conflict with local identities. In some cases, this may lead to a calm, but firm distancing; in others it may almost lead to a revolt. This depends on how strong the sense of identity, or the sense of the self, is. As said before, articulation always leads to a rejection of forest regulations. This practice may also result in a situation of normative pluriformity and selective adherence to different institutional features (Von Benda-Beckman and Van Meijl, 1999; Cleaver, 2002). The process of articulation then results in a situation resembling a clash, a friction or a discord between the different types of institutions.

When identities are felt very strongly and considered to be traditionally rooted, articulation practices become visible. This was the case in an indigenous community in Ecuador. Identity was formed by asserting community traditions and by people's perceptions that the community was very independent and proud. This identity was also shaped by community members' dislike of government interventions of any kind. Throughout history, various governments had tried to change their culture and traditions by converting them to Catholicism, forcing them to make use of the modern health care system and obliging them to send their children to an Ecuadorian school. These activities led to a growing dislike of the state, which the indigenous community felt was threatening its original identity. This resulted in practices of articulation in which the community *called upon their tradition* and forest regulations for forest management plans were collectively rejected. As forest use held economic, social-cultural and spiritual values, the community had a very clear image of the *right way of doing things* which did not correspond at all with the new forest regulations. In this process of articulation, the traditional culture as an independent community *leaked* from a social-political context to a forest-use context. Old symbols – such as their warfare practices and status as fierce fighters – were used to justify the rejection of forest regulations. However, the idea of what is 'traditional' lost part of its meaning in this process since the traditional identity did not always coincided with current needs for community development and new *moral rationalities* such as conflict avoidance. What remained was a rather confusing situation in which identity articulation only happened under certain circumstances and when the timing was right. It mostly occurred during traditional festivities and political gatherings. In other instances, articulation was much less evident.

17.6 Conclusions

The examples of community forestry from Bolivia and Ecuador show the diversity in key elements and practices of institutional bricolage. The often assumed direct relationship between well-designed community level institutions on the one hand and well-managed forest and improved livelihoods on the other does not hold in practice. The existence of processes of aggregation, alteration and articulation show that formal institutions do not necessarily function in the way they are intended. On the contrary, they often result in a multitude of possible and unexpected outcomes. The identified local practices align well with the key elements of institutional bricolage outlined earlier in the chapter, but their particular configuration differs according to context and time. Nonetheless, some more general observations on the relationship between key elements and social practices can be made. For example, practices of *aggregation* can be linked to the creation of *multipurpose institutions. Alteration* often involves *innovation* and *necessary improvisation*, and *articulation* mostly relates to the *invention of tradition* and *leakage of meaning*.

However, this still leaves us with many questions about the scope of institutional bricolage and the extent to which these processes can be managed. The challenge of studying and analysing institutional bricolage processes is that they are not confined to conscious, well-thought-out strategies in response to introduced institutions, such as for community forestry. The responses of local actors can appear calculated and conscious, but are also a much more natural, think-on-your-feet, almost instinctive reaction. Furthermore, local institutions are so deeply embedded in local livelihoods that it is difficult for actors to consciously draw upon them. This can be particularly visible in traditional communities with a strong social and cultural identity.

To summarise, we argue that the concept of bricolage, underpinned by the idea that institutions are animated by socially located people, offers a useful way to understand the interface between structure and agency. We have outlined some key characteristics of processes of bricolage and identified some of the practices through which people shape institutions. The concept of institutional bricolage offers a number of insights while simultaneously raise questions which point to areas for further research. These include:

- The numerous dimensions of plurality in resource management (of actors, scale, uses and values) mean that the institutional arrangements for community forestry are multi-stranded, overlapping and imbued with a variety of meanings and interests. This plurality raises questions of how far processes of bricolage can be managed, and to what extent institutions elude design.
- Recognising the interplay between the creative exercise of agency, the constraining effects of social relationships and designed rules should enable us to track just how much room for manoeuvre specific institutions offer to different actors.
- Institutional analysis from a bricolage perspective is centrally concerned with tracking the effects of arrangements on equity as well as resource optimisation. This usefully situates institutional analysis within the broader processes of governance and raises questions about where the boundaries of institutional analysis for natural resource management should be drawn.
- A bricolage perspective reveals the centrality of power relations to both the functioning and the outcomes of institutional processes and so moves beyond over-instrumental 'technical' approaches to natural resource management, which are bound to fail. Illuminating the operation of power in everyday relationships and through authority exercised by the state poses further challenges for those concerned with promoting better resource management. To what extent can this power be channelled to effect equitable and sustainable natural resource management without reproducing entrenched inequalities?

The examples from community forestry and the questions raised reflect the need for further research on institutional functioning. The bringing together of different key elements and practices of institutional bricolage is a first step towards illuminating the diversity of responses to external interventions, such as community forest management systems. However, it also raises further questions about the predictability, manageability and scope of institutional bricolage processes. To answer these questions and to further identify central aspects of institutional bricolage, more studies are necessary. These studies should not only focus on mapping out the wider institutional landscape, but also track the institutional processes over time. These aspects of 'breadth' and 'depth' can be best achieved by zooming in on the local practices of social actors, but also placing these within wider frameworks of governance. This calls for longitudinal and comparative studies in which it becomes possible to compare and examine the processes of institutional bricolage in community forestry and other natural resource management sectors across difference contexts and over time.

References

Brundtland Commission (1987). Our common future. Oxford University Press, Oxford, UK.

Campbell, B.A., Mandondo, N., Nemarundwe, B., Sithole, W., De Jong, M., Luckert, F. and Matose, F. (2001). Challenges to proponents of common property resource systems: despairing voices from the social forests of Zimbabwe. World Development 29(4): 589-600.

Chase Smith, R., Pinedo, D., Summers, P.M. and Almeyda, A. (2001). Tropical rhythms and collective action community-based fisheries management in the face of Amazonian unpredictability. IDS Bulletin 32(4): 36-46.

Cleaver, F. (2001). Institutional bricolage, conflict and cooperation in Usangu, Tanzania. IDS Bulletin 32(4): 26-35.

Cleaver, F. (2002). Reinventing institutions: bricolage and the social embeddedness of natural resource management. European Journal of Development Research 14(2): 11-30.

Cleaver, F. (2005). The inequality of social capital and the reproduction of chronic poverty. World Development 33(6): 893-906.

Cleaver, F. (2012). Development through bricolage: institutions and natural resource management. Earthscan, London, UK.

De Koning, J. (2011). Reshaping institutions: bricolage processes in smallholder forestry in the Amazon. PhD thesis, Wageningen University, Wageningen, the Netherlands.

Galvan, D. (1997). The market meets sacred fire: land pawning as institutional syncretism in inter-war Senegal. African Economic History 25: 9-41.

Kant, S, and Berry, R.A. (2005). Sustainability, institutions and forest management. In: Kant, S. and Berry, R.A. (eds.). Institutions, sustainability and natural resources: institutions for sustainable forest management. Springer, Dordrecht, the Netherlands, pp. 1-20.

Karky, B.S. and Skutsch, M. (2010). The cost of carbon abatement through community forest management in Nepal Himalaya. Ecological Economics 93(3): 666-672.

Lanzara, G. F. (1999) Why is it so hard to build institutions. Desarrorollo Economico-Revista de Ciencias Sociales 38(152): 925-952.

Leach, M., Mearns, R. and Scoones, I. (1999). Environmental entitlements: dynamics and institutions in community-based natural resource management. World Development 27(2): 225-247.

Lund, C. (2006). Twilight Institutions: public authority and local politics in Africa. Development and Change 37(4): 685-705.

Matose, F. (2008). Institutional configurations around forest reserves in Zimbabwe: the challenge of nested institutions for resource management. International Journal of Justice and Sustainability 13(5): 393-404.

Mehta, L., Leach, M., Newell, P., Scoones, I., Silvaramakrishnan, Way, S.A. (1999). Exploring understandings of institutions and uncertainty: new directions in naturual resource management. IDS Discussion Papers 372, Institute of Development Studies, Brighton, UK.

North, D.C. (1990). Institutions, institutional change and economic performance. Cambridge University Press, Cambridge, UK.

Osei-Kufuor, P. (2010). Does institutionalising decentralisation work? Rethinking agency, institutions and authority in local governance: a case study Ntonaboma in Kwahu-North Disrict, Ghana. PhD thesis, University of Bradford, Bradford, UK.

Ostrom, E. (1990). Governing the commons: the evolution of institutions for collective action. Cambridge University Press, Cambridge, UK.

Ostrom, E. (1991). Managing our common resources: introduction. Nature and Resources 27(4): 2-3.

Ostrom, E. (2005). Understanding institutional diversity. Princeton University Press, Princeton, NJ, USA.

Paavola, J. (2007). Institutions and environmental governance: a reconceptualization. Ecological Economics 63(1): 93-103.

Rametsteiner, E. and Simula, M. (2003). Forest certification – An instrument to promote sustainable forest management? Journal of Environmental Management 67(1): 87-98.

Ribot, J.C., Lund, J.F. and Treue, T. (2010). Democratic decentralization in Sub-Saharan Africa: its contribution to forest management, livelihoods, and enfranchisement. Environmental Conservation 37(1): 35-44.

Ros-Tonen, M.A.F., Van Andel, T., Morsello, C., Otsuki, K., Rosendo, S. and Scholz, I. (2008). Forest-related partnerships in Brazilian Amazonia: there is more to sustainable forest management than reduced impact logging. Forest Ecology and Management 256(7): 1482-1497.

Sandstrom, E. (2008). Reinventing the commons: exploring the emergence of local natural resource management arrangements. Doctoral thesis, Swedish University of Agricultural Sciences, Uppsala, Sweden.

Schanz, H. (2004). Sustainable forest management. In: Burley, J., Evans, J. and Youngquist, J. (eds.) Encyclopaedia of forest sciences. Elsevier, Oxford, UK.

Sehring, J. (2009). Path dependencies and institutional bricolage in post-soviet water governance. Water Alternatives 2(1): 61-81.

Von Benda-Beckman, F. and Van Meijl, T. (1999). Property rights and economic development: land and natural resources in Southeast Asia and Oceania. Colombia University Press, New York, NY, USA.

Wiersum, K.F. (1995) 200 years of sustainability in forestry: lessons from history. Environmental Management 19(3): 321-329.

Wiersum, K.F. (1997). Indigenous exploitation and management of tropical forest resources: an evolutionary continuum in forest-people interactions. Agriculture, Ecosystems and Environment 63(1): 1-16.

Wong, S. (2010). Elite capture or capture elites? Lessons from the 'counter-elite' and 'co-opt-elite' approaches in Bangladesh and Ghana. UNU-WIDER vol. 82. Available at: http://www.wider.unu.edu/publications/working-papers/2010/en_GB/wp2010-82/_files/83816057260277885/default/wp2010-82.pdf.

18. Forest market governance: exploring a practice-based approach

Marjanke A. Hoogstra
Forest and Nature Conservation Policy group, Wageningen University, P.O. Box 47, 6700 AA Wageningen, the Netherlands; marjanke.hoogstra@wur.nl

Abstract

Market governance - the use of the market mechanism in governance processes - has become highly popular. The enthusiasm for market-oriented approaches has become so overwhelming that governments even promote new markets in some areas of policy, including for forest conservation. However, doubts exist about how sustainable this mechanism is for conservation. But how can we analyse and evaluate forest markets when several disciplines see markets as something different? As several scientists indicate, a detailed, all-encompassing specification of the market concept is urgently needed. This chapter reviews the different conceptualisations of markets in different disciplines and how they are used in forest market research. A revised conceptualisation of markets as practices is proposed. During the past twenty years practice-based approaches have received growing interest within the social sciences. Recently they have also been introduced in the study of markets as they offer a richer conceptual tool, and therefore a better understanding of the market than other methods. Using the example of the certified timber market, the potential for using a practice-based approach to analyse forestry markets is explored.

Keywords: forest, governance, markets, practices

18.1 Introduction

The growing public concern about deforestation and forest degradation has pressured governments into developing responses to the wide diversity of forestry problems (Stavins, 1995). Until recently, the predominant approach to forest protection by governments around the world has been to use 'command-and-control' instruments. The key characteristic of command-and-control regulation is that the regulator specifies what individuals and individual firms can and cannot do, enforced by the threat of penalties for non-compliance (Tollefson, 1998). Although these approaches have sometimes been effective, they have also been very costly. At a time when public sectors face huge challenges due to budget cuts, which limit the possibility of preserving forests (Kroeger and Casey, 2007), alternative (cheaper) approaches to protection are clearly attractive.

In recent years, serious attention has been given to forest 'market governance', the use of the market mechanism in governance processes. Relying more on market-based approaches offers two important advantages: (1) they guide private behaviour; and (2) they provide mechanisms for sharing the costs of forest management and protection between governmental and non-governmental actors (Landell-Mills, 2002). The enthusiasm for market-based approaches has become so great that governments even promote the creation of new markets, the development of markets for carbon sequestration being by far the most ambitious to date. Other markets created include those for the supply of clean water, the protection of threatened species, and the avoided disruption of forest and watershed functions (Landell-Mills, 2002; Wunder, 2007).

However, several scientists cast doubt on this new form of governance. Crook and Clapp (1998), for example, warned of an excessive faith in markets as a way to efficiently protect forests 'as they are likely to have contradictory, and largely negative, effects' on forests. Arnold and Perez (2001) also expressed their fear for markets as a solution to efficiently conserve forests, especially due to the selective nature of market demand and the uneven distribution of resources within forests. Different scientists also state that introducing a more market-oriented way of protecting forests inescapably leads to a transition of many subsistence-based use systems to market-oriented production systems, with negative effects such as loss of biodiversity and loss of livelihoods (Bennett and Robinson, 2000; Rico-Gray *et al.*, 1990). Moreover, as market prices seldom reflect the 'true' value of the benefits, markets will fail, leading to overexploitation and the disappearance of valuable benefits (Fa *et al.*, 1995; Hansis, 1998; Vasquez and Gentry, 1989; Witkowski *et al.*, 1994).

In general, forest markets are considered to be an efficient solution to sustainability problems in forestry, although some view these markets as having contradictory, negative effects. This difference in view clearly reflects the lack of knowledge on forest markets. As several authors have already stated, only few scholars really understand how these markets work and how well (or how badly) they operate (Landell-Mills and Porras, 2002; Veeman, 2002). The problem already arises with the concept of 'market'. Even though one might expect some agreement on what is meant by the word market, the reality is that there is a wide range of different conceptualisations across and within different scientific disciplines, especially economics and sociology (Depeyre and Dumez, 2010; Law and Hassard, 1999; Rosenbaum, 2000). But how can we analyse and evaluate forest markets when several disciplines understand markets in different ways? As Rosenbaum (2000) stated, markets cannot be evaluated or tested unless we are able to explain what markets are; the need for an extensive, all-encompassing specification of the market concept is 'particularly pressing'.

A promising, innovative and multidisciplinary approach to analysing forest markets is a practice-based approach. Practice-based approaches have received growing interest within the social sciences during the past two decades (see e.g. Latour,

1987; Schatzki *et al.*, 2001; Wenger, 1998). They have recently also been introduced in the study of markets (see e.g. Andersson *et al.*, 2008; Araujo, 2007; Kjellberg and Helgesson, 2007a,b). According to Kjellberg and Helgesson (2006), such an approach offers a richer conceptual tool and therefore a better understanding of the market than hitherto.

This chapter reviews attempts by scientists from different disciplines to define the concept of a market and discusses how these conceptualisations are used in forest market research. I propose a revised conceptualisation of markets based on practice theory and argue that this has the potential to enrich our understanding of how forest markets come about and how they work.

18.2 Setting the stage: markets from a disciplinary view

The term 'market' was introduced into the English language in the 12th century (or possibly even earlier) and comes from the Latin word *mercatus*, which means 'trade' or 'place to trade'. The word soon acquired three distinct meanings, all related to the locational aspect: (1) a physical marketplace; (2) the gathering at such a place; and (3) the legal right to hold a meeting at a marketplace. It is therefore not surprising that some scientists see the market as either a marketplace or a geographical area: the location where exchanges of a certain commodity take place (Swedberg, 1994). Lipsey (1983), for instance, defined a market as 'an area over which buyers and sellers negotiate the exchange of a well-defined commodity'. In different studies on forest markets, such as studies on local markets for non-timber forest products, the physical marketplace is still the focus of analysis. The study by Dold and Cocks (2002), for example, describes indigenous medicinal plants traded in medicinal markets around the Eastern Cape in South Africa. In many cases, however, the area in question is now apparently much larger than the square around the church or the site of a fair (Rosenbaum, 2000). One example is the market for forest carbon storage, which is not limited to a geographical area.

The reference to the locational aspect is lost in definitions which see the markets as equal to 'exchange'. Jevons (1871, quoted in Hodgson, 1988), for example, defines the market as 'any body of persons who are in intimate business relations and carry on extensive transactions in any commodity'. However, the core of this definition remains the market as something that can be observed (i.e. an exchange is taking place). An example of this aspect of the markets can be found in the study by Hart (1978), who described the transition of a group of net-hunting Mbuti (a nomadic society of the Ituri Forest of Zaire) from subsistence to market hunting. The central elements are the material and non-material exchanges, with different relations over time.

Several scientists have abandoned this observational definition and focused on what the market *does* rather than on what the market *is*: the function of the market

(Rosenbaum, 2000). Most prominent in this respect is the neoclassical economics perspective. In neoclassical economics, which dominates the literature on markets, the market is an abstract concept that is of interest as an allocation mechanism (Swedberg, 1994). Markets are thought of as a way to bring those that want a certain product together with those that are willing to provide this product. The market price is essentially determined by the relation between the quantity supplied and the quantity demanded (see Figure 18.1). This 'equilibrium point' also indicates the optimal market allocation of the particular product.

The neoclassical perspective is also prominent in the literature on forest markets. Most forestry economics handbooks (such as the one by Pearse (1990)) will contain an explanation of markets as allocation mechanisms. Such an explanation was recently used in a study by Trømborg and Solberg (2010) to calculate the consequences of changes in supply and demand on the market for forest biomass.

Different forest scientists have criticised this view on the market. Mantau (1981), for example, stated that the neoclassical model fails to describe the reality of markets, as it ignores psychological and sociological determinants. After analysing different studies on timber markets, both Borowski (1996) and Lückge (2000) also concluded that the neoclassical model is inadequate to describe (the development of) markets as it ignores the social aspects. These critiques of the neoclassical market model in forestry coincided with critiques of the classical market model by several scientists from other disciplines (Keister, 2002). For example, Granovetter (1992) stated that 'the anonymous market of neoclassical models is virtually nonexistent in economic life'. White (1981) even wrote that a neoclassical theory of the market does not exist; it is only a pure theory of exchange.

Two important lines of research within economics that distance themselves from at least some of the premises of the neoclassical model of the market are information

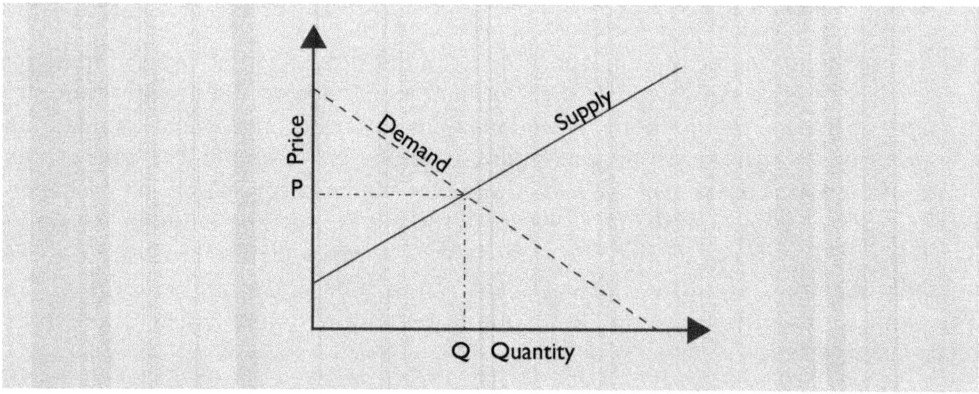

Figure 18.1. Market as an allocation

economics and the new institutional economics (Beckert, 2008). Both approaches consider markets as institutional arrangements that make exchange possible.

In the neoclassical view on the market, market actors have perfect knowledge. In informational economics, however, the focus is on informational asymmetry between the market actors. A classic paper in this field is Akerlof's (1970) 'The Market for Lemons'. In this article, he used the market for used cars as an example to illustrate how the difference between knowledge of the seller and the potential buyer leads to market failure. In his example, he distinguished between good used cars (which he called 'cherries') and defective used cars (called 'lemons'). The difference is the result of several not-always-traceable variables, such as the owner's driving style, quality and frequency of maintenance and accident history. The potential buyer of a used car knows less about how good or how bad the car for sale is than the seller. He therefore also knows that there is a probability that he will buy a lemon. The price he is willing to pay is therefore less than he would pay if he were certain that he was buying a cherry. This lower price may result in owners of good cars not selling them. In such situations, it is impossible for an efficient market to develop and thus market failure ensues. A solution to this problem is the introduction of safeguarding institutions, such as guarantees on used cars, which reduce the risk to buyers of buying a 'lemon' and increase their willingness to purchase. In this case the market is less efficient than it would be if all parties had the same information, but at least markets can exist.

Several forest scientists draw a parallel between forestry markets and Akerlof's market for 'lemons'. Rametsteiner (2002) and Costa and Ibanez (2007), for example, explain certification as a way to remedy informational asymmetry: as consumers cannot differentiate sustainable forest management from regular forest management, forest certification should be considered as an attribute of reliability for consumers who lack the required knowledge. Nunes and Riyanto (2005) even state that the presence of an informational problem is in fact the cornerstone of any certification and ecolabelling policy instrument.

The new institutional economics takes a more radical point of view than information economics. It assumes that factors like incomplete information, a limited mental capacity of market actors to process information, and lack of trust lead to uncertainty. This might be the cause for market failure. Institutions permit the stabilisation of market actors' expectations by guarding against possible negative consequences (such as antitrust laws), thus helping to make markets possible (Beckert, 2007). Which institutions are developed is explained by the interests of the individual market actors. Market actors want to reduce this uncertainty before becoming involved in an exchange, and to do so they have to incur costs. These costs are called transaction costs. Actors will tend to organise an exchange so as to minimise transaction costs. The main kinds of transaction costs are:

- search and information costs – the costs of finding out what products are on offer, whether the required product is available on the market, and which of the sellers offers the lowest price, etc.;
- negotiation costs – the costs of coming to an acceptable agreement with the other party to the transaction, drawing up an appropriate contract, etc.;
- enforcement costs – the costs of making sure the other party adheres to the terms of the contract, and taking appropriate action (often through the legal system) if this turns out not to be the case.

The new institutional economics' approach to forest markets has received considerable attention in the forest literature. Arts and Kerwer (2007), for example, referred in their analysis of timber certification to certification as a regulatory mechanism to reduce search costs. Benneker (2008) looked into the market transaction costs of community forest enterprises to explain the performance of these enterprises. Galik and Jackson (2009) studied the lack of supply of forest carbon offsets by small private forest holdings due to the high transaction costs.

Both informational economics and the new institutional economics emphasise the importance of institutional regulatory mechanisms for the very existence of markets. In both cases the emergence of institutional regulations is explained in terms of the interests of the participating actors, which means that they retain the individualist ontology of markets (Beckert, 2007). Such an individualistic view is rejected in the sociology of markets, a field that has received considerable attention in recent decades (Fligstein and Dauter, 2007). However, although there is a great deal of agreement that markets are social structures characterised by extensive social relations between different market actors, different perspectives have emerged at the theoretical level (Dobbin, 2004; Fligstein and Dauter, 2007). The perspectives are often divided into three theory groups according to the mechanisms explaining the emergence and ongoing dynamics of markets: (1) the networks approach; (2) the institutional approach; and (3) the performativity approach (Fligstein and Dauter, 2007).

The networks approach focuses on social relations as a determining factor of market action. Emphasis is placed on how economic activity comes to be coordinated by groups of people instead of being carried out by individuals (Granovetter, 1992). In other words, it focuses on the structures of the social relations and the positions the individual market actors hold within these market structures (Granovetter, 1985; White, 2001). Often, a division is made between horizontal and vertical relations. Analysis of the horizontal (or non-hierarchical) relations often leads to discussions on topics like 'trust' and 'solidarity', which lead to cooperation. The topic of 'power' is central in the vertical (hierarchical) relations (Granovetter and Swedberg, 2001). The networks perspective has received some attention in connection with forest markets. Recently, Murphy and Lawhon (2011), for example, studied the possibility of trusting partnerships in Bolivia's forest products sector. In a study of small-scale private forest owners in Bavaria (Germany), Schlüter and Koch (2009) found the enormous importance of trust among the people within the network.

The institutional approach focuses on the way exchange is determined by institutional settings (e.g. Fligstein, 1990). But in contrast to the new institutional economics' view on markets, market actors only play a limited role. They are incorporated as actors trying to change the structures within which they have to function in an attempt to enhance their market position. However, these structures are the result of a historical development (the evolution of a specific market) and are considered far too complex to be directly steered by the individual actors (Beckert, 2002). An example of a study on forest markets using this approach is McNichol's (2000) study analysing British efforts to set up new markets for NGO-certified sustainable wood products.

The third group focuses on performativity. Performativity is a concept that primarily expresses the idea that phenomena exist only through the processes of creating them; nothing exists without being continually performed (MacKenzie, 2003, 2007). Markets are usually presented as a mysterious force which specific enterprises and individuals have no means of affecting. Seeing markets in a performative manner, however, means that they consist of and are constantly performed by actors (Kortelainen, 2008). The performativity perspective has received only limited attention in the literature on forest markets, an exception being the research of Kortelainen (2008). He showed in his research on green markets in the Russian forest sector that these green markets did not come about and develop through the demand for green products, but were (and still are) performed by a diversified group of actors, including environmental NGOs, publishing companies, certification agencies, market researchers and critical citizens.

This section has shown how a wide array of scientific disciplines have developed their own view on the market. Although the interest in markets from different disciplines has provided important insights into the functioning of markets, many scientists (in economics as well as in other sciences) still feel discontented with the current state of knowledge about markets as they sense that much (or even most) of the work remains to be done (Krippner, 2001; Swedberg, 2003). Several scientists (see e.g. Krippner, 2001) state that markets should not be conceived within a single concept, but as complex combinations of multiple dimensions. This is in sharp contrast to the studies so far, which have all taken a one-dimensional, mostly disciplinary view. This may provide insights, especially when taken to the extreme, but at the same time can stand in the way of gaining an overall view on markets. And it is exactly this overall view on the market that is of utmost importance: if there is one thing the different studies have taught us, it is that a market is many things at once; it is complex, multifaceted and sometimes even paradoxical. Such an understanding demands approaches which are not limited to a certain theoretical perspective, as in the predominantly disciplinary studies on markets carried out so far (e.g. economics, sociology), but which cross disciplinary boundaries.

18.3 Towards a new approach: markets as practices

Such a promising, innovative and multidisciplinary way to analyse forest markets can be found in exploring markets as *practices*. Following the strong call for a practice-based approach from different social scientists (e.g. Bourdieu, 1977; De Certeau, 1984; Giddens, 1984; Schatzki *et al.*, 2001), in recent years a number of scientists working on markets and marketing have also proposed using a practice-based approach to explain the functioning of markets (e.g. Aurojo *et al.*, 2008; Kjellberg and Helgesson, 2007a,b; Smith, 2007a,b).

In order to explain what a practice-based market approach is, I turn first to the question of what a practice-based approach in itself is. Although the term 'practice' is used frequently, no unifying theory exists. A growing number of researchers from several fields, such as organisation studies, sociology, anthropology, philosophy, and science and technology studies have developed and adopted a range of different practice approaches. What they all have in common is that they place practices at the centre of understanding social phenomena. Giddens (1984) was one of the social scientists who argued that social sciences 'is neither the experience of the individual actor, nor the existence of any form of social totality, but social practices ordered across space and time'. In other words, a practice approach stands in opposition to both the individualistic view, in which society is explained as the result of the actions of individual agents (as, for example, economics does), and the structured view, in which society is understood as the result of underlying institutions (as, for example, in sociology).

What then are practices? Practices are based on the idea that in the continual flow of activities it is possible to identify clusters or blocks of activities (Røpke, 2009). In Schatzki's (2002) words, practices are organised constellations of human activity in social life. An organised set of activities is considered to be a practice when it is recognisable across time and space: a relatively enduring, relatively recognisable entity (Shove *et al.*, 2007). However, Schatzki (2002) separates out the spatial arrangement and the material elements of practices (Jarzabkowski and Spee, 2009). In contrast, other practice scholars, such as Orlikowski (2007) and Reckwitz (2002), advocate an explicit incorporation of the material. As Orlikowski (2007) states, there is an 'absence of any considered treatment or theorising of the material artefacts, bodies, arrangements, and infrastructures through which practices are performed'. Orlikowski thus conceptualises practices not only as a doing, but also as the material arrangements in which doing occurs. The two cannot be separated. This is also the way Reckwitz (2002) defined practices, as:

> a routinized type of behavior which consists of several elements, interconnected to one another: forms of bodily activities, forms of mental activities, 'things' and their use, a background knowledge in the form of understanding, know-how, states of emotion and motivational knowledge. A practice – a way of cooking, of consuming, of working,

> of investigating, of taking care of oneself or of others, etc. – forms so to speak a 'block' whose existence necessarily depends on the existence and specific interconnectedness of these elements, and which cannot be reduced to one of these elements.

Sandberg and Dall'Alba (2009) conceptualise practices in a similar way, as 'specific worlds in which members dwell, made up of an array of activities, people, knowledge, equipment, concerns and so on'.

Practices are thus coordinated entities, but they also require performance for their existence. To make the distinction between the entity and the enactment clear, Schatzki (2002) applies two different notions of practice: practice as a coordinated entity (in the following: practice-as-entity) and practice as performance (in the following: practice-as-performance). The first notion is of 'practice as a temporally unfolding and spatially dispersed nexus of doings and sayings. Examples are cooking practices, voting practices, industrial practices, recreational practices, and correctional practices' (Schatzki, 1996). The second notion, practice-as-performance, refers to the carrying out of practices, the performing of the doings and sayings which 'actualizes and sustains practices in the sense of nexuses' (Schatzki, 2002). Coordinated entities can only exist when the activities involved are performed by people – not just a few individuals, but shared by more. As Reckwitz (2002) put it: 'a practice represents a pattern which can be filled out by a multitude of single and often unique actions reproducing the practice'. Individuals face practices-as-entities, as these are formed historically as a collective achievement. Through their own practice-as-performance, individuals reproduce and transform the entities over time. Individuals thus act as 'carriers' of practices (Røpke, 2009). In fact, individuals are the carriers of many different practices at the same time; at any time, an individual actor can engage in different practices. Practices are also not singular and unitary, but multiple (Sandberg and Dall'Alba, 2009). Practices such as cooking and voting vary considerably between regions, countries and cultural contexts.

During the last ten years, several scientists have started to develop thinking about markets as practices. In the literature on markets and practices, one can find two different 'streams'. One stream is based on the work of Kjellberg and Helgesson (2006, 2007a,b), Andersson *et al.* (2008) and Araujo *et al.* (2008). Inspired by the ideas of Callon (1998) on the sociology of translation, they suggested that markets are performed as actors engage in 'market practices', which is taken to mean all activities that contribute to perform markets. Kjellberg and Helgesson (2006, 2007a,b) developed a conceptual model in which a distinction is made between three types of practices: (1) exchange practices; (2) normalising practices; and (3) representational practices. Exchange practices relate to the concrete activities related to the exchanges of a specific commodity. Representational practices include activities that contribute to depicting markets and/or how they work. Normalising practices are the activities that contribute to establish guidelines for how a market should be shaped and reshaped or should work according to a certain actor or group of actors. Kjellberg and Helgesson (2006, 2007a,b) argue that markets can be understood as emergent orders constituted by ongoing exchange, normalising and representational practices, and the interlinked translations between these involving intermediaries such as rules, tools, measures and measurements.

Kjellberg and Helgesson (2006) have applied this framework in different situations, for example to study the change from full service to self-service in grocery retailing. Their framework has also been used in other studies. Veal and Mouzas (2008), for example, used the framework to analyse barriers to market formation, and Rohracher (2009) used this model to analyse the reframing of electricity markets as a strategically oriented non-state governance activity of intermediary organisations.

A somewhat different stream is Smith's (2007a,b) practice-based view on markets. He frames markets as 'evolving social practices'. In contrast to the aforementioned scholars on markets and practices, Smith understands markets themselves as 'unfolding practices' and not as *consisting of* market practices that perform markets. Smith presents markets as social entities in which actors define themselves and their activities by generating shared meaning. Taking a snapshot of a market, one can see that most markets exhibit fairly well-defined notions of their who (the actors involved), what (the commodity that is being exchanged) and how (the way of exchange). These who, what and how factors, however, not only emerge in different ways, but are also not stable givens. 'Participants not only come and go, but their interests and styles change. The same can be said for the items exchanged and the rules governing these exchanges' (Smith, 2007a,b).

The differences between the two streams on markets as practices imply that there is still a need to continue the theoretical developments to develop a better understanding of the what, how and when of practices shaping and making markets. Nevertheless, the practice-based view on markets clearly shows that markets are not one-dimensional (as in the disciplinary views); they are something people do and say, with specific tools and rules. Markets involve different elements as described in different disciplines, such as markets being performed (sociology) and being an ongoing accomplishment over time (evolutionary economics). The practice-based view on markets also dissolves the dichotomy in the markets discussion between structure (as in sociology) and agency (which, for example, is the focus in economics), thereby offering an overarching coherence.

18.4 Implications of researching markets as practices

Despite the fact that the theory of markets as practices still has to be developed further, a valid question is what exactly needs to be studied empirically. Although it is not the intention of this article to discuss an extensive framework for analysing forest markets, this section briefly explores different elements that have been used in other fields of practice-based research (especially strategy research) which could form a starting point for developing a frame of analysis for forest markets. This is discussed using the certified timber market as an example.

Different scholars distinguish between different elements when analysing practices empirically. According to Rasche and Chia (2009), one can distinguish four main elements:
1. practices as routinised bodily doings and sayings;
2. practices as based on objects and artefacts;
3. practices as routinised subjective interpretation;
4. practices as the (ongoing) constitution of the practitioner.

Practices as routinised bodily doings and sayings

In most of the definitions of practices, the focus is on the routinised type of behaviour (see e.g. Reckwitz, 2002; Schatzki, 2002). Research on markets as practices therefore does not deal with 'market activity' per se, but with 'the patterns of bodily doings and sayings' that market actors perform (Rasche and Chia, 2009). 'Bodily doings' means activities such as walking, cycling, gesturing, keeping to time, etc. 'Bodily sayings' refers to speech and the use of language. In the certified timber market, 'bodily doings' is illustrated by Kiker and Putz (1997). They draw attention to the fact that certified timber means an ecologically and socially acceptable way of production and therefore involves specific activities that occur on different spatial levels (from the forest stand to international centres of trade) on different time scales (ranging from day-to-day activities of loggers in the forest to the traders of the timber). 'Bodily sayings' is described by Fletcher and Hansen (1999), who state that certification has given rise to an own (shared) language. For more than a decade, a range of concepts and terms has been used to elaborate aspects involved in the certification process (Rametsteiner and Simula, 2003).

Practices as based on objects

As described in the second section, practices are more than just doings; they also include material arrangements (Orlikowski, 2007; Reckwitz, 2002). Just as playing tennis requires a racket, a ball, a special outfit and special shoes, the practice of forest markets involves objects which are vital for markets to be performed. Besides the 'hardware' of forests, trees and timber, brands like the tick-tree of the Forest Stewardship Council (FSC) label and the deciduous and coniferous trees of the PEFC logo are examples of objects essential in the process of legitimising the certified timber market by guaranteeing that the timber products come from sustainably managed forests. Another example is given by Rametsteiner and Simula (2003), who describe some tools that are commonly used in certification, for example the criteria and indicators (C&I) for sustainable forest management (SFM).

Practices as routinised subjective interpretation

Practices also include a shared understanding of the world based on shared knowledge. Different scholars use different terms to describe this. Foucault (1990, 1992), for example, talks about 'codes' and Bourdieu (1990) uses the concept of 'habitus'. 'Frames' is the word Goffman (1977) uses, while 'background understanding' is the term coined by Taylor (1985). Such shared knowledge transcends the individual level of understanding and is part of the practice as a whole. This is exactly what Eden (2009) meant when writing about FSC: 'The FSC ... through a complex credibility alliance across the science-policy interface ... reflects and uses socially distributed knowledge.' A certification standard means, among other things, a shared understanding of what the sustainable management of forest and the sustainable production of timber is. As Pattberg (2007) stated, actors in the certified timber market share norms of behaviour.

Practices as the (ongoing) constitution of the practitioner

Practice theory emphasises that identity plays an important role in practices; identity is embedded in practices, including markets. The identity of an actor is not a given, but the result of an ongoing construction, realised through the engagement of the actor in social practices (Foucault, 1992). In Crow and Danks' (2010) study on certification in community-based enterprises, they found that community forest enterprises were willing to certify their forests because it helped them to tell their story. In other words, they could realise their identity by taking part in the practices, and by performing the practices, their identity is continuously reshaped.

These four elements seem to reflect some of the main elements of what practices are, and they therefore provide a good starting point for research. But other elements may be added when further developing a practice-based approach. For example, a fifth element might be 'practices as performance'. As explained above, practices are performed continuously; they are not stable givens. As Eden (2009) explained for the standards in the certified timber market, standards vary across time, because of changing knowledge and views. Hence, standards 'are always subject to renegotiation in the light of future events' (Busch, 2000).

18.5 Discussion and conclusion

The practice-based approach to markets has not previously been addressed in the forestry literature. This chapter is by no means an attempt to present a complete framework and should be viewed as exploratory. It offers an illustration of how markets may be defined from a practices perspective. What it has shown is that such an approach has the potential to contribute to the study of forest markets in terms of enriching our understanding of how forest markets come about and how they 'work'. It also shows that some further theoretical development of this approach is necessary.

'Markets as practices' not only has a theoretical agenda, but a methodological one as well. As Miettinen *et al.* (2009) stated, developing a practice-based approach not only means developing a theory, but also an empirical approach. So far, however, only the minority of current practice theory has, as Bueger (2011) recently stated, 'been practical in the sense of thinking about how to actually conduct practice research'. Bueger (2011) therefore even advocated using a specific term to signify the importance of methodology: 'praxiography' – the practice of doing practice-theory-driven research. This praxiography is a combination of methodological considerations derived from ethnomethodology, ethnography, activity theory and actor-network theory. In his review of research strategies and research tools in practice research, Bueger (2011) looked at three strategies (investigating site, investigating controversies, and following concepts, objects and technologies) and three different tools (participant observation, the analysis of narratives of practitioners and the analysis of experts). His conclusion is that it is often not an 'either or' question, but an 'and and' question. Clearly, studying living practices 'here and now' and relating them to their histories and larger social contexts is not easy (Miettinen *et al.*, 2009) and presents a challenge for the future.

References

Akerlof, G.A. (1970). The market for 'lemons': quality uncertainty and the market mechanism. The Quarterly Journal of Economics 84: 488-500.

Andersson, P., Aspenberg, K. and Kjellberg, H. (2008). The configuration of actors in market practice. Marketing Theory 8: 67-80.

Arnold, J-E.M. and Perez, M.R. (2001). Can non-timber forest products match tropical forest conservation and development objectives? Ecological Economics 39: 437-447.

Araujo, L. (2007). Markets, market making and marketing. Marketing Theory 7: 211-226.

Araujo, L., Kjellberg, H. and Spencer, R. (2008). Market practices and forms: Introduction to the special issue. Marketing Theory 8: 5-14.

Arts, B. and Kerwer, D. (2007). Beyond legalisation? How global standards work. In: Lehmkuhl, D. and Brütsch, C. (eds.) Law and legalization in transnational relations. Routledge, London, UK, pp. 144-165.

Beckert, J. (2002). Beyond the market: the social foundations of economic efficiency. Princeton University Press, Princeton, NJ, USA.

Beckert, J. (2007). The social order of markets. MPIFG Discussion Paper, Max-Planck-Institut für Gesellschaftsforschung, Cologne, Germany.

Beckert, J. (2008). Introduction. Socio-Economic Review 6: 515-516.

Benneker, C. (2008). Dealing with the state, the market and NGOs: the impact of institutions on the constitution and performance of community forest enterprises (CFE) in the lowlands of Bolivia. PhD thesis, Wageningen University, Wageningen, the Netherlands.

Bennet, E.L. and Robinson, J.G. (2000). Hunting for sustainability in tropical forests. Columbia University Press, New York, NY, UK.

Borowski, S. (1996). Marketingstrategien von Forstbetrieben. Schriften aus dem Institut für Forstökonomie, Band 7, Institut für Forstökonomie der Universität Freiburg, Freiburg, Germany.

Bourdieu, P. (1977). Outline of a theory of practice. Cambridgue University Press, Cambridge, UK.

Bourdieu, P. (1990). The logic of practice. Cambridgue University Press, Cambridge, UK.

Bueger, C. (2011). Praxiography in international relations: methodological implications of the practical turn. Draft paper prepared for presentation at the third Conference of the World International Studies Committee, Porto, Portugal, August 2011. Available at: http://www.bueger.info/documents/Praxiography%20in%20International%20Relations.pdf.

Busch, L. (2000). The moral economy of grades and standards. Journal of Rural Studies 16: 273-283.

Callon, M. (1998). Introduction: the embeddedness of economic markets in economics. In: Callon, M. (ed.) The laws of the markets. Blackwell Publishers/The Sociological Review, Oxford, UK, pp. 1-57.

Costa, S. and Ibanez, L. (2007). Certification of sustainable forest management: differentiation strategies and asymmetric information. International Journal of Agricultural Resources, Governance and Ecology 6: 393-414.

Crook, C. and Clapp, R.A. (1998). Is market-oriented forest conservation a contradiction in terms? Environmental Conservation 25: 131-145.

Crow, S. and Danks, C. (2010). Why certify? Motivations, outcomes and the importance of facilitating organizations in certification of community-based forestry initiatives. Small-Scale Forestry 9: 195-211.

De Certeau, M. (1984). The practice of everyday life. University of California Press, Berkeley, CA, USA.

Depeyre, C. and Dumez, H. (2009). The role of architectural players on coopetition: the case of the US defense industry. In: Yami, S., Castaldo, S., Dagnino, G.B. and Le Roy, F. (eds.) Coopetition: winning strategies for the 21st century. Edward Elgar, Cheltenham, UK, pp. 124-140.

Dobbin, F. (2004). Introduction: the sociology of the economy. In: Dobbin, F. (ed.) The sociology of the economy. Russell Sage Foundation, New York, NY, USA, pp. 1-25.

Dold, A.P. and Cocks, M.L. (2002). The trade in medicinal plants in the Eastern Cape Province, South Africa. South African Journal of Science 98: 589-597.

Eden, S. (2009). The work of environmental governance networks: traceability, credibility and certification by the Forest Stewardship Council. Geoforum 40: 383-394.

Fa, J.E., Juste, J., Perez del Val, J. and Castroviejo, J. (1995). Impact of market hunting on mammal species in Equatorial Guinea. Conservation Biology 9: 1107-1115.

Fletcher, R. and Hansen, E. (1999). Forest certification trends in North America and Europe. New Zealand Journal of Forestry 44: 4-6.

Fligstein, N.D. (1990). The transformation of the corporate control. Harvard University Press, Cambridge, MA, USA.

Fligstein, N.D. and Dauter, L. (2007). The sociology of markets. Annual Review of Sociology 33: 105-128.

Foucault, M. (1990). The order of things. Routledge, London, UK.

Foucault, M. (1992). Language, counter-memory, practice. Cornell University Press, New York, NY, USA.

Galik, C.S. and Jackson, R.B. (2009). Risks to forest carbon offset projects in a changing climate. Forest Ecology and Forest Management 257: 2209-2216.

Giddens, A. (1984). The constitution of society: an outline of the theory of structuration. University of California Press, Berkeley, CA, USA.

Goffman, E. (1977). Frame analysis: an essay on the organization of experience. Harvard University Press, Cambridge, MA, USA.

Granovetter, M. (1985). Economic action and social structure: the problem of embeddedness. American journal of sociology 91: 481-510.

Granovetter, M. (1992). Economic institutions as social constructions: a framework for analysis. Acta Sociologica 35: 3-11.

Granovetter, M. and Swedberg, R. (2001). The sociology of economic life. Westview Press, Boulder, CO, USA.

Hansis, R. (1998). A political ecology of picking: non-timber forest products in the Pacific Northwest. Human Ecology 26: 67-86.

Hart, J.A. (1978). From subsistence to market: a case study of the Mbuti net hunters. Human Ecology 6(3): 325-353.

Hodgson, G.M. (1988). Economics and institutions: a manifesto for a modern institutional economics. Polity Press, Cambridge, MA, USA.

Jarzabkowski, P. and Spee, A.P. (2009). Strategy-as-practice: a review and future directions for the field. International Journal of Management Reviews 11: 69-95.

Keister, L.A. (2002). Financial markets, money, and banking. Annual Review of Sociology 28: 39-61.

Kiker, C.F. and Putz, F.E. (1997). Ecological certification of forest products: economic challenges. Ecological Economics 20: 37-52.

Kjellberg, H. and Helgesson, C.-F. (2006). Multiple versions of markets: multiplicity and performativity in market practice. Industrial Marketing Management 35: 839-855.

Kjellberg, H. and Helgesson, C.-F. (2007a). On the nature of markets and their practices. Marketing Theory 7: 137-162.

Kjellberg, H. and Helgesson, C.-F. (2007b). The mode of exchange and shaping of markets: distributor influence in the Swedish post-war food industry. Industrial Marketing Management 36: 861-878.

Kortelainen, J. (2008). Performing the green market – creating space: emergence of the green consumer in the Russian woodlands. Environment and Planning A 40: 1294-1311.

Krippner, G.R. (2001). The elusive market: embeddedness and the paradigm of economic sociology. Theory and Society 30: 775-810.

Kroeger, T. and Casey, F. (2007). An assessment of market-based approaches to providing ecosystem services on agricultural lands. Ecological Economics Ecological Economics 64: 321-32.

Landell-Mills, N. (2002). Developing markets for forest environmental services: An opportunity for promoting equity while securing efficiency? Philosophical Transactions of the Royal Society of London A 360: 1817-1825.

Landell-Mills, N. and Porras, T.I. (2002). Silver bullet or fools' gold? A global review of markets for forest environmental services and their impact on the poor. Instruments for sustainable private sector forestry series, International Institute for Environment and Development, London, UK.

Latour, B. (1987). Science in action: how to follow scientists and engineers through society. Harvard University Press, Cambridge, MA, USA.

Law, J. and Hassard, J. (1999). Actor network theory and after. Blackwell/Sociological Review, Oxford, UK.

Lipsey, R.G. (1983). Positive economics. Weidenfeld and Nicolson, London, UK.

Lückge, F.J. (2000). Modelle des Angebotsverhaltens deutscher Forstbetriebe – Stand und Perspektiven der Forschung. Allgemeine Forst- und Jagdzeitschrift 1: 15-19.

MacKenzie, D. (2003). An equation and its worlds: bricolage, exemplars, disunity, and performativity in financial economics. Social Studies of Science 33: 831-868.

MacKenzie, D. (2007). Do economists make markets? On the performativity of economics. Princeton University Press, Princeton, NJ, USA.

McNichol, J. (2000). Contesting governance in the global marketplace: a sociological assessment of British efforts to build new markets for NGO-certified sustainable wood products. Working Paper no. 2000-06, University of California, Center for Culture, Organization, and Politics, Berkeley, CA, USA.

Miettinen, R., Samra-Fredericks, D. and Yanow, D. (2009). Re-turn to practice: an introductory essay. Organization Studies 30: 1309-1327.

Murphy, J.T. and Lawhon, M. (2011). Market intermediaries and rural people in Bolivia's forest products sector: are trusting partnerships possible? Singapore Journal of Tropical Geography 32: 203-219.

Nunes, P.A.L.D. and Riyanto, Y.E. (2005). Information as a regulatory instrument to price biodiversity benefits: certification and ecolabeling policy practices. Biodiversity and Conservation 14: 35-52.

Orlikowski, W.J. (2007). Sociomaterial practices: exploring technology at work. Organization Studies 28: 1435-1448.

Pattberg, P.H. (2007). Partnerships for sustainability: an analysis of transnational environmental regimes. In: Glasbergen, P., Biermann, F. and Mol, A.P.J. (eds.) Partnerships, governance and sustainable development: reflections on theory and practice. Edward Elgar, Cheltenham, UK, pp. 173-193.

Pearse, P.H. (1990). Introduction to forestry economics. University of British Columbia Press, Vancouver, Canada.

Rametsteiner, E. (2002). The role of governments in forest certification – A normative analysis based on new institutional economics theories. Forest Policy and Economics 4: 163-173.

Rametsteiner, E. and Simula, M. (2003). Forest certification – An instrument to promote sustainable forest management? Journal of Environmental Management 67: 87-98.

Rasche, A. and Chia, R. (2009). Researching strategy practices: a genealogical social theory perspective. Organization Studies 30: 713-734.

Reckwitz, A. (2002). Toward a theory of social practices'. European Journal of Social Theory 5: 243-263.

Rico-Gray, V., Garcia-Franco, J.G., Chemas, A., Puch, A. and Sima, P. (1990). Economic Botany 44: 470-487.

Rohracher, H. (2009). Intermediaries and the governance of choice: The case of green electricity labelling. Environment and Planning 41: 2014-2028.

Røpke, I. (2009). Theories of practice – New inspiration for ecological economic studies on consumption. Ecological Economics 68: 2490-2497.

Rosenbaum, E.F. (2000). What is a market? On the methodology of a contested concept. Review of Social Economy LVIII: 455-482

Sandberg, J. and Dall'Alba, G. (2009). Returning to practice anew: a life-world perspective. Organization Studies 30: 1349-1368.

Schatzki, T.R. (1996). Social practices: a Wittgensteinian approach to human activity and the social. Cambridge University Press, Cambridge, UK.

Schatzki, T.R. (2002). The site of the social: a philosophical account of the constitution of social life and change. Pennsylvania State University Press, University Park, PA, USA.

Schatzki, T.R., Knorr-Cetina, K. and Von Savigny, E. (2001). The practice turn in contemporary theory. Routledge, New York, NY, USA.

Schlüter, A. and Koch, M. (2009). Institutional change in the forest sector: trust and mental models. European Journal of Forest Research 3: 383-393.

Shove, E., Watson, M., Hand, M. and Ingram, J. (2007). The design of everyday life. Berg, Oxford, UK.

Smith, C.W. (2007a). Continuities in markets as definitional practices: a response to Aspers, Knorr Cetina, and Prus. Canadian Journal of Sociology 32: 503-510.

Smith, C.W. (2007b). Markets as definitional practices. Canadian Journal of Sociology 32: 1-39.

Stavins, R.N. (1995). Transaction costs and tradeable permits. Journal of Environmental Economics and Management 29: 133-148.

Swedberg, R. (1994). Markets as social structures. In: Smelser, N.J. and Swedberg, R. (eds.) The handbook of economic sociology. Princeton University Press, Princeton, NJ, USA, pp. 255-282.

Swedberg, R. (2003). Economic and sociological approaches to markets. In: Swedberg, R. (ed.) Principles of economic sociology. Princeton University Press, Princeton, NJ, USA.

Taylor, C. (1985). Human agency and language: Philosophical papers I. Cambridge University Press, Cambridge, UK.

Tollefson, C. (1998). The wealth of forests: markets, regulation, and sustainable forestry. UBC Press, Vancouver, Canada.

Trømborg, E. and Solberg, B. (2010). Forest sector impacts of the increased use of wood in energy production in Norway. Forest Policy and Economics 12: 39-47.

Vasquez, R. and Gentry, A.H. (1989). Use and misuse of forest-harvested fruits in the Iquitos area. Conservation Biology 3: 350-361.

Veal, G. and Mouzas, S. (2008). Barriers to market formation. Representing multiplicity in markets – multiple representations of markets. Paper for the 26th IMP Conference, Corvinus University of Budapest, Hungary. Available at: http://impgroup.org/uploads/papers/7525.pdf.

Veeman, M. (2002). Understanding local and regional markets for forest products. In: Campbell, B.M. and Luckert, M.K. (eds.) Uncovering the hidden harvest: valuation methods for woodland and forest resources. Earthscan Publications, London, UK, pp. 66-102.

Wenger, E. (1998). Communities of practice: learning, meaning, and identity. Cambridge University Press, Cambridge, UK.

White, H.C. (1981). Where do markets come from? The American Journal of Sociology 87: 517-547.

White, H.C. (2001). Markets from networks. Socioeconomic models of production. Princeton University Press, Princeton, NJ, USA.

Witkowski, E.T.F., Lamont, B.B. and Obbens, F.J. (1994). Commercial picking of *Banksia hookeriana* in the wild reduces subsequent shoot, flower and seed production. Journal of Applied Ecology 31: 508-520.

Wunder, S. (2007). The efficiency of payments for environmental services in tropical conservation. Conservation Biology 21: 48-58.

Index

D